From a Molecule to a Drug: Chemical Features Enhancing Pharmacological Potential

From a Molecule to a Drug: Chemical Features Enhancing Pharmacological Potential

Editors

Giovanni Ribaudo
Laura Orian

MDPI • Basel • Beijing • Wuhan • Barcelona • Belgrade • Manchester • Tokyo • Cluj • Tianjin

Editors
Dr. Giovanni Ribaudo, PhD
Dipartimento di Medicina
Molecolare e Traslazionale,
Università di Brescia, Viale
Europa 11, 25123 Brescia, Italy

Prof. Laura Orian, PhD
Dipartimento di Scienze Chimiche,
Università degli Studi di Padova,
Via Marzolo, 1,
35131 Padova, Italy

Editorial Office
MDPI
St. Alban-Anlage 66
4052 Basel, Switzerland

This is a reprint of articles from the Special Issue published online in the open access journal *Molecules* (ISSN 1420-3049) (available at: http://www.mdpi.com).

For citation purposes, cite each article independently as indicated on the article page online and as indicated below:

LastName, A.A.; LastName, B.B.; LastName, C.C. Article Title. *Journal Name* **Year**, *Volume Number*, Page Range.

ISBN 978-3-0365-4753-4 (Hbk)
ISBN 978-3-0365-4754-1 (PDF)

Cover image courtesy of Giovanni Ribaudo and Laura Orian.

© 2022 by the authors. Articles in this book are Open Access and distributed under the Creative Commons Attribution (CC BY) license, which allows users to download, copy and build upon published articles, as long as the author and publisher are properly credited, which ensures maximum dissemination and a wider impact of our publications.

The book as a whole is distributed by MDPI under the terms and conditions of the Creative Commons license CC BY-NC-ND.

Contents

About the Editors vii

Giovanni Ribaudo and Laura Orian
From a Molecule to a Drug: Chemical Features Enhancing Pharmacological Potential
Reprinted from: *Molecules* **2022**, 27, 4144, doi:10.3390/molecules27134144 1

Tsun-Thai Chai, Clara Chia-Ci Wong, Mohamad Zulkeflee Sabri and Fai-Chu Wong
Seafood Paramyosins as Sources of Anti-Angiotensin-Converting-Enzyme and Anti-Dipeptidyl-Peptidase Peptides after Gastrointestinal Digestion: A Cheminformatic Investigation
Reprinted from: *Molecules* **2022**, 27, 3864, doi:10.3390/molecules27123864 5

Karolina Wanat and Elżbieta Brzezińska
Statistical Methods in the Study of Protein Binding and Its Relationship to Drug Bioavailability in Breast Milk
Reprinted from: *Molecules* **2022**, 27, 3441, doi:10.3390/molecules27113441 27

Etimad Huwait, Dalal A. Al-Saedi and Zeenat Mirza
Anti-Inflammatory Potential of Fucoidan for Atherosclerosis: In Silico and In Vitro Studies in THP-1 Cells
Reprinted from: *Molecules* **2022**, 27, 3197, doi:10.3390/molecules27103197 47

Elena Alba Álvaro-Alonso, Ma Paz Lorenzo, Andrea Gonzalez-Prieto, Elsa Izquierdo-García, Ismael Escobar-Rodríguez and Antonio Aguilar-Ros
Physicochemical and Microbiological Stability of Two Oral Solutions of Methadone Hydrochloride 10 mg/mL
Reprinted from: *Molecules* **2022**, 27, 2812, doi:10.3390/molecules27092812 63

Ivan Yu. Torshin, Olga A. Gromova, Konstantin S. Ostrenko, Marina V. Filimonova, Irina V. Gogoleva, Vladimir I. Demidov and Alla G. Kalacheva
Lithium Ascorbate as a Promising Neuroprotector: Fundamental and Experimental Studies of an Organic Lithium Salt
Reprinted from: *Molecules* **2022**, 27, 2253, doi:10.3390/molecules27072253 77

Daniel Muñoz-Reyes, Alfredo G. Casanova, Ana María González-Paramás, Ángel Martín, Celestino Santos-Buelga, Ana I. Morales, Francisco J. López-Hernández and Marta Prieto
Protective Effect of Quercetin 3-*O*-Glucuronide against Cisplatin Cytotoxicity in Renal Tubular Cells
Reprinted from: *Molecules* **2022**, 27, 1319, doi:10.3390/molecules27041319 97

Hoang Thai Ha, Dang Xuan Cuong, Le Huong Thuy, Pham Thanh Thuan, Dang Thi Thanh Tuyen, Vu Thi Mo and Dinh Huu Dong
Carrageenan of Red Algae *Eucheuma gelatinae*: Extraction, Antioxidant Activity, Rheology Characteristics, and Physicochemistry Characterization
Reprinted from: *Molecules* **2022**, 27, 1268, doi:10.3390/molecules27041268 117

Amin Osman Elzupir
Molecular Docking and Dynamics Investigations for Identifying Potential Inhibitors of the 3-Chymotrypsin-like Protease of SARS-CoV-2: Repurposing of Approved Pyrimidonic Pharmaceuticals for COVID-19 Treatment
Reprinted from: *Molecules* **2021**, 26, 7458, doi:10.3390/molecules26247458 133

Everaldo F. Krake and Wolfgang Baumann
Selective Oxidation of Clopidogrel by Peroxymonosulfate (PMS) and Sodium Halide (NaX) System: An NMR Study
Reprinted from: *Molecules* **2021**, *26*, 5921, doi:10.3390/molecules26195921 . . . . **151**

Giovanni Ribaudo, Marco Bortoli, Erika Oselladore, Alberto Ongaro, Alessandra Gianoncelli, Giuseppe Zagotto and Laura Orian
Selenoxide Elimination Triggers Enamine Hydrolysis to Primary and Secondary Amines: A Combined Experimental and Theoretical Investigation
Reprinted from: *Molecules* **2021**, *26*, 2770, doi:10.3390/molecules26092770 . . . . **163**

Amalia Stefaniu, Lucia Pirvu, Bujor Albu and Lucia Pintilie
Molecular Docking Study on Several Benzoic Acid Derivatives against SARS-CoV-2
Reprinted from: *Molecules* **2020**, *25*, 5828, doi:10.3390/molecules25245828 . . . . **177**

Sebastián A. Cuesta and Lorena Meneses
The Role of Organic Small Molecules in Pain Management
Reprinted from: *Molecules* **2021**, *26*, 4029, doi:10.3390/molecules26134029 . . . . **191**

Giuseppe Zagotto and Marco Bortoli
Drug Design: Where We Are and Future Prospects
Reprinted from: *Molecules* **2021**, *26*, 7061, doi:10.3390/molecules26227061 . . . . **211**

About the Editors

Dr. Giovanni Ribaudo, PhD

Dr. Giovanni Ribaudo, PhD, carries out his research activity at the Department of Molecular and Translational Medicine of the University of Brescia. Taking advantage of a combination of synthetic and analytical (HPLC, NMR, mass spectrometry) tools and computational medicinal chemistry, his main research topics consist of the design and screening of small molecules interacting with peculiar DNA arrangements and the study of nature-inspired phosphodiesterase (PDE) inhibitors targeting the central nervous system.

Prof. Laura Orian, PhD

Prof. Laura Orian, PhD, is conducting her research at the Department of Chemical Sciences of the University of Padova. Her main research interest is the theoretical rationalization of the chemical reactivity, particularly of metal-based catalysts, biological and bioinspired systems. Her studies are focused on the comprehension of the reaction mechanisms of different systems, ranging from small-model molecules to enzymes. The major goal of her research is the rational design of functional molecules assisted by the computer, aiming at reactivity prediction in advance of the experiment. Her computational investigations are applied in different scientific areas, including redox biology, medicine and toxicology, and are systematically based on a rigorous approach to chemical structure and reactivity shared with numerous national and international collaborators. LO is also active in scientific divulgation in the schools for pupils and in STEM activities for secondary school students and shares numerous chemistry projects with teachers.

Editorial

From a Molecule to a Drug: Chemical Features Enhancing Pharmacological Potential

Giovanni Ribaudo [1] and Laura Orian [2,*]

[1] Dipartimento di Medicina Molecolare e Traslazionale, Università degli Studi di Brescia, Viale Europa 11, 25123 Brescia, Italy; giovanni.ribaudo@unibs.it
[2] Dipartimento di Scienze Chimiche, Università degli Studi di Padova, Via Marzolo 1, 35131 Padova, Italy
* Correspondence: laura.orian@unipd.it

Citation: Ribaudo, G.; Orian, L. From a Molecule to a Drug: Chemical Features Enhancing Pharmacological Potential. *Molecules* 2022, 27, 4144. https://doi.org/10.3390/molecules27134144

Received: 17 June 2022
Accepted: 27 June 2022
Published: 28 June 2022

Publisher's Note: MDPI stays neutral with regard to jurisdictional claims in published maps and institutional affiliations.

Copyright: © 2022 by the authors. Licensee MDPI, Basel, Switzerland. This article is an open access article distributed under the terms and conditions of the Creative Commons Attribution (CC BY) license (https://creativecommons.org/licenses/by/4.0/).

Health is a fundamental human right and is a global goal to which extensive research effort is devoted in all fields. Chemistry plays a key role in understanding the mechanisms ruling health and disease conditions at the molecular level, as well as in discovering substances with pharmacological potential which can restore health status or mitigate pathology-related damage. One of the major challenges is to understand, rationalize, and control those molecular features which are crucial for a specific drug action. This problem is rooted in the well-known chemical ambition of establishing structure–activity relationships of general validity, although other relevant aspects must be considered, such as solubility, targeting efficiency, and toxicity.

Stitching to the first essential aspect, we assist the continuous evolution of the chemical design approach, which was mainly based on the expensive 'trial and error' method only few decades ago. It is commonly accepted that the trials can be efficiently delegated to computers. Machine-assisted drug design has gained importance with the implementation of different methodologies, ranging from quantum chemistry to classic and continuum approaches, and, more recently, with the application of artificial intelligence algorithms. Despite the fact that there is plenty of room for improvement, large-scale screenings, protein–ligand and protein–protein docking, simulations, and molecular- and multi-scale mechanistic studies play an important role in research progress and receive a large consensus in health sciences.

When we conceived this Special Issue, it became apparent for us to choose a topic and title which reflect our different background in medicinal and theoretical computational chemistry and is close to our joint collaboration. Combining our complementary expertise, we recently developed a project repurposing or better redesigning a popular antidepressant drug molecule, i.e., fluoxetine, which is better known by its commercial name, Prozac. We designed in silico a series of selenoderivatives of fluoxetine and assessed their enhanced antioxidant capacity through chemical and computational protocols [1,2], and, finally, we demonstrated in vivo that selenofluoxetine maintains its SSRI antidepressant action [3]. These outcomes paved the route to our contribution on this Special Issue, in which we report on a new ability of these selenofluoxetine derivatives, i.e., a novel strategy to selectively release bioactive molecules within a selenoxide elimination-triggered enamine hydrolysis [4].

The Special Issue collected contributions from researchers all over the world, demonstrating the flourishing interest of the international scientific community towards the above-mentioned aims and scopes. Amalia Stefaniu and colleagues reported a computer-aided screening of benzoic acid derivatives and semisynthetic alkyl gallates against SARS-CoV-2 main protease [5]. Furthermore, the paper from Amin Osman Elzupir focuses on the SARS-CoV-2 outbreak, but a different mechanism was considered, as the author presented an in silico evaluation of pyrimidonic pharmaceuticals against papain-like protease [6]. In their review article, Sebastián A. Cuesta and Lorena Menes provided an overview on the evolution of analgesic and anti-inflammatory drugs, including theories on novel mechanisms of action [7]. Everaldo F. Krake and Wolfgang Baumann used NMR to investigate the reactivity of clopidogrel towards reactive halogen species [8]. Giuseppe Zagotto and Marco Bortoli provided a perspective on the evolution of medicinal chemistry, which nowadays faces novel challenges in the context of precision medicine and advanced drug delivery [9]. This aspect was also approached by Karolina Wanat and Elżbieta Brzezińska, who studied the effects of protein binding on drug bioavailability by means of statistical methods related to molecular and chromatographic descriptors [10], and by Tsun-Thai Chai and colleagues, who predicted pharmacokinetic and pharmacodynamic properties of seafood paramyosins peptides though computational tools [11]. Hoang Thai Ha and colleagues presented a comprehensive study on the extraction, characterization, and evaluation of antioxidant activity of carrageenan from *Eucheuma gelatinae* [12]. Daniel Muñoz-Reyes and colleagues described a novel application for a known molecule, investigating the role of quercetin 3-*O*-glucuronide against cisplatin cytotoxicity in renal tubular cells [13]. Ivan Yu Torshin and colleagues provided novel insights on the use of a known therapeutic agent, as they reported their study on lithium salts with reduced toxicity as neuroprotective agents [14]. In the context of neuroprotection, Etimad Huwait, Dalal A. Al-Saedi, and Zeenat Mirza presented a combined in silico and in vitro study assessing the potential of fucoidan against atherosclerosis [15]. In their analytical chemistry-oriented contribution, Elena Alba Álvaro-Alonso focused their study on the investigation of physicochemical and microbiological of oral solutions of methadone in different storage conditions [16].

As a conclusive note as Guest Editors, we would like to sincerely thank all the authors for choosing our Special Issue to share the results of their research work, as well as the reviewers and the assistant editors for their valuable support.

Author Contributions: Conceptualization, G.R. and L.O.; Writing—Original Draft Preparation, G.R. and L.O.; Writing—Review & Editing, G.R. and L.O. All authors have read and agreed to the published version of the manuscript.

Funding: This research was funded by Università degli Studi di Brescia and Università degli Studi di Padova.

Conflicts of Interest: The authors declare no conflict of interest.

References

1. Ribaudo, G.; Bortoli, M.; Ongaro, A.; Oselladore, E.; Gianoncelli, A.; Zagotto, G.; Orian, L. Fluoxetine Scaffold to Design Tandem Molecular Antioxidants and Green Catalysts. *RSC Adv.* **2020**, *10*, 18583–18593. [CrossRef]
2. Muraro, C.; Polato, M.; Bortoli, M.; Aiolli, F.; Orian, L. Radical Scavenging Activity of Natural Antioxidants and Drugs: Development of a Combined Machine Learning and Quantum Chemistry Protocol. *J. Chem. Phys.* **2020**, *153*, 114117. [CrossRef] [PubMed]
3. Ribaudo, G.; Bortoli, M.; Witt, C.E.; Parke, B.; Mena, S.; Oselladore, E.; Zagotto, G.; Hashemi, P.; Orian, L. ROS-Scavenging Selenofluoxetine Derivatives Inhibit In Vivo Serotonin Reuptake. *ACS Omega* **2022**, *7*, 8314–8322. [CrossRef]
4. Ribaudo, G.; Bortoli, M.; Oselladore, E.; Ongaro, A.; Gianoncelli, A.; Zagotto, G.; Orian, L. Selenoxide Elimination Triggers Enamine Hydrolysis to Primary and Secondary Amines: A Combined Experimental and Theoretical Investigation. *Molecules* **2021**, *26*, 2770. [CrossRef] [PubMed]
5. Stefaniu, A.; Pirvu, L.; Albu, B.; Pintilie, L. Molecular Docking Study on Several Benzoic Acid Derivatives against SARS-CoV-2. *Molecules* **2020**, *25*, 5828. [CrossRef] [PubMed]
6. Elzupir, A.O. Molecular Docking and Dynamics Investigations for Identifying Potential Inhibitors of the 3-Chymotrypsin-like Protease of SARS-CoV-2: Repurposing of Approved Pyrimidonic Pharmaceuticals for COVID-19 Treatment. *Molecules* **2021**, *26*, 7458. [CrossRef]
7. Cuesta, S.A.; Meneses, L. The Role of Organic Small Molecules in Pain Management. *Molecules* **2021**, *26*, 4029. [CrossRef]
8. Krake, E.F.; Baumann, W. Selective Oxidation of Clopidogrel by Peroxymonosulfate (PMS) and Sodium Halide (NaX) System: An NMR Study. *Molecules* **2021**, *26*, 5921. [CrossRef] [PubMed]
9. Zagotto, G.; Bortoli, M. Drug Design: Where We Are and Future Prospects. *Molecules* **2021**, *26*, 7061. [CrossRef] [PubMed]
10. Wanat, K.; Brzezińska, E. Statistical Methods in the Study of Protein Binding and Its Relationship to Drug Bioavailability in Breast Milk. *Molecules* **2022**, *27*, 3441. [CrossRef] [PubMed]

11. Chai, T.-T.; Wong, C.C.-C.; Sabri, M.Z.; Wong, F.-C. Seafood Paramyosins as Sources of Anti-Angiotensin-Converting-Enzyme and Anti-Dipeptidyl-Peptidase Peptides after Gastrointestinal Digestion: A Cheminformatic Investigation. *Molecules* **2022**, *27*, 3864. [CrossRef] [PubMed]
12. Ha, H.T.; Cuong, D.X.; Thuy, L.H.; Thuan, P.T.; Tuyen, D.T.T.; Mo, V.T.; Dong, D.H. Carrageenan of Red Algae Eucheuma Gelatinae: Extraction, Antioxidant Activity, Rheology Characteristics, and Physicochemistry Characterization. *Molecules* **2022**, *27*, 1268. [CrossRef] [PubMed]
13. Muñoz-Reyes, D.; Casanova, A.G.; González-Paramás, A.M.; Martín, Á.; Santos-Buelga, C.; Morales, A.I.; López-Hernández, F.J.; Prieto, M. Protective Effect of Quercetin 3-O-Glucuronide against Cisplatin Cytotoxicity in Renal Tubular Cells. *Molecules* **2022**, *27*, 1319. [CrossRef] [PubMed]
14. Torshin, I.Y.; Gromova, O.A.; Ostrenko, K.S.; Filimonova, M.V.; Gogoleva, I.V.; Demidov, V.I.; Kalacheva, A.G. Lithium Ascorbate as a Promising Neuroprotector: Fundamental and Experimental Studies of an Organic Lithium Salt. *Molecules* **2022**, *27*, 2253. [CrossRef] [PubMed]
15. Huwait, E.; Al-Saedi, D.A.; Mirza, Z. Anti-Inflammatory Potential of Fucoidan for Atherosclerosis: In Silico and In Vitro Studies in THP-1 Cells. *Molecules* **2022**, *27*, 3197. [CrossRef] [PubMed]
16. Álvaro-Alonso, E.A.; Lorenzo, M.P.; Gonzalez-Prieto, A.; Izquierdo-García, E.; Escobar-Rodríguez, I.; Aguilar-Ros, A. Physico-chemical and Microbiological Stability of Two Oral Solutions of Methadone Hydrochloride 10 mg/mL. *Molecules* **2022**, *27*, 2812. [CrossRef] [PubMed]

Article

Seafood Paramyosins as Sources of Anti-Angiotensin-Converting-Enzyme and Anti-Dipeptidyl-Peptidase Peptides after Gastrointestinal Digestion: A Cheminformatic Investigation

Tsun-Thai Chai [1,2,*], Clara Chia-Ci Wong [1], Mohamad Zulkeflee Sabri [3] and Fai-Chu Wong [1,2]

[1] Department of Chemical Science, Faculty of Science, Universiti Tunku Abdul Rahman, Kampar 31900, Malaysia; clara2000genesis@1utar.my (C.C.-C.W.); wongfc@utar.edu.my (F.-C.W.)
[2] Center for Agriculture and Food Research, Universiti Tunku Abdul Rahman, Kampar 31900, Malaysia
[3] Green Chemistry and Sustainable Technology Cluster, Bioengineering Section, Malaysian Institute of Chemical and Bioengineering Technology, Universiti Kuala Lumpur, Lot 1988, Bandar Vendor Taboh Naning, Alor Gajah 78000, Malaysia; mzulkeflee@unikl.edu.my
* Correspondence: chaitt@utar.edu.my; Tel.: +60-5-468-8888

Abstract: Paramyosins, muscle proteins occurring exclusively in invertebrates, are abundant in seafoods. The potential of seafood paramyosins (SP) as sources of anti-angiotensin-converting-enzyme (ACE) and anti-dipeptidyl-peptidase (DPP-IV) peptides is underexplored. This in silico study investigated the release of anti-ACE and anti-DPP-IV peptides from SP after gastrointestinal (GI) digestion. We focused on SP of the common octopus, Humboldt squid, Japanese abalone, Japanese scallop, Mediterranean mussel, Pacific oyster, sea cucumber, and Whiteleg shrimp. SP protein sequences were digested on BIOPEP-UWM, followed by identification of known anti-ACE and anti-DPP-IV peptides liberated. Upon screening for high-GI-absorption, non-allergenicity, and non-toxicity, shortlisted peptides were analyzed via molecular docking and dynamic to elucidate mechanisms of interactions with ACE and DPP-IV. Potential novel anti-ACE and anti-DPP-IV peptides were predicted by SwissTargetPrediction. Physicochemical and pharmacokinetics of peptides were predicted with SwissADME. GI digestion liberated 2853 fragments from SP. This comprised 26 known anti-ACE and 53 anti-DPP-IV peptides exhibiting high-GI-absorption, non-allergenicity, and non-toxicity. SwissTargetPrediction predicted three putative anti-ACE (GIL, DL, AK) and one putative anti-DPP-IV (IAL) peptides. Molecular docking found most of the anti-ACE peptides may be non-competitive inhibitors, whereas all anti-DPP-IV peptides likely competitive inhibitors. Twenty-five nanoseconds molecular dynamics simulation suggests the stability of these screened peptides, including the three predicted anti-ACE and one predicted anti-DPP-IV peptides. Seven dipeptides resembling approved oral-bioavailable peptide drugs in physicochemical and pharmacokinetic properties were revealed: AY, CF, EF, TF, TY, VF, and VY. In conclusion, our study presented in silico evidence for SP being a promising source of bioavailable and safe anti-ACE and anti-DPP-IV peptides following GI digestions.

Keywords: anti-ACE; anti-DPP-IV; gastrointestinal digestion; in silico; molecular docking; molecular dynamics; paramyosin; pharmacokinetics; seafood; target fishing

Citation: Chai, T.-T.; Wong, C.C.-C.; Sabri, M.Z.; Wong, F.-C. Seafood Paramyosins as Sources of Anti-Angiotensin-Converting-Enzyme and Anti-Dipeptidyl-Peptidase Peptides after Gastrointestinal Digestion: A Cheminformatic Investigation. *Molecules* **2022**, *27*, 3864. https://doi.org/10.3390/molecules27123864

Academic Editors: Giovanni Ribaudo, Laura Orian and Chojiro Kojima

Received: 9 May 2022
Accepted: 14 June 2022
Published: 16 June 2022

Publisher's Note: MDPI stays neutral with regard to jurisdictional claims in published maps and institutional affiliations.

Copyright: © 2022 by the authors. Licensee MDPI, Basel, Switzerland. This article is an open access article distributed under the terms and conditions of the Creative Commons Attribution (CC BY) license (https://creativecommons.org/licenses/by/4.0/).

1. Introduction

Bioactive peptides, especially those derived from dietary sources, are short fragments of food proteins that exert physiologically relevant activities. Such peptides, frequently 2–20 residues in length, could be liberated from food proteins by means of chemical or enzymatic hydrolysis, microbial fermentation, but also naturally in vivo during gastrointestinal (GI) digestion. The past ten years have seen a drastic surge in research exploring food-derived bioactive peptides. Such investigations have led to the discovery of numerous peptides that exert diverse bioactivities, encompassing antihypertension, antidiabetic, antioxidant, and anticancer activities. A key driver behind such intensive explorations is the recognition that that such peptides could have potential applications as nutraceuticals/functional food ingredients and therapeutic/prophylactic agents [1–4].

Traditionally, bioactive peptide discovery is mainly driven by wet-lab research, often involving the time-consuming process of isolating proteins from chosen samples, release of peptides from food proteins, bioactivity-guided purification of protein hydrolysates, mass spectrometric identification of peptides, synthesis of peptides, and lastly, validation of peptide bioactivity [1,5]. However, the in silico approach is increasingly embraced by researchers in bioactive peptide discovery due to its low cost and efficiency in peptide screening. Some studies have focused on only an in silico approach; others have integrated in silico analysis into their wet-lab experimentations. The toolbox for in silico bioactive peptide discovery encompasses, among others, various online servers, cheminformatics tools, simulation and visualization software, and bioactive peptide databases [6,7]. In this computational study, we adopted the in silico approach to screen for anti-angiotensin-converting-enzyme (ACE) and anti-dipeptidyl peptidase IV (DPP-IV) peptides released from seafood paramyosins following in silico GI digestion. ACE is a key player in the renin-angiotensin system, a pathway for the regulation of blood pressure in vivo. ACE inhibitors (e.g., Captopril) can help maintain normal blood pressure and thus can be used as antihypertensive drugs [8,9]. On the other hand, DPP-IV inhibitors improve the control of blood sugar levels in type 2 diabetes mellitus [10]. Inhibitors of DPP-IV (e.g., Anagliptin) can be used as oral antidiabetic drugs [11]. In this study, we also attempted to screen for bifunctional peptides exhibiting both anti-ACE and anti-DPP-IV activities. Such bifunctional peptides are valuable particularly in addressing complex pathological conditions (e.g., co-occurrence of high blood pressure in patients experiencing type 2 diabetes mellitus) [12].

Paramyosins are muscle proteins that occur exclusively in invertebrates, absent in vertebrate muscles. Paramyosins are enriched with about 20% glutamic acid residues. Paramyosin contents in scallop, squid, and oysters are 3, 14, and 19%, respectively. Notably, in the white adductor muscle of oysters and clams, paramyosins may comprise 38–48% of the total myofibrillar protein [13]. Despite their uniqueness and abundance in seafood invertebrates, there is very little information about seafood paramyosins as sources of bioactive peptides. A recent in silico investigation on the Portuguese oyster (*Crassostrea angulata*) found that paramyosin isoform X2 of the species could be a source of hundreds of anti-ACE (294) and anti-DPP-IV (517) peptides [14]. Thus, we hypothesized that other seafood paramyosins may also be sources of anti-ACE and anti-DPP-IV. In this in silico study, we focused on the paramyosins of eight species: the common octopus, Humboldt squid, Japanese abalone, Japanese scallop, Mediterranean mussel, Pacific oyster, sea cucumber, and Whiteleg shrimp, which are widely consumed worldwide. By virtually screening for anti-ACE and anti-DPP-IV peptides liberated from the paramyosins, we aimed to not only fill gaps of knowledge in the literature. Importantly, promising paramyosins that can be prioritized in future research as sources of nutraceuticals/drug candidates targeting hypertension and diabetes would be pinpointed. Mechanistic information on peptide-enzyme interactions as well as pharmacokinetics and drug-likeness of the peptides would also be explored.

2. Results and Discussion
2.1. Seafood Paramyosins

Nine paramyosin protein sequences were retrieved from UniProtKB (Table 1). One paramyosin sequence was found for each seafood species, except for the common octopus (CO), for which two isoforms (CO-X1 and CO-X2) were found. The paramyosins of the seafoods ranged from 516 residues (CO) to 934 residues (Japanese scallop, JS). Similarly, paramyosin isoform X2 of the common octopus (CO-X2) has the smallest molecular mass (59 kDa), whereas paramyosin of JS has the largest (107.5 kDa).

Table 1. Length and molecular masses of paramyosin proteins of eight seafood species.

Seafood	Accession Number	Number of Residues	Mass (Da)
Common octopus (CO-X1)	A0A6P7TIV8 (isoform X1)	523	59,847
Common octopus (CO-X2)	A0A7E6FQ28 (isoform X2)	516	59,026
Humboldt squid (HS)	A0A1Y1DCG9	880	102,476
Japanese abalone (JA)	A0A286QYA2	860	99,648
Japanese scallop (JS)	A0A210R0B2	934	107,548
Mediterranean mussel (MM)	O96064	864	99,573
Pacific oyster (PO)	K1QTC1	851	97,876
Sea cucumber (SC)	A0A2G8LGY5	727	83,851
Whiteleg shrimp (WS)	A0A3R7QCP1	828	96,537

2.2. In Silico GI Digestion

The in silico GI digestion of the nine paramyosins in Table 1 resulted in the release of 2853 peptide fragments. The outcome of the in silico hydrolysis is presented in Figure 1. Among the 2853 fragments liberated by the nine paramyosins, 1706 of them comprised two or more residues. In this study, we paid special attention to short peptides of several residues rather than the free amino acids since digested proteins are absorbed predominantly in the form of di- and tri-peptides, rather than individual amino acids [15–18]. More than 300 peptide fragments were liberated from each protein, except for the two paramyosin isoforms of the common octopus (CO-X1; CO-X2). The paramyosin of JS, which has the largest number of residues (Table 1), liberated the largest number of fragments (367). CO-X2, the shortest among the nine paramyosins, released the lowest number of fragments (223). With the exceptions of CO-X1 and CO-X2, the other seafood paramyosins each potentially liberated more than 100 peptide fragments collectively as di- and tripeptides. Numerous such short peptides are known for ready uptake by the human intestinal cells, a process mediated by PepT1, a proton-coupled oligopeptide cotransporter [19]. Thus, whether such peptides exhibit any health-promoting effects, particularly anti-ACE and anti-DPP-IV activities, is of great interest. Dipeptides consistently formed the major group of short peptide fragments released from the nine paramyosins, ranging from 27.3% in JS to 22.4% in CO-X2. The paramyosin of JS also released the largest number of dipeptides (100) following in silico GI digestion, whereas CO-X2 released the fewest (50) (Figure 1). In contrast to the dipeptide group, peptide fragments > 4 residues long formed only 8.3% to 16.5% of the total pool of fragments released from the seafood paramyosins. The longest peptide fragment released was an 18-residue peptide originating from JS (Data not shown). Overall, our observations agree with that previously reported [20] where more peptide fragments were liberated from housefly larval proteins of larger peptide lengths. Our results also suggests that among the nine seafood paramyosins, the one from JS likely has the greatest number of pepsin-, trypsin-, and chymotrypsin cleavage sites in its sequence.

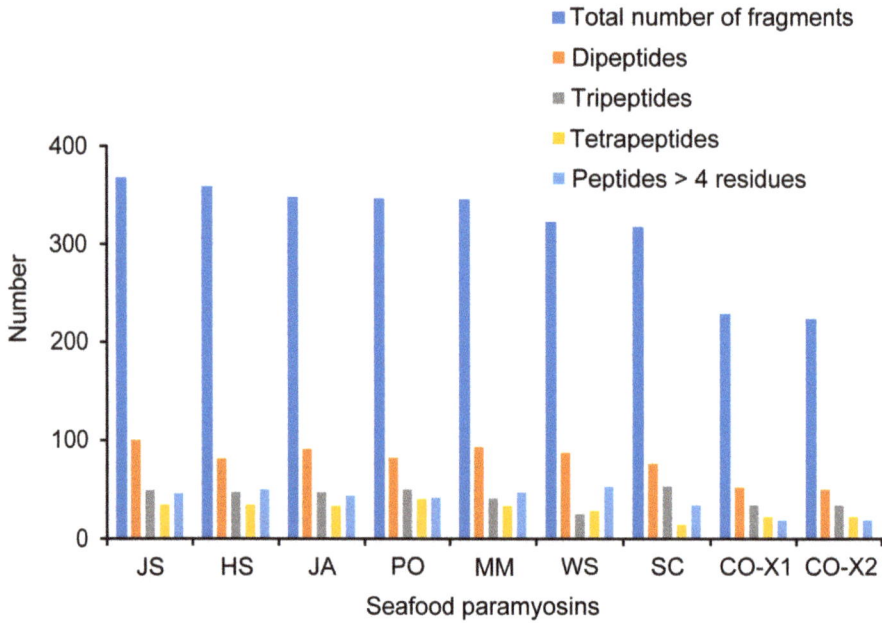

Figure 1. Distribution of peptides of different lengths released by in silico GI digestion of seafood paramyosins. An individual amino acid released from in silico GI digestion was counted as one fragment.

2.3. Screening for Anti-ACE and Anti-DPP-IV Peptides

A search against the BIOPEP-UWM database where experimentally validated bioactive peptides were deposited [21] revealed that 92 and 174 seafood paramyosin-derived peptides are known anti-ACE and anti-DPP-IV peptides, respectively. This implies that when ingested and digested in the GI tract, the seafood paramyosins are better as sources of anti-DPP-IV peptides than as sources of anti-ACE peptides. Nevertheless, not all such peptides are promising bioavailable anti-ACE and anti-DPP-IV agents. For example, ASL, ITF, and IVR are three known anti-ACE peptides released from the paramyosins following in silico GI digestion. The three peptides were predicted as having only low GI absorption by SwissADME (data not shown). Furthermore, molluscan paramyosins such as those from the common octopus and the Mediterranean mussel have been connected to food allergies [22]. Therefore, whether the peptide fragments released from the paramyosins following GI digestion is safe and easily absorbed by the GI tract is a pertinent issue. Thus, to better explore the potential of the nine seafood paramyosins as sources of orally available anti-ACE and anti-DPP-IV agents, we virtually screened the 266 known anti-ACE and anti-DPP-IV peptides for high GI absorption, non-allergenicity and non-toxicity.

2.4. Screening for High GI Absorption, Non-Allergenicity, and Non-Toxicity

Twenty-six of the ninety-two known anti-ACE peptides (28.3%) liberated from the nine paramyosins through in silico GI digestion are potentially highly-absorbed by the human GI tract, non-allergenic, and non-toxic (Table 2). By contrast, 53 of the 174 known anti-DPP-IV peptides (30.5%) released from the paramyosins were likely to exhibit high GI absorption, non-allergenicity, and non-toxicity. By comparison, the nine paramyosins are a more promising source of potentially bioavailable and safe anti-DPP-IV peptides versus anti-ACE peptides. As shown in Table 2, WS was the most promising source of high-GI-absorption, non-allergenic, and non-toxic anti-DPP-IV peptides. At the individual paramyosin's level, repeated sequences are found. For example, the 12 WS-derived anti-DPP-IV peptides can be consolidated into six unique sequences (IL, SL, TF, TY, VL, and VY). Across the different seafood species, identical dipeptides were also found, suggesting sequence similarity between the paramyosins. VY, a known anti-ACE peptide [23], was found in seafood paramyosins of five species (i.e., SC, WS, JA, MM, and PO). On the other hand, SL, a known anti-DPP-IV peptide [24], was found in seafood paramyosins of seven species, except for SC. Meanwhile, identical sets of anti-ACE and anti-DPP-IV peptides were found for both CO-X1 and CO-X2 as well as for both MM and PO (Table 2). Interestingly, some dipeptides in Table 2 (i.e., AY, IL, TF, VF, and VY) are bi-functional peptides. For instance, VY derived from five species (SC, WS, JA, MM and PO) has been reported to have both anti-ACE [23] and anti-DPP-IV [25] activity. One peptide sequence having bifunctional anti-ACE/anti-DPP-IV activities (VF) was released by HS paramyosin, whereas three bifunctional peptide sequences were found for WS (IL, TF, and VY) and JA (AY, IL, and VY). Taken together, when the number of peptides and bifunctionality are considered, the WS paramyosin was the most outstanding. It liberated three anti-ACE and 12 anti-DPP-IV peptides predicted as high-GI-absorbable, non-allergenic and non-toxic, among which three were bifunctional anti-ACE/anti-DPP-IV peptides. It should be noted that although other seafood proteins may also release similar high-GI-absorbable, non-allergenic and non-toxic anti-ACE and anti-DPP-IV peptides as paramyosins following GI digestions, we hypothesize based on our in silico scientific evidence that any anti-ACE and anti-DPP-IV effects of consumed seafoods could be attributed, at least partly, to paramyosins.

Table 2. The numbers of high-GI-absorption, non-allergenic, and non-toxic peptides with known anti-ACE and anti-DPP-IV activities.

Seafood	Anti-ACE Peptides		Anti-DPP-IV Peptides	
	Number	Unique Sequences [a]	Number	Unique Sequences [a]
CO-X1	5	AY, CF, EF, VF	6	AY, SL, VF, VL
SC	5	EF, GM, IL, VY	4	IL, VL, VY
CO-X2	4	AY, CF, EF, VF	6	AY, SL, VF, VL
WS	3	IL, TF, VY	12	IL, SL, TF, TY, VL, VY
JA	3	AY, IL, VY	5	AY, IL, SL, VY
MM	2	TF, VY	5	SL, TF, TY, VY
PO	2	TF, VY	5	SL, TF, TY, VY
JS	1	TF	6	SL, TF, TY, VL
HS	1	VF	4	SL, TY, VF
Total	26		53	

[a] Bifunctional dipeptides with both anti-ACE and anti-DPP-IV activities are underlined.

2.5. Predicting Anti-ACE and Anti-DPP-IV Peptides with SwissTargetPrediction

To further explore the possible presence of paramyosin-derived peptides which are novel, or not documented in the BIOPEP-UWM database, we adopted a target-fishing strategy. All paramyosin-derived peptides (2587) not recognized as anti-ACE and anti-DPP-IV peptides by BIOPEP-UWM were first screened for high GI absorption, non-allergenicity, and non-toxicity. This reduced the number of peptides to 64. Altogether, they can be narrowed down to four unique sequences: AK, DL, GIL and IAL (Table 3). Among the nine paramyosins investigated, the paramyosin of JA was the one from which these four high-GI-absorption, non-allergenic, and non-toxic peptides without any known anti-ACE and anti-DPP-IV activities can be found. On the other hand, AK can be consistently found in all nine paramyosins. DL can be found from all paramyosins, except for that of SC. This set of four peptides were analyzed with SwissTargetPrediction tool to predict potential anti-ACE and anti-DPP-IV peptides, based on their structural similarity to drug or compounds known to be ligands to ACE and to DPP-IV. This step led to the discovery of three putative anti-ACE peptides (GIL, DL, AK) and one putative anti-DPP-IV peptide (IAL) (Table 4).

Table 3. The numbers of high-GI-absorption, non-allergenic, and non-toxic peptides without known anti-ACE and anti-DPP-IV activities.

Seafood	Number	Unique Sequences
JA	12	AK, DL, GIL, IAL
JS	10	AK, DL
PO	9	AK, DL
WS	8	AK, DL
MM	8	AK, DL
SC	6	AK, IAL
HS	5	AK, DL
CO-X1	3	AK, DL
CO-X2	3	AK, DL
Total	64	

Table 4. Peptide sequences having ACE or DPP-IV as potential target as predicted by SwissTargetPrediction.

Peptide	Potential Target	Probability	Known Actives (3D/2D)	ChEMBL ID of Known Active Compound with Top Similarity to Peptide */IC$_{50}$
GIL	ACE	0.5345	167/189	CHEMBL128399/4200 nM
DL	ACE	0.0580	33/130	CHEMBL358439/2400 nM
AK	ACE	0.0524	2/183	CHEMBL430554/7 nM
IAL	DPP-IV	0.5776	167/362	CHEMBL214381/2530 nM

* Based on 3D structure comparison to known anti-ACE/anti-DPP-IV compounds stored in ChEMBLE database.

2.6. Molecular Docking

In order to clarify the mechanisms of interactions between the known/predicted anti-ACE/anti-DPP-IV peptides and their target enzymes, we analyzed them by performing molecular docking. Overall, the 26 known anti-ACE peptides liberated from the paramyosins through in silico GI digestion (Table 2) can be narrowed down to eight unique sequences of anti-ACE peptides: AY, CF, EF, GM, IL, TF, VF, and VY. While the anti-ACE activity of the eight peptides were previously demonstrated, their mechanisms of inhibition have not been elucidated for all of them. Molecular docking represents a fast and economical in silico tools to clarify the potential mechanisms of action of the eight peptides side-by-side in the same study. On the other hand, comparison of the ACE-binding modes of the predicted anti-ACE peptides with those of the eight known anti-ACE peptides may also provide hints on the former's anti-ACE potential. As shown in Table 5, the eight known anti-ACE peptides ranged between −112.800 (VY) to −75.728 (GM) in their docking scores. These scores are clearly inferior to the score of bradykinin potentiating peptide b (BPPb) (−376.180), which is the co-crystallized inhibitor in the human ACE crystal 4APJ. Underlying these weaker scores in the eight anti-ACE peptides could be their fewer interactions with ACE, in contrast with those formed between BPPb and ACE. Our results suggest that even at binding stability weaker than of BPPb, it is still possible for a peptide to be an effective ACE inhibitor. Based on comparison with the known anti-ACE peptides, it could be deduced that among the three predicted peptides, GIL was likely a potential anti-ACE peptide. This is due to the fact that it could bind to ACE with the strongest binding stability, which also falls within the score range exhibited by the eight known anti-ACE peptides.

Table 5. Docking scores and intermolecular interactions between ACE and known/predicted anti-ACE peptides.

	Peptide	Docking Score	Interaction with ACE [b,c]		
			Hydrogen Bond	Hydrophobic Interaction	Salt Bridge
	BPPb [a]	−376.180	Lys118, Asp121, Gln281, Ala356(2), Tyr360, Glu403, Lys511, His513, Ser516, Ser517, Tyr520, Tyr523	Trp59, Ile88, Lys118, Asp121, Glu123, Gln281, His353, Ala354, Ser355, Ala356, Trp357, Tyr360, His387, Glu403, Glu411, Phe457, Lys511, Phe512, His513, Ser516, Ser517, Val518, Tyr520, Tyr523	Glu403
Indicated by BIOPEP-UWM	VY	−112.800	Glu123	Tyr51, Trp59, Tyr62, Ala63, Ile88, Lys118, Glu123, Tyr360	-
	CF	−108.762	Tyr62, Leu122, Glu123, Ala125	Trp59, Tyr62, Thr92, Glu123, Arg124, Ala125, Tyr360	-
	AY	−108.695	Glu123, Arg124, Tyr135, Asn211, Ser517	Glu123, Arg124, Tyr135, Leu139, Ile204, Ala207, Ala208, Ser219, Trp220, Ser517	-
	VF	−107.589	Glu123	Trp59, Tyr62, Ile88, Thr92, Leu122, Glu123, Arg124, Tyr360	-
	TF	−103.827	Tyr51, Glu123	Tyr51, Trp59, Ile88, His91, Thr92, Lys118, Asp121, Glu123	-
	EF	−103.021	Glu123, Arg124, Tyr135	Glu123, Arg124, Tyr135, Leu139, Ile204, Ala207, Ser219, Trp220, Ser517, Val518, Pro519, Arg522	Arg522(4)
	IL	−79.044	Tyr62, Asn85	Trp59, Tyr62, Asn85, Ile88, Ala89, Arg124, Leu132	-
	GM	−75.728	Tyr146, Phe512	Tyr146, Leu161, Glu162, Trp279, His353, Lys511, Phe512, His513	-
Predicted by Swiss Target Prediction	GIL	−103.475	Asp121, Glu123	Trp59, Tyr62, Ile88, Ala89, Thr92, Asp121, Leu122, Glu123, Arg124, Ala125	-
	AK	−64.629	Glu162, Lys511(2), His513	Tyr146, Leu161, Glu162, Trp279, Gln281, His353, Lys511, Phe512, His513	-
	DL	−60.501	Tyr62, Glu123, Arg124	Tyr62, Asn85, Ile88, Ala89, Glu123, Arg124	Arg124

[a] Bradykinin potentiating peptide b, co-crystalized inhibitor of ACE in 4APJ crystal. [b] Residues in the active site of ACE are underlined. [c] The number in bracket indicates the number of hydrogen bonds or salt bridges formed with the same residues of ACE.

The active site of the ACE enzyme includes the inhibitor binding site (His383, His387, and Glu411), the S1 pocket (Ala354, Glu384, and Tyr523), the S2 pocket (Gln281, His353, Lys511, His513, and Tyr520), and S1' (Glu162). An inhibitor that binds to ACE through other than the aforementioned active site is a non-competitive inhibitor [26]. Thus, our results in Table 5 implies that the eight known anti-ACE peptides derived from seafood paramyosins are mostly non-competitive ACE inhibitors (i.e., AY, CF, EF, IL, TF, VF, and VY). The role of VY as a non-competitive inhibitor of ACE was reported [27]. Our findings imply that the putatively high-GI-absorption, non-allergenic, and non-toxic anti-ACE peptides in all paramyosin sources, except SC, are all putatively non-competitive ACE inhibitors. By contrast, a combination of competitive (GM) and non-competitive (EF, IL, and VY) peptides could be derived from SC. Among the three anti-ACE peptides predicted by SwissTargetPrediction tool, two (GIL and DL) are possible non-competitive ACE inhibitors. Taken together, this prevalence of possible non-competitive ACE inhibitors is not unusual. In our previous in silico investigation of anti-ACE peptides from calpain 2-digested silkworm cocoon proteins, all four shortlisted dipeptides (AF, IL, PG and AG) were also deduced to be non-competitive ACE inhibitors [28].

The 53 known anti-DPP-IV peptides generated from the paramyosins through in silico GI digestion (Table 2) can be consolidated into eight unique sequences: AY, IL, SL, TF, TY, VF, VL, and VY. As shown in Table 4, IAL was the only putative anti-DPP-IV peptide predicted by SwissTargetPrediction. As presented in Table 6, the docking scores of TF, TY, VY and VF (-134.788 to -122.342) are clearly superior to the score of diprotin A (-115.228), which is the co-crystallized peptide inhibitor in the human DPP-IV crystal 1WCY. The stronger scores of the four anti-DPP-IV peptides may be partly accounted by their more frequent interactions with the active site residues of DPP-IV, in comparison with diprotin A. The four dipeptides formed 9–12 hydrophobic interactions with the residues in the DPP-IV active site, whereas diprotin A formed only seven. The active site of DPP-IV consists of a catalytic triad (Ser630, Asn710, and His740), a hydrophobic S1 pocket (Tyr631, Val656, Trp659, Tyr662, Tyr666, Val711), and a S2 pocket (Arg125, Glu205, Glu206, Ser209, Phe357, Arg358) [29]. Our molecular docking revealed that the eight known and one predicted anti-DPP-IV peptides could bind to at least one residue (Ser630) of the catalytic triad through hydrophobic interactions (Table 6). TF, TY, and AY were predicted to engage all three residues in the catalytic triad. Five known (TF, TY, AY, and IL) and one predicted (IAL) anti-DPP-IV peptides was also predicted to interact with His740 by hydrogen bonds. By contrast, diprotin A only bound to Ser630 of the DPP-IV catalytic triad via hydrophobic interaction (Table 6). Based on their modes of binding to the active site of DPP-IV, the paramyosin-derived peptides listed in Table 6 are all potentially competitive inhibitors. Our interpretation agrees with a previous report of SL being a competitive inhibitor of DPP-IV [30].

2.7. Molecular Dynamics

To further dissect the dynamics of interactions between the aforementioned anti-ACE and anti-DPP-IV peptides and their respective target proteins, we have performed 25 ns molecular dynamics (MD) simulations on selected anti-ACE and anti-DPP-IV peptides from both BIOPEP-UWM and SwissTargetPrediction results. Structural parameters RMSD, Radius of gyration (Rg), intermolecular H-bonds and protein-ligand distance were examined to determine the stability, dynamical behavior, and the compactness of the protein-ligand complexes.

Table 6. Docking scores and intermolecular interactions between DPP-IV and known/predicted anti-DPP-IV peptides.

	Peptide	Docking Score	Interaction with DPP-IV [b]	
			Hydrogen Bond [c]	Hydrophobic Interaction
	Diprotin A [a]	−115.228	Arg125(2), Glu205, Glu206(2), Tyr547, Tyr662	Arg125, Glu205, Glu206, Phe357, Tyr547, Ser630, Tyr631, Tyr662, Tyr666
Indicated by BIOPEP-UWM	TF	−134.788	Glu205(2), Glu206, Tyr662, His740	Arg125, Glu205, Glu206, Tyr547, Ser630, Tyr631, Val656, Trp659, Tyr662, Tyr666, Asn710, Val711, His740
	TY	−130.756	Glu205(2), Glu206, Tyr662, His740	Arg125, Glu205, Glu206, Tyr547, Ser630, Tyr631, Val656, Trp659, Tyr662, Tyr666, Asn710, Val711, His740
	VY	−125.108	Arg125, Glu205(2), Tyr547	Arg125, Glu205, Glu206, Ser209, Tyr547, Ser630, Tyr631, Val656, Trp659, Tyr662, Tyr666
	VF	−122.342	Glu205(2), Glu206	Arg125, Glu205, Glu206, Phe357, Tyr547, Ser630, Tyr631, Tyr662, Tyr666, Asn710
	AY	−114.150	Arg125, Tyr547, Ser630, His740	Arg125, Tyr547, Ser630, Tyr631, Val656, Trp659, Tyr662, Tyr666, Asn710, Val711, His740
	IL	−86.409	Glu205, Glu206, Tyr547, Ser630, His740	Glu206, Phe357, Tyr547, Ser630, Tyr631, Val656, Trp659, Tyr662, Tyr666, His740
	SL	−85.505	Glu205, Glu206(2), Tyr547, Tyr631, Tyr662(2)	Arg125, Glu205, Glu206, Tyr547, Ser630, Tyr631, Val656, Trp659, Tyr662, Tyr666, Val711
	VL	−84.356	Arg125, Glu205(3)	Arg125, Glu205, Glu206, Tyr547, Ser630, Tyr631, Tyr662, Tyr666, Arg669
Predicted by Swiss Target Prediction	IAL	−109.567	Arg125, Glu205, Glu206, Tyr547, Tyr662, His740	Arg125, Glu205, Glu206, Tyr547, Trp629, Ser630, Tyr662, Tyr666, Val711, His740

[a] Bound ligand of DPP-IV in the crystal (PDB ID: 1WCY). [b] Residues in the active site of DPP-IV are underlined. [c] The number in the brackets indicates the number of hydrogen bonds formed with the same residues of DPP-IV.

2.7.1. Root Mean Square Deviation

The root mean square deviation (RMSD) of the all-atom protein structure and the peptide ligand was employed to analyze each protein and ligand stability in complex. The mean RMSD value of protein-ligand that is lower than of the BPPb for ACE, and diprotin A for DPP-IV indicates a stable complex formation [31]. For ACE inhibitory peptide, VY and AK from BIOPEP-UWM and GIL from SwissTargetPrediction were subjected to 25 ns MD. The mean RMSD value of ACE complexed with BPPb was 1.75 ± 0.16 Å, while for VY, AK and GIL were 1.88 ± 0.19 Å, 1.96 ± 0.23 Å, and 1.81 ± 0.15 Å, respectively. While the mean RMSD value of complexed-ACE for the selected peptides were slightly higher than the BPPb, Figure 2a exhibit the overall stability of the RMSD for the whole 25 ns duration. The ACE-VY complex RMSD value was slightly fluctuated early at 7 ns, while for ACE-AK complex the value increased during 16–20 ns. The mean RMSD of free ACE protein at 1.88 ± 0.20 Å was higher than of the BPPb, VY and GIL complexes, and its graph shows gradual increase during the 25 ns course (Figure 2a). The low RMSD of protein-ligand complex in comparison with free protein also indicates a stable protein-ligand complex formation [32].

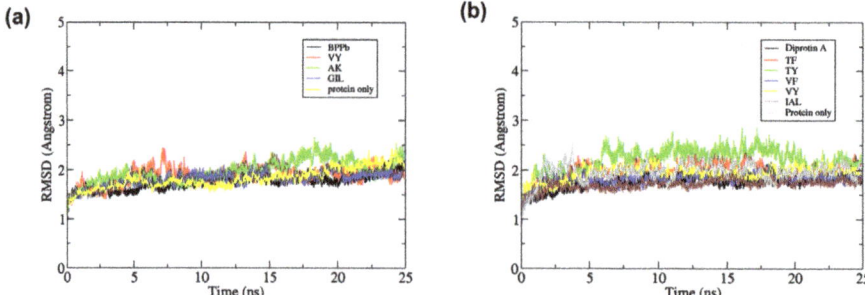

Figure 2. All-atom root mean square deviation (RMSD) of (a) Free ACE and ACE-peptide complexes (b) Free DPP-IV and DPP-IV-peptide complexes during 25 ns of the molecular dynamics simulation.

Similarly, the RMSD analysis for anti DPP-IV and peptide complexes was also conducted. Here, the inhibitory peptide candidates TF, TY, VF, VY, and IAL which form complexes with DPP-IV, and DPP-IV- diprotin A complex were subjected to 25 ns MD simulation. The all-atom protein mean RMSD value for DPP-IV-diprotin A complex was lower (1.76 ± 0.15 Å) compared to all dipeptides subjected in the study (TF = 2.00 ± 0.15 Å, TY = 2.20 ± 0.25 Å, VF = 1.82 ± 0.12 Å, VY = 1.93 ± 0.15 Å). However, the tripeptide IAL (1.71 ± 0.11 Å) showed a considerably low mean RMSD value against Diprotin A, which suggested a stable protein-ligand binding [33]. In comparison, the mean RMSD of free DPP-IV was higher at 1.96 ± 0.19 Å. The all-atom protein RMSD plotted in Figure 2b shows DPP-IV that formed complexes with each diprotin A and IAL were stable during 25 ns compared to free DPP-IV which was highly fluctuated, especially at the few first ns of the simulation. In addition, DPP-IV-TY complex gave a high all-atom protein RMSD fluctuation during 25 ns, even though it had a lower docking score predicted by the BIOPEP-UWM server compared to diprotin A.

To ensure the binding stability of inhibitory peptides in the active site of ACE and DPP-IV, the ligand positional all-atom RMSD was also calculated. This is also important as from the docking result, the screened peptides were suggested to be the non-competitive inhibitors as it binds to the site other than the inhibitor binding site, the S1 pocket, the S2 pocket and S1'. Figure 3a shows that each BPPb, VY, and AK gave a stable all-atom ligand RMSD when complexed with ACE protein with the mean RMSD value of 1.45 ± 0.24 Å, 1.65 ± 0.33 Å and 1.46 ± 0.17 Å, respectively. In comparison, GIL mean RMSD value was a magnitude lower at 0.89 ± 0.16 Å, although the plot shows that its RMSD fluctuated for the whole 25 ns. Therefore, five snapshots of ACE-peptide complexes were downloaded in the interval of 5 ns during the entire simulation, as in Figure 4a–d. It was observed that docked GIL had moved considerably at the active site during the intervals, while BPPb, VY and AK remained firmly bound at the active site of ACE. BPPb occupied the most active site regions due to its large decapeptide structure which gave a spatial space around the binding pockets. While VY and GIL ligand shared the similar binding region of alpha-helices domain where Glu123 and residues Tyr51, Trp59, Tyr62, and Ala63 were located, AK tends to bind on the different region where residues Glu162, Cys352 and Lys511 were located and highly interacted with the N-terminal and O-terminal of the peptide.

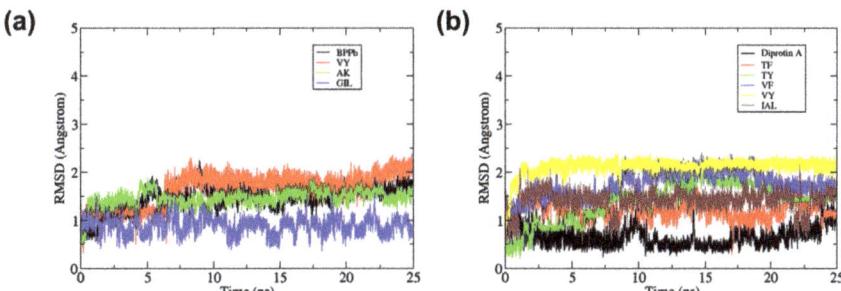

Figure 3. All-atom root mean square deviation (RMSD) of (**a**) ACE-docked peptide complexes (**b**) DPP-IV-docked peptide complexes during 25 ns of the molecular dynamics simulation.

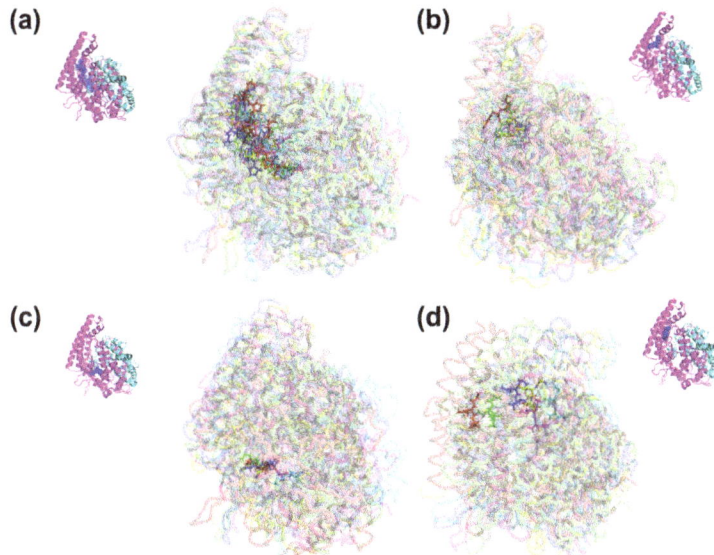

Figure 4. Snapshot of the superimposed structures of ACE in complex with peptide (**a**) BPPb, (**b**) VY, (**c**) AK and (**d**) GIL. Structures were obtained from the trajectory file in the interval of 5 ns for 25 ns MD.

As the GIL inhibitory peptide predicted from SwissTargetPrediction gave a lower ligand RMSD compared to the positive control peptide (BPPb) as the ACE inhibitor candidate, the same cannot be said for the peptide ligand against DPP-IV. DPP-IV peptide inhibitor diprotin A gave a relatively low ligand RMSD value as observed in the plot Figure 3b with the mean RMSD of 0.64 ± 0.19 Å, compared to the dipeptide inhibitors suggested by BIOPEP-UWM which gave a higher mean RMSD values (TF = 1.22 ± 0.17 Å, TY = 1.39 ± 0.41 Å, VF = 1.74 ± 0.28 Å and VY = 2.11 ± 0.18 Å). However, IAL peptide obtained from SwissTargetPrediction gave a considerably low ligand RMSD with the mean value of 1.41 ± 0.22 Å when compared to the dipeptides obtained from BIOPEP-UWM. The snapshots in Figure 5a–f shows that the docked region of diprotin A and other inhibitor peptides candidates shared the similar binding residues of Arg125, Glu205 and the pockets residues around Glu206, Tyr547, Trp629, Ser630, and His740.

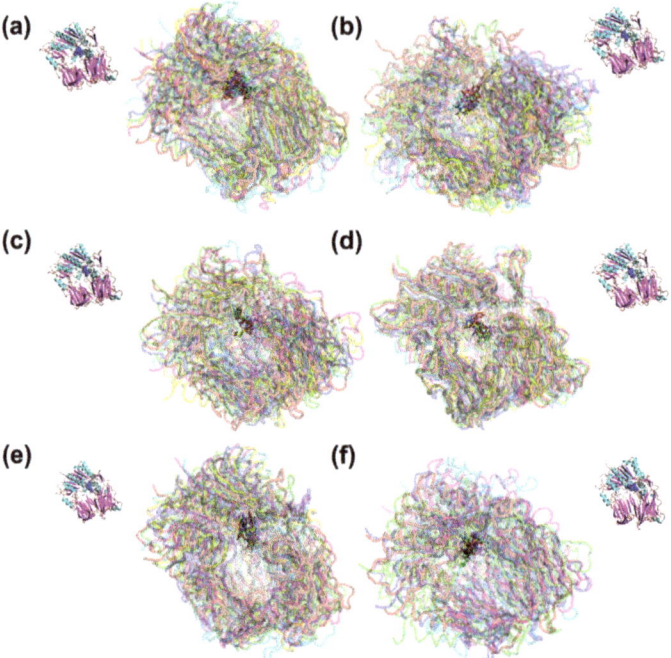

Figure 5. Snapshot of the superimposed structures of DPP-IV in complex with peptide (**a**) Diprotin A, (**b**) TF, (**c**) TY, (**d**) VF, (**e**) VY and (**f**) IAL. Structures were obtained from the trajectory file in the interval of 5 ns for 25 ns.

2.7.2. Radius of Gyration

Radius of gyration (Rg) measures the compactness of a protein which allows the understanding of protein folding properties [34]. While the Gromacs tool 'gmx_gyrate' is applicable to compute the protein radius of gyration, the equation can be written as below;

$$R_g^2 = \frac{1}{M} \sum_{i=1}^{N} m_i (r_i - R)^2 \qquad (1)$$

where $M = \sum_{i=1}^{N} m_i$ is the total mass, and $R = N^{-1} \sum_{i=1}^{N} r_i$ is the center of mass of the protein consisting of N atoms. A small Rg values shows that the protein is in a tight packing with a relatively steady value of Rg, while high Rg values indicate a floppy packing of protein with lack of compactness. In addition, a stable Rg value of protein-ligand complex during the time frame indicates that the ligand holds the folding behavior of protein whilst high Rg fluctuations might denotes the protein-ligand folding instability over time [35].

The free ACE protein (mean Rg = 2.42 nm) was shown to have similar compactness with the ACE-BPPb complex (mean Rg = 2.42 nm) and ACE-GIL (mean Rg = 2.42 nm), while the Rg value was slightly higher against two dipeptides VY (mean Rg = 2.44 nm) and AK (mean Rg = 2.43 nm) after 25 ns. The gyration of ACE-VY and ACE-AK complexes were observed to be fluctuated for the whole 25 ns compared to free ACE, as observed in Figure 6a. The gyration of ACE-GIL complex was shown to be more stable during the period. In comparison, free DPP-IV protein and each of its peptide ligand complex are slightly less compact, with the free DPP-IV Rg mean value of 2.72 nm, followed by diprotin A (mean Rg = 2.70 nm), TF (mean Rg = 2.72 nm), TY (mean Rg = 2.72 nm), VF (mean Rg = 2.71 nm), VY (mean Rg = 2.72 nm) and IAL (mean Rg = 2.70 nm), as observed in

Figure 6b. The high fluctuations of free DPP-IV and DPP-IV peptide complexes was mainly caused by higher number of residues in the protein and wider space for protein-ligand spatial interaction in the protein active site [36]. The Rg kept constant with no abrupt fluctuations through the time in all of these complexes, indicating that VY, AK, and GIL maintain the folding behavior as similarly as ACE-BPPb complex, while TF, TY, VF, VY, and IAL maintains DPP-IV folding as similarly from the complex formation with diprotin A (Figure 6).

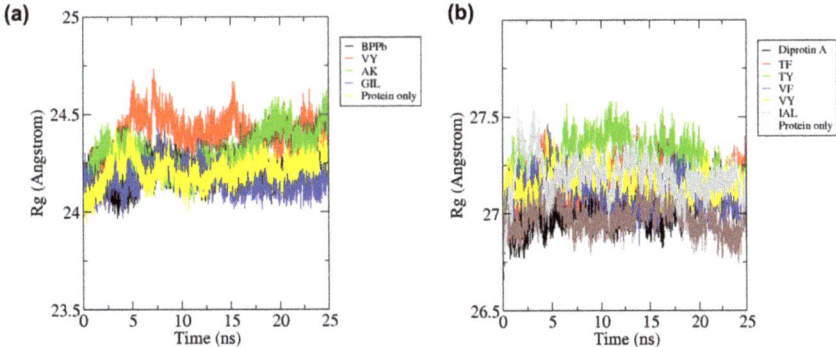

Figure 6. All-atom radius of gyration of (**a**) ACE and (**b**) DPP-IV as free proteins and forming complexes with the inhibitor peptides during 25 ns of the molecular dynamics simulation period.

2.7.3. Hydrogen Bonds and Protein-Ligand Distance

Hydrogen bonds (H-bonds) are non-covalent bonds that provide most of the directional interactions underneath the formation of secondary and tertiary structure protein motifs where it satisfies the hydrogen-bonding potential between carbonyl oxygen and amide nitrogen in the hydrophobic core of protein [37]. Close proximity of the polar atoms in protein and its ligand, with acceptor-donor distance between 2.0 and 2.5 Å and its geometric angle of less than 120° also provides a directionality and specificity of the H-bond interaction. In addition, it also explains the binding affinity of a ligand towards the protein target in the molecular dynamics simulation. Therefore, a higher number of intermolecular H-bonds can be translated to stronger interactions between the complex and smaller protein-ligand intermolecular distance [38].

Figure 7a shows that the ACE protein complexed with BPPb provides the highest number of intermolecular H-bonds with mean seven bonds, while AK and GIL provide the mean H-bonds of three bonds followed by VY with only one bond during 25 ns simulation. Number of H-bonds formed by ACE-BPPb complex was high while H-bonds between ACE-GIL complex seems to increase within 25 ns. Figure 7b shows that these were translated to the protein-ligand intermolecular distance where ACE-BPPb has the smallest distance of 2.0 Å, ACE-VY complex distance slowly increased to more than 3.0 Å after 15 ns and was highly fluctuated. On the other hand, ACE-GIL complex tends to stabilize after 10 ns and the distance decrease to less than 2.0 Å until the end of the duration.

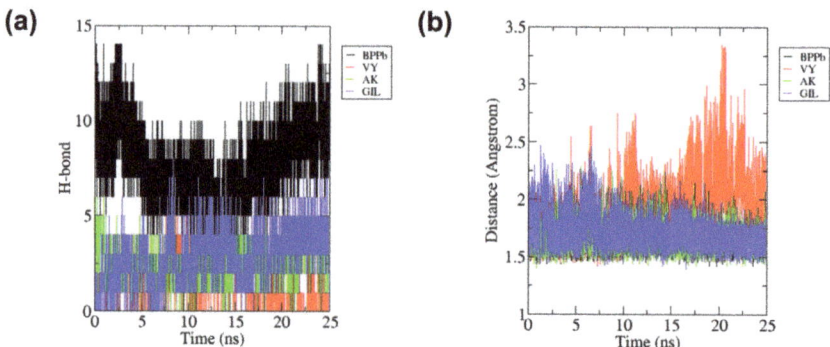

Figure 7. (a) Total number of hydrogen bonds interactions between ACE and each peptide and (b) intermolecular distance of ACE with peptides during 25 ns of the molecular dynamics simulation period.

Similarly, for anti-DPP-IV peptides, DPP-IV-diprotin A complex formed the most H-bonds with mean four bonds, followed by TF, TY and VY with the mean three bonds each (Figure 8a). IAL and DPP-IV complex has the lowest intermolecular H-bond with only one bond, where the amide group on the N-terminal of IAL bonded with either residues Glu205 or Glu206 of the DPP-IV during the simulation (not shown). Due to the pocketed position of the peptide ligand which is in the close proximity with the surrounded alpha-helices domains, the peptides were also supported by strong hydrophobic interactions and tend to stabilize. This also contributed to the intermolecular distance of protein-ligand for each complex were low and less than 2.5 Å each (Figure 8b).

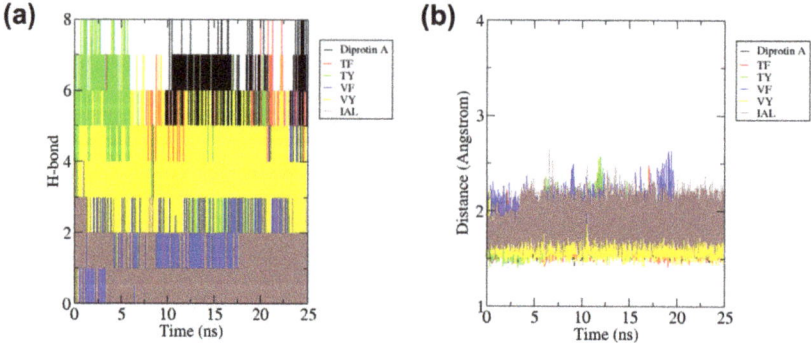

Figure 8. (a) Total number of hydrogen bonds interactions between DPP-IV and each peptide and (b) intermolecular distance of DPP-IV with peptides during 25 ns of the molecular dynamics simulation period.

2.8. SwissADME Analysis

SwissADME is a free web tool used in some in silico studies to assess the physicochemical properties, pharmacokinetics and drug-likeness of bioactive peptides [39,40]. Based on the predicted information, it may be possible to compare the peptides under investigation with approved peptide drugs [39]. Table 7 shows the physicochemical properties of the 15 unique sequences of high-GI-absorption, non-allergenic, non-toxic anti-ACE and/or anti-DPP-IV peptides derived from the seafood paramyosins. The physicochemical properties of the 15 peptides are generally comparable to those of Captopril and Anagliptin. Captopril is an ACE inhibitor used as an oral antihypertension drug [8]. Anagliptin is an anti-DPP-IV agent used as an oral antidiabetic drug [11]. The 15 peptides range between 217 g/mol (AK) and 315 g/mol (IAL) in molecular weights, not remarkably different from the two small-molecule drugs Captopril and Anagliptin. Meanwhile, the majority of the United States Food and Drug Administration (FDA)-approved, orally available peptide drugs have the following characteristics: fraction Csp3 up to 0.55; rotatable bonds (RB) up to 20; number of H-bond acceptors (HBA) up to 50; number of H-bond donors (HBD) up to 25; TPSA up to 400 Å^2; and lipophilicity between −5 and 8 [41]. It is clear from Table 7 that the RB, HBA, HBD, TPSA, and lipophilicity values of all of the 15 peptides are comparable to those found in orally available peptide drugs. However, only seven peptides (AY, CF, EF, TF, TY, VF, and VY) have fraction Csp3 values smaller than 0.55. Taken together, based on the physicochemical descriptors in Table 7, only the seven aforementioned dipeptides resemble FDA-approved, orally available peptide drugs.

Table 7. Physicochemical properties of peptides having known and potential anti-ACE/anti-DPP-IV activities, in comparison with Captopril (antihypertension drug) and Anagliptin (antidiabetic drug).

Peptide [a]	MW (g/mol)	Fraction Csp3	RB	HBA	HBD	TPSA (Å^2)	Lipophilicity (Consensus Log $P_{o/w}$)
AK	217.27	0.78	8	5	4	118.44	−0.96
AY	252.27	0.33	6	5	4	112.65	−0.54
CF	268.33	0.33	7	4	3	131.22	−0.02
DL	246.26	0.70	8	6	4	129.72	−0.93
EF	294.30	0.36	9	6	4	129.72	−0.21
GIL	301.38	0.79	11	5	4	121.52	0.05
GM	206.26	0.71	7	4	3	117.72	−0.82
IAL	315.41	0.80	11	5	4	121.52	0.58
IL	244.33	0.83	8	4	3	92.42	0.49
SL	218.25	0.78	7	5	4	112.65	−0.80
TF	266.29	0.38	7	5	4	112.65	−0.52
TY	282.29	0.38	7	6	5	132.88	−0.97
VF	264.32	0.43	7	4	3	92.42	0.44
VL	230.30	0.82	7	4	3	92.42	0.26
VY	280.32	0.43	7	5	4	112.65	0.01
Captopril	217.29	0.78	4	3	1	96.41	0.62
Anagliptin	383.45	0.53	8	6	2	115.42	0.73

[a] MW, molecular weight; fraction Csp3, the ratio of sp^3 hybridized carbons over the total carbon count of the molecule; RB, number of rotatable bonds; HBA, number of H-bond acceptors; HBD, number of H-bond donors; TPSA, topological polar surface area.

Table 8 shows the predicted pharmacokinetic properties, drug-likeness and lead-likeness of the 15 selected anti-ACE/anti-DPP-IV peptides shortlisted from seafood paramyosins in this study. All 15 peptides were predicted to be non-substrates of P-glycoprotein (P-gp). P-gp is one of the drug transporters that regulate the update and efflux of drugs in the body. It is known to reduce the oral bioavailability of its substrates. Furthermore, P-gp substrates may potentially act as inducers or inhibitors of P-gp. This could enhance the risk of drug-drug interactions, particularly involving drugs acting on P-gp [42]. That the 15 peptides were predicted as non-P-gp-substrates is therefore desirable, implying that their oral bioavailability would not be compromised by P-gp, neither would the peptides bring about risks of drug-drug interactions. On the other hand, the 15 peptides were predicted as non-inhibitors of all five cytochrome P450 (CYP) isozymes (CYP1A2, CYP2C19, CYP2C9, CYP2D6 and CYP3A4) (Table 8). The aforementioned enzymes play an important role in Phase I biotransformation. Inactivation of the CYP enzymes by a drug or other molecules may cause bioaccumulation and, consequently, toxicity [43]. The predicted non-inhibition of the CYP isozymes by the 15 peptides is therefore consistent with their predicted non-toxicity by Toxinpred in this study. The 15 peptides are comparable to the anti-ACE antihypertension drug Captopril based on the peptides' status as non-P-gp substrates and non-inhibitors of CYP isozymes. Theoretically, the 15 peptides are also less likely to raise risk of drug-drug interactions relative to the anti-DPP-IV antidiabetic drug Anagliptin, a P-gp substrate. In terms of drug-likeness, all 15 paramyosin-derived peptides were predicted to comply with the Lipinski's rule-of-five, with zero violations each. The Abbot bioavailability score estimates the probability that a compound has at least 10% oral bioavailability in the rat or measurable Caco-2 permeability [44]. Similar scores were predicted among the 15 peptides and the two oral drugs we used for comparison. The observation suggests that all 15 peptides can be considered as oral drug candidates [45] that can be further tested for their in vivo effects. In the context of drug development, six of the 15 paramyosin-derived peptides stood out in terms of lead-likeness: AY, CF, TF, TY, VF, and VY. These six dipeptides are also among the seven peptides resembling FDA-approved, orally available peptide drugs based on the physicochemical descriptors in Table 7. Theoretically, the six dipeptides are considered suitable to be subjected to further chemical modifications for lead optimization [45].

Table 8. Pharmacokinetics, drug-likeness and lead-likeness of peptides having known and potential anti-ACE and anti-DPP-IV activities, in comparison with Captopril (antihypertension drug) and Anagliptin (antidiabetic drug).

Peptide	Pharmacokinetics							Drug-Likeness		Lead-Likeness (Number of Violations)
	P-gp Substrate	CYP1A2 Inhibitor	CYP2C19 Inhibitor	CYP2C9 Inhibitor	CYP2D6 Inhibitor	CYP3A4 Inhibitor		Lipinski (Number of Violations)	Abbot Bioavailability Score	
AK	No	No	No	No	No	No		Yes (0)	0.55	No (2)
AY	No	No	No	No	No	No		Yes (0)	0.55	Yes (0)
CF	No	No	No	No	No	No		Yes (0)	0.55	Yes (0)
DL	No	No	No	No	No	No		Yes (0)	0.56	No (2)
EF	No	No	No	No	No	No		Yes (0)	0.56	No (1)
GIL	No	No	No	No	No	No		Yes (0)	0.55	No (1)
GM	No	No	No	No	No	No		Yes (0)	0.55	No (1)
IAL	No	No	No	No	No	No		Yes (0)	0.55	No (1)
IL	No	No	No	No	No	No		Yes (0)	0.55	No (2)
SL	No	No	No	No	No	No		Yes (0)	0.55	No (1)
TF	No	No	No	No	No	No		Yes (0)	0.55	Yes (0)
TY	No	No	No	No	No	No		Yes (0)	0.55	Yes (0)
VF	No	No	No	No	No	No		Yes (0)	0.55	Yes (0)
VL	No	No	No	No	No	No		Yes (0)	0.55	No (1)
VY	No	No	No	No	No	No		Yes (0)	0.55	Yes (0)
Captopril	No	No	No	No	No	No		Yes (0)	0.56	No (1)
Anagliptin	Yes	No	No	No	No	No		Yes (0)	0.55	No (2)

In summary, based on our SwissADME analysis, we found seven putative drug-like peptides (AY, CF, EF, TF, TY, VF, and VY) resembling FDA-approved oral peptide drugs (Figure 9). Among these seven, AY, CF, EF, TF, VF, and VY were demonstrated anti-ACE peptides, whereas AY, TF, TY, VF, and VY were demonstrated anti-DPP-IV peptides. These two sets could be consolidated into four bifunctional anti-ACE/anti-DPP-IV peptides: AY, TF, VF and VY. Notably, all seven peptides have an aromatic residue (F or Y) in their C-termini. Overall, our results suggest that the nine paramyosins investigated can serve as sources of bioavailable, safe, single/dual-function anti-ACE and anti-DPP-IV peptides upon oral ingestion and GI digestion. Considering the in silico/theoretical nature of this study, the actual pool of anti-ACE and anti-DPP-IV peptides liberated from the paramyosins as well as the activity of the three putative anti-ACE peptides (GIL, DL, AK) and one putative anti-DPP-IV peptide (IAL) must be validated in future wet lab experiments. Notably, this virtual screening study has pinpointed promising candidates that can be prioritized in future investigations. Among the nine paramyosins investigated, the two paramyosins isoforms of the common octopus appear to be the most frequent sources of drug-like peptides exhibiting either only anti-ACE activity (CF and EF) or anti-ACE + anti-DPP-IV activities (AY and VF) (Figure 9). Thus, for future discovery of food-derived nutraceuticals or drug candidates targeting hypertension and/or diabetes, the common octopus paramyosins represent a desirable raw material.

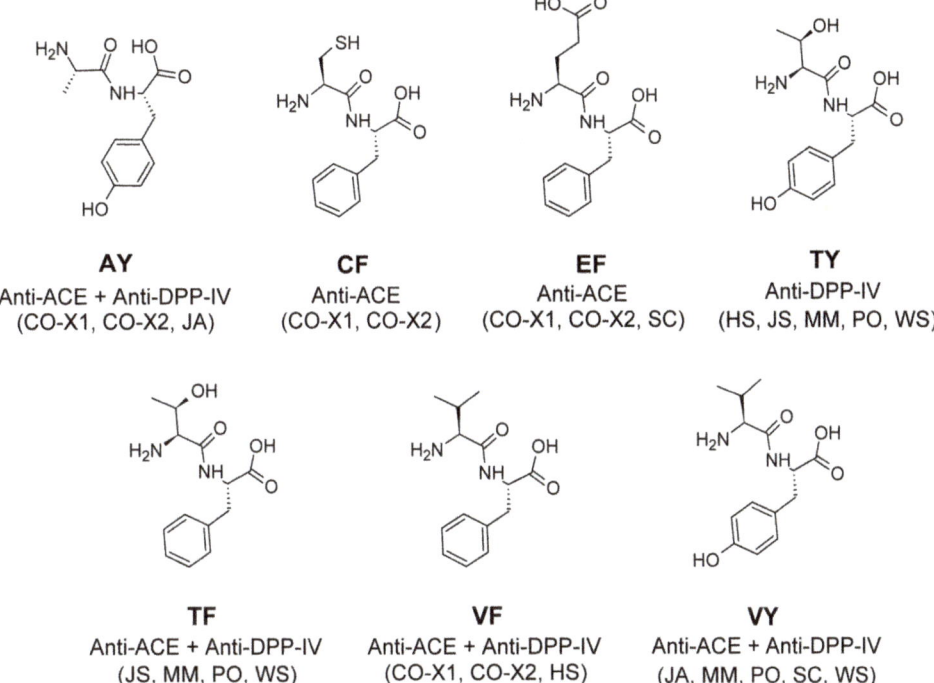

Figure 9. Seven putative drug-like peptides derived from seafood paramyosins.

3. Materials and Methods

3.1. Paramyosin Protein Sequences

The protein sequences of paramyosins of eight seafood species were retrieved from the UniProt Knowledgebase (UniProtKB) (https://www.uniprot.org/, accessed on 28 August 2021) [46] in the FASTA format. The eight species were the common octopus (*Octopus vulgaris*), Humboldt squid (*Dosidicus gigas*), Japanese abalone (*Haliotis discus hannai*), Japanese scallop (*Mizuhopecten yessoensis*), Mediterranean mussel (*Mytilus galloprovincialis*), Pacific oyster (*Crassostrea gigas*), sea cucumber (*Stichopus japonicus*), and Whiteleg shrimp (*Penaeus vannamei*). The number of residues and molecular mass of each paramyosin protein were recorded. In this study, the protein sequences retrieved from UniProtKB were used in in silico GI digestion. Fragments liberated from the digestion were used in in silico screening for high-GI-absorption, non-allergenic, and non-toxic anti-ACE and anti-DPP-IV peptides, as depicted in Figure 10.

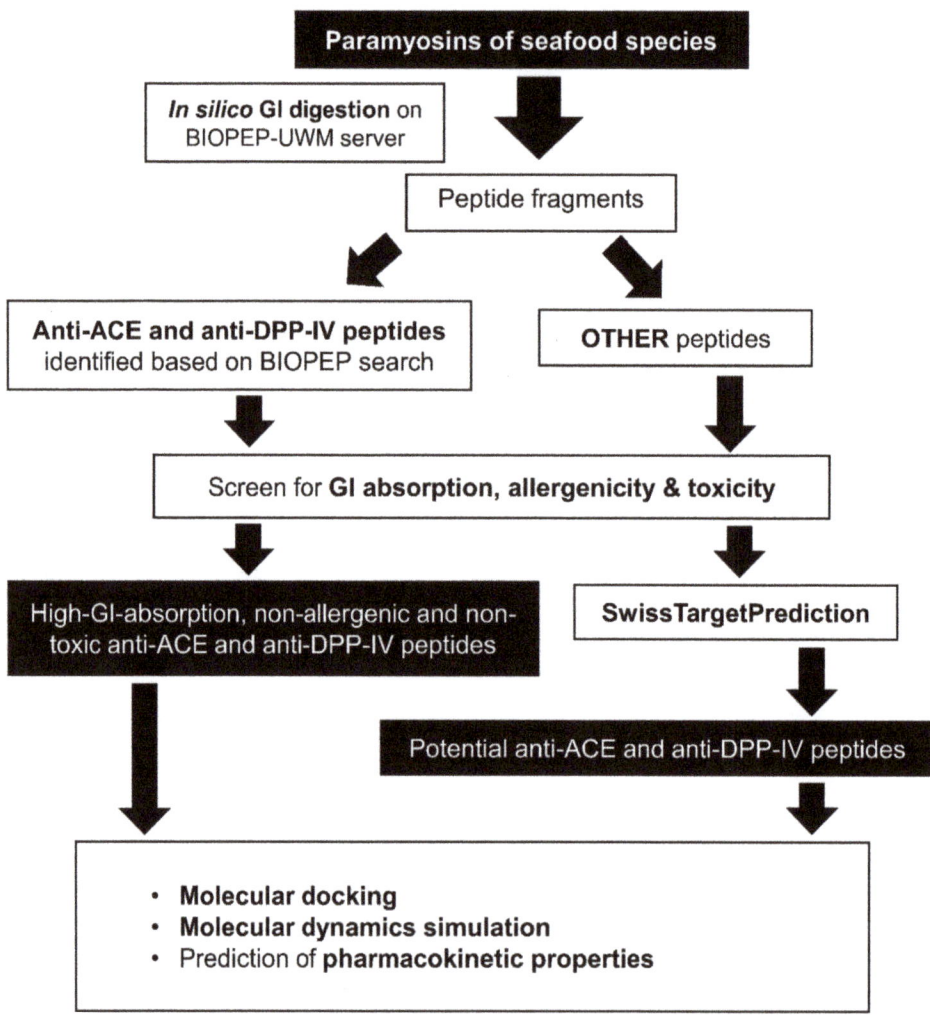

Figure 10. An overview of the study.

3.2. In Silico GI Digestion of Paramyosins

The paramyosin sequences were subjected to in silico GI digestion on the BIOPEP-UWM server (https://biochemia.uwm.edu.pl/en/start/, accessed on 8 September 2021) [21] using the "enzyme(s) action" tool. Chymotrypsin A (EC 3.4.21.1), trypsin (EC 3.4.21.4), and pepsin (pH 1.3) (EC 3.4.23.1) were used for in silico GI digestion as previously reported [20]. The peptide fragments released from each protein were recorded and divided into separate groups: two residues, three residues, four residues, and >four residues. Peptides with previously demonstrated anti-ACE and anti-DPP-IV activities were identified by using the "Search for active fragments" tool in BIOPEP-UWM.

3.3. Prediction of GI Absorption, Allergenicity, and Toxicity of Peptides

Peptides released from the in silico GI digestion of paramyosins were screened for GI absorption in SwissADME (http://www.swissadme.ch/, accessed on 17 September 2021) [45]. The conversion of peptide sequences into the Simplified Molecular Input Line

Entry System (SMILES) format was conducted with the "SMILES" tool of BIOPEP-UWM; the SMILES strings were then submitted to SwissADME as input for analysis. Peptide allergenicity was screened with AllerTOP v.2.0 (https://www.ddgpharmfac.net/AllerTOP/, accessed on 17 September 2021) [47]. Toxicity was screened with ToxinPred (https://webs.iiitd.edu.in/raghava/toxinpred/index.html, accessed on 17 September 2021) [48].

3.4. Ligand-Based In Silico Target Fishing

High-GI-absorption, non-allergenic, and non-toxic peptides not indicated as having anti-ACE and anti-DPP-IV activities based on BIOPEP-UWM search were further analysed with SwissTargetPrediction (http://www.swisstargetprediction.ch/, accessed on 20 September 2021). SwissTargetPrediction is a free, web-based tool that can be used to predict putative human protein targets of any small molecules. Through reverse screening, target prediction is accomplished by matching the structures of query compounds to similar two-dimensional (2D) and three-dimensional (3D) structures of compounds experimentally active on human protein targets [49]. For this analysis, the "Select a species" was set to "*Homo sapiens*". Peptide sequences in the SMILES format were generated with BIOPEP-UWM as described in Section 3.3 (accessed on 20 September 2021). Output of prediction was ranked based on the "Known actives (3D/2D)" parameter. Peptides whose top predicted target was ACE or DPP-IV were recorded, along with the probability of the prediction.

3.5. Molecular Docking Analysis

The docking of peptides onto ACE and DPP-IV was accomplished with HPEPDOCK (http://huanglab.phys.hust.edu.cn/hpepdock/, accessed on 21 December 2021) [50]. The crystal structures of ACE and DPP-IV were downloaded from RCSB Protein Data Bank (https://www.rcsb.org/, accessed on 21 December 2021) [51]. The crystal of the human ACE was complexed with bradykinin potentiating peptide b (BPPb)(PDB ID: 4APJ) [52], whereas the human DPP-IV was complexed with diprotin A (PDB ID: 1WCY) [53]. Upon removal of the bound ligands, the receptors (ACE and DPP-IV) were subjected to energy minimization in the Swiss-PdbViewer 4.0 software [54] prior to docking with HPEPDOCK. To ensure suitability of docking procedure, redocking of co-crystalized ligands (BPPb and diprotin A) to their respective crystals were performed with HPEPDOCK. Peptides were submitted to HPEPDOCK in the PDB format. The 3D structures of peptides were retrieved from Mendeley Data. (https://data.mendeley.com/datasets/z8zh5rpthg/1, accessed on 26 September 2021) [55] and converted into the PDB format by using BIOVIA Discovery Studio Visualizer (BIOVIA, Dassault Systèmes, BIOVIA Discovery Studio Visualizer, Version 20.1.0.192, San Diego: Dassault Systèmes, 2020). The top (most negative) docking score for each peptide-ACE or peptide-DPP-IV docking, as reported by HPEPDOCK, was recorded. BIOVIA Discovery Studio Visualizer was used for the visualization of the 3D structures of the docked models generated by HPEPDOCK. LigPlot+ v.2.2 was used for the 2D visualization and analysis of intermolecular interactions between a peptide and the target proteins [56,57].

3.6. Molecular Dynamics Analysis

Molecular dynamics (MD) was performed in GROMACS 2020 using AMBER99SB-ILDN force field. MD simulation was performed on free proteins (ACE and DPP-IV), docked peptide-ACE and peptide-DPP-IV complexes. Similarly, MD of ACE- BPPb complex and DPP-IV-diprotin A complex were also performed as control. In the MD, each complex was solvated in a cubic box with a distance of 1.2 nm between the complex and each side of the solvated box, and sodium and chloride ions were added to neutralize the total charge of the system [58]. The complex was then energy-minimized using the steepest descent algorithm. The simulation conditions were set at room temperature (300 K) and atmospheric pressure (1 bar) to mimic the general experiment conditions. The fully temperature and pressure equilibrated system was treated as the minimization step for the complex and used as the initial configuration for the MD production dynamic analysis. All simulations were conducted for 25 ns using a 2 fs time step. The results then were analyzed using common GROMACS functions RMSD and RMSF, while the formation of the intermolecular hydrogen bonds in the complex were analyzed using 'gmx_hbond' function. Radius of gyration for free and protein-ligand complexes were also analyzed. The intermolecular distance between each ACE and DPP-IV and their peptide ligand was measured using the 'gmx_pairdist' function.

3.7. Prediction of Physicochemical and Pharmacokinetic Properties

The physicochemical and pharmacokinetic properties of selected peptides were assessed using SwissADME (http://www.swissadme.ch/, accessed on 27 January 2022) [45]. Peptide sequences in the SMILES format were generated as described in Section 3.3 (access date: 27 January 2022). Physicochemical properties, as well as other predicted information concerning the pharmacokinetics, drug-likeness, and lead-likeness of the peptides was recorded. The 2D structures of selected peptides were drawn by using the ACD/ChemSketch freeware (ACD/ChemSketch, version 2019.2.1, Advanced Chemistry Development, Inc., Toronto, ON, Canada, www.acdlabs.com, 2019).

Author Contributions: Conceptualization, T.-T.C. and F.-C.W.; methodology, T.-T.C. and M.Z.S.; software, T.-T.C. and M.Z.S.; validation, C.C.-C.W. and M.Z.S.; formal analysis, C.C.-C.W. and M.Z.S.; investigation, T.-T.C. and M.Z.S.; resources, T.-T.C. and F.-C.W.; data curation, C.C.-C.W. and F.-C.W.; writing—original draft preparation, M.Z.S. and F.-C.W.; writing—review and editing, F.-C.W. and T.-T.C.; visualization, C.C.-C.W. and M.Z.S.; supervision, F.-C.W.; project administration, T.-T.C. All authors have read and agreed to the published version of the manuscript.

Funding: This research received no external funding.

Institutional Review Board Statement: Not applicable.

Informed Consent Statement: Not applicable.

Data Availability Statement: The data presented in this study are available on request from the corresponding author.

Conflicts of Interest: The authors declare no conflict of interest.

Sample Availability: Not applicable.

References

1. Chai, T.-T.; Law, Y.-C.; Wong, F.-C.; Kim, S.-K. Enzyme-assisted discovery of antioxidant peptides from edible marine invertebrates: A review. *Mar. Drugs* **2017**, *15*, 42. [CrossRef] [PubMed]
2. Chai, T.-T.; Ee, K.-Y.; Kumar, D.T.; Manan, F.A.; Wong, F.-C. Plant bioactive peptides: Current status and prospects towards use on human health. *Protein Pept. Lett.* **2021**, *28*, 623–642. [CrossRef] [PubMed]
3. Islam, M.S.; Wang, H.; Admassu, H.; Sulieman, A.A.; Wei, F.A. Health benefits of bioactive peptides produced from muscle proteins: Antioxidant, anti-cancer, and anti-diabetic activities. *Process Biochem.* **2022**, *116*, 116–125. [CrossRef]
4. Apostolopoulos, V.; Bojarska, J.; Chai, T.-T.; Elnagdy, S.; Kaczmarek, K.; Matsoukas, J.; New, R.; Parang, K.; Lopez, O.P.; Parhiz, H.; et al. A global review on short peptides: Frontiers and perspectives. *Molecules* **2021**, *26*, 430. [CrossRef]
5. Wong, F.-C.; Xiao, J.; Wang, S.; Ee, K.-Y.; Chai, T.-T. Advances on the antioxidant peptides from edible plant sources. *Trends Food Sci. Technol.* **2020**, *99*, 44–57. [CrossRef]
6. Daroit, D.J.; Brandelli, A. In vivo bioactivities of food protein-derived peptides–A current review. *Curr. Opin. Food Sci.* **2021**, *39*, 120–129. [CrossRef]
7. Agyei, D.; Tsopmo, A.; Udenigwe, C.C. Bioinformatics and peptidomics approaches to the discovery and analysis of food-derived bioactive peptides. *Anal. Bioanal. Chem.* **2018**, *410*, 3463–3472. [CrossRef]
8. Kaya, A.; Tatlisu, M.A.; Kaplan Kaya, T.; Yildirimturk, O.; Gungor, B.; Karatas, B.; Yazici, S.; Keskin, M.; Avsar, S.; Murat, A. Sublingual vs. oral captopril in hypertensive crisis. *J. Emerg. Med.* **2016**, *50*, 108–115. [CrossRef]
9. Majumder, K.; Wu, J. Molecular targets of antihypertensive peptides: Understanding the mechanisms of action based on the pathophysiology of hypertension. *Int. J. Mol. Sci.* **2014**, *16*, 256–283. [CrossRef]
10. Nongonierma, A.B.; FitzGerald, R.J. Features of dipeptidyl peptidase IV (DPP-IV) inhibitory peptides from dietary proteins. *J. Food Biochem.* **2019**, *43*, e12451. [CrossRef]
11. Nishio, S.; Abe, M.; Ito, H. Anagliptin in the treatment of type 2 diabetes: Safety, efficacy, and patient acceptability. *Diabetes Metab. Syndr. Obes.* **2015**, *8*, 163–171. [CrossRef] [PubMed]

12. de Boer, I.H.; Bangalore, S.; Benetos, A.; Davis, A.M.; Michos, E.D.; Muntner, P.; Rossing, P.; Zoungas, S.; Bakris, G. Diabetes and hypertension: A position statement by the American Diabetes Association. *Diabetes Care* **2017**, *40*, 1273–1284. [CrossRef] [PubMed]
13. Venugopal, V. *Marine Products for Healthcare: Functional and Bioactive Nutraceutical Compounds from the Ocean*; CRC Press: Boca Raton, FL, USA, 2009; pp. 1–527.
14. Gomez, H.L.R.; Peralta, J.P.; Tejano, L.A.; Chang, Y.W. In silico and in vitro assessment of portuguese oyster (*Crassostrea angulata*) proteins as precursor of bioactive peptides. *Int. J. Mol. Sci.* **2019**, *20*, 5191. [CrossRef] [PubMed]
15. Adibi, S.A. The oligopeptide transporter (Pept-1) in human intestine: Biology and function. *Gastroenterology* **1997**, *113*, 332–340. [CrossRef]
16. Cheng, H.M.; Mah, K.K.; Seluakumaran, K. Protein absorption. In *Defining Physiology: Principles, Themes, Concepts. Volume 2: Neurophysiology and Gastrointestinal Systems*; Cheng, H.M., Mah, K.K., Seluakumaran, K., Eds.; Springer International Publishing: Cham, Switzerland, 2020; pp. 71–73.
17. Mathews, D.M.; Adibi, S.A. Peptide absorption. *Gastroenterology* **1976**, *71*, 151–161. [CrossRef]
18. Leibach, F.H.; Ganapathy, V. Peptide transporters in the intestine and the kidney. *Annu. Rev. Nutr.* **1996**, *16*, 99–119. [CrossRef]
19. Wang, B.; Xie, N.; Li, B. Influence of peptide characteristics on their stability, intestinal transport, and in vitro bioavailability: A review. *J. Food Biochem.* **2019**, *43*, e12571. [CrossRef]
20. Koh, J.-A.; Ong, J.-H.; Abd Manan, F.; Ee, K.-Y.; Wong, F.-C.; Chai, T.-T. Discovery of bifunctional anti-DPP-IV and anti-ACE peptides from housefly larval proteins after in silico gastrointestinal digestion. *Biointerface Res. Appl. Chem.* **2022**, *12*, 4929–4944. [CrossRef]
21. Minkiewicz, P.; Iwaniak, A.; Darewicz, M. BIOPEP-UWM database of bioactive peptides: Current opportunities. *Int. J. Mol. Sci.* **2019**, *20*, 5978. [CrossRef]
22. Khora, S.S. Seafood-associated shellfish allergy: A comprehensive review. *Immunol. Investig.* **2016**, *45*, 504–530. [CrossRef]
23. Saito, Y.; Wanezaki, K.; Kawato, A.; Imayasu, S. Structure and activity of angiotensin I converting enzyme inhibitory peptides from sake and sake lees. *Biosci. Biotechnol. Biochem.* **1994**, *58*, 1767–1771. [CrossRef] [PubMed]
24. Nongonierma, A.B.; Mooney, C.; Shields, D.C.; FitzGerald, R.J. Inhibition of dipeptidyl peptidase IV and xanthine oxidase by amino acids and dipeptides. *Food Chem.* **2013**, *141*, 644–653. [CrossRef] [PubMed]
25. Lan, V.T.T.; Ito, K.; Ohno, M.; Motoyama, T.; Ito, S.; Kawarasaki, Y. Analyzing a dipeptide library to identify human dipeptidyl peptidase IV inhibitor. *Food Chem.* **2015**, *175*, 66–73. [CrossRef]
26. Xue, L.; Yin, R.; Howell, K.; Zhang, P. Activity and bioavailability of food protein-derived angiotensin-I-converting enzyme–inhibitory peptides. *Compr. Rev. Food Sci. Food Saf.* **2021**, *20*, 1150–1187. [CrossRef]
27. Matsufuji, H.; Matsui, T.; Seki, E.; Osajima, K.; Nakashima, M.; Osajima, Y. Angiotensin I-converting enzyme inhibitory peptides in an alkaline protease hydrolyzate derived from sardine muscle. *Biosci. Biotechnol. Biochem.* **1994**, *58*, 2244–2245. [CrossRef]
28. Ong, J.-H.; Koh, J.-A.; Siew, Y.-Q.; Manan, F.-A.; Wong, F.-C.; Chai, T.-T. In silico discovery of multifunctional bioactive peptides from silkworm cocoon proteins following proteolysis. *Curr. Top. Pept. Protein Res.* **2021**, *22*, 47–57.
29. Juillerat-Jeanneret, L. Dipeptidyl peptidase IV and its inhibitors: Therapeutics for type 2 diabetes and what else? *J. Med. Chem.* **2014**, *57*, 2197–2212. [CrossRef]
30. Nongonierma, A.B.; FitzGerald, R.J. Dipeptidyl peptidase IV inhibitory and antioxidative properties of milk protein-derived dipeptides and hydrolysates. *Peptides* **2013**, *39*, 157–163. [CrossRef]
31. Nagasundaram, N.; Zhu, H.; Liu, J.; Karthick, V.; George Priya Doss, C.; Chakraborty, C.; Chen, L. Analysing the effect of mutation on protein function and discovering potential inhibitors of CDK4: Molecular modelling and dynamics studies. *PLoS ONE* **2015**, *10*, e0133969. [CrossRef]
32. Liu, K.; Watanabe, E.; Kokubo, H. Exploring the stability of ligand binding modes to proteins by molecular dynamics simulations. *J. Comput.-Aided Mol. Des.* **2017**, *31*, 201–211. [CrossRef]
33. Shao, Q.; Zhu, W. Exploring the ligand binding/unbinding pathway by selectively enhanced sampling of ligand in a protein–ligand complex. *J. Phys. Chem. B* **2019**, *123*, 7974–7983. [CrossRef] [PubMed]
34. Lobanov, M.Y.; Bogatyreva, N.S.; Galzitskaya, O.V. Radius of gyration as an indicator of protein structure compactness. *Mol. Biol.* **2008**, *42*, 623–628. [CrossRef]
35. Khan, R.J.; Jha, R.K.; Amera, G.M.; Jain, M.; Singh, E.; Pathak, A.; Singh, R.P.; Muthukumaran, J.; Singh, A.K. Targeting SARS-CoV-2: A systematic drug repurposing approach to identify promising inhibitors against 3C-like proteinase and 2′-O-ribose methyltransferase. *J. Biomol. Struct. Dyn.* **2021**, *39*, 2679–2692. [CrossRef] [PubMed]
36. Lee, M.S.; Olson, M.A. Calculation of absolute protein-ligand binding affinity using path and endpoint approaches. *Biophys. J.* **2006**, *90*, 864–877. [CrossRef] [PubMed]
37. Hubbard, R.E.; Haider, M.K. Hydrogen bonds in proteins: Role and strength. In *Encyclopedia of Life Sciences (ELS)*; John Wiley & Sons, Ltd.: Chichester, UK, 2010.
38. Menéndez, C.A.; Accordino, S.R.; Gerbino, D.C.; Appignanesi, G.A. Hydrogen bond dynamic propensity studies for protein binding and drug design. *PLoS ONE* **2016**, *11*, e0165767. [CrossRef]
39. Wong, F.-C.; Ong, J.-H.; Kumar, D.T.; Chai, T.-T. In silico identification of multi-target anti-SARS-CoV-2 peptides from quinoa seed proteins. *Int. J. Pept. Res. Ther.* **2021**, *27*, 1837–1847. [CrossRef]

40. Ji, D.; Xu, M.; Udenigwe, C.C.; Agyei, D. Physicochemical characterisation, molecular docking, and drug-likeness evaluation of hypotensive peptides encrypted in flaxseed proteome. *Curr. Res. Food Sci.* **2020**, *3*, 41–50. [CrossRef]
41. Santos, G.B.; Ganesan, A.; Emery, F.S. Oral administration of peptide-based drugs: Beyond Lipinski's rule. *ChemMedChem* **2016**, *11*, 2245–2251. [CrossRef]
42. Finch, A.; Pillans, P. P-glycoprotein and its role in drug-drug interactions. *Aust. Prescr.* **2014**, *37*, 137–139. [CrossRef]
43. Sychev, D.A.; Ashraf, G.M.; Svistunov, A.A.; Maksimov, M.L.; Tarasov, V.V.; Chubarev, V.N.; Otdelenov, V.A.; Denisenko, N.j.P.; Barreto, G.E.; Aliev, G. The cytochrome P450 isoenzyme and some new opportunities for the prediction of negative drug interaction in vivo. *Drug Des. Dev. Ther.* **2018**, *12*, 1147–1156. [CrossRef]
44. Martin, Y.C. A bioavailability score. *J. Med. Chem.* **2005**, *48*, 3164–3170. [CrossRef] [PubMed]
45. Daina, A.; Michielin, O.; Zoete, V. SwissADME: A free web tool to evaluate pharmacokinetics, drug-likeness and medicinal chemistry friendliness of small molecules. *Sci. Rep.* **2017**, *7*, 42717. [CrossRef] [PubMed]
46. Consortium, T.U. UniProt: The universal protein knowledgebase in 2021. *Nucleic Acids Res.* **2021**, *49*, D480–D489. [CrossRef] [PubMed]
47. Dimitrov, I.; Bangov, I.; Flower, D.R.; Doytchinova, I. AllerTOP v.2—A server for in silico prediction of allergens. *J. Mol. Model.* **2014**, *20*, 2278. [CrossRef] [PubMed]
48. Gupta, S.; Kapoor, P.; Chaudhary, K.; Gautam, A.; Kumar, R.; Open Source Drug Discovery, C.; Raghava, G.P.S. In silico approach for predicting toxicity of peptides and proteins. *PLoS ONE* **2013**, *8*, e73957. [CrossRef]
49. Daina, A.; Michielin, O.; Zoete, V. SwissTargetPrediction: Updated data and new features for efficient prediction of protein targets of small molecules. *Nucleic Acids Res.* **2019**, *47*, W357–W364. [CrossRef]
50. Zhou, P.; Jin, B.; Li, H.; Huang, S.Y. HPEPDOCK: A web server for blind peptide-protein docking based on a hierarchical algorithm. *Nucleic Acids Res.* **2018**, *46*, W443–W450. [CrossRef]
51. Burley, S.K.; Bhikadiya, C.; Bi, C.; Bittrich, S.; Chen, L.; Crichlow, G.V.; Christie, C.H.; Dalenberg, K.; Di Costanzo, L.; Duarte, J.M.; et al. RCSB Protein Data Bank: Powerful new tools for exploring 3D structures of biological macromolecules for basic and applied research and education in fundamental biology, biomedicine, biotechnology, bioengineering and energy sciences. *Nucleic Acids Res.* **2021**, *49*, D437–D451. [CrossRef]
52. Masuyer, G.; Schwager, S.L.U.; Sturrock, E.D.; Isaac, R.E.; Acharya, K.R. Molecular recognition and regulation of human angiotensin-I converting enzyme (ACE) activity by natural inhibitory peptides. *Sci. Rep.* **2012**, *2*, 717. [CrossRef]
53. Hiramatsu, H.; Yamamoto, A.; Kyono, K.; Higashiyama, Y.; Fukushima, C.; Shima, H.; Sugiyama, S.; Inaka, K.; Shimizu, R. The crystal structure of human dipeptidyl peptidase IV (DPPIV) complex with diprotin A. *Biol. Chem.* **2004**, *385*, 561–564. [CrossRef]
54. Guex, N.; Peitsch, M.C. SWISS-MODEL and the Swiss-Pdb Viewer: An environment for comparative protein modeling. *Electrophoresis* **1997**, *18*, 2714–2723. [CrossRef] [PubMed]
55. Prasasty, V.D.; Istyastono, E.P. Data of small peptides in SMILES and three-dimensional formats for virtual screening campaigns. *Data Brief* **2019**, *27*, 104607. [CrossRef]
56. Laskowski, R.A.; Swindells, M.B. LigPlot+: Multiple ligand–protein interaction diagrams for drug discovery. *J. Chem. Inf. Modeling* **2011**, *51*, 2778–2786. [CrossRef] [PubMed]
57. Wallace, A.C.; Laskowski, R.A.; Thornton, J.M. LIGPLOT: A program to generate schematic diagrams of protein-ligand interactions. *Protein Eng. Des. Sel.* **1995**, *8*, 127–134. [CrossRef] [PubMed]
58. Oyewusi, H.A.; Huyop, F.; Wahab, R.A.; Hamid, A.A.A. In silico assessment of dehalogenase from Bacillus thuringiensis H2 in relation to its salinity-stability and pollutants degradation. *J. Biomol. Struct. Dyn.* **2021**. [CrossRef]

Article

Statistical Methods in the Study of Protein Binding and Its Relationship to Drug Bioavailability in Breast Milk

Karolina Wanat * and Elżbieta Brzezińska

Department of Analytical Chemistry, Faculty of Pharmacy, Medical University of Lodz, 90-419 Łódź, Poland; elzbieta.brzezinska@umed.lodz.pl
* Correspondence: karolina.wanat@umed.lodz.pl; Tel.: +48-42-677-92-11

Abstract: Protein binding (PB) is indicated as the factor most severely limiting distribution in the organism, reducing the bioavailability of the drug, but also minimizing the penetration of xenobiotics into the fetus or the body of a breastfed child. Therefore, PB is an important aspect to be analyzed and monitored in the design of new drug substances. In this paper, several statistical analyses have been introduced to find the relationship between protein binding and the amount of drug in breast milk and to select molecular descriptors responsible for both pharmacokinetic phenomena. Along with descriptors related to the physicochemical properties of drugs, chromatographic descriptors from TLC and HPLC experiments were also used. Both methods used modification of the stationary phase, using bovine serum albumin (BSA) in TLC and human serum albumin (HSA) in HPLC. The use of the chromatographic data in the protein binding study was found to be positive —the most effective application of normal-phase TLC and $HPLC_{HSA}$ data was found. Statistical analyses also confirmed the prognostic value of affinity chromatography data and protein binding itself as the most important parameters in predicting drug excretion into breast milk.

Keywords: protein binding; breast milk; M/P ratio; statistical modeling; molecular descriptors; chromatographic descriptors; affinity chromatography

1. Introduction

Excretion of drugs into breast milk is an important aspect to be considered in the pharmacotherapy of breastfeeding women. Due to ethical considerations, in vivo studies are very rare and it is difficult to obtain the milk-to-plasma (M/P) ratio of many active pharmaceutical compounds (APIs). A mathematical model capable of calculating M/P values using the available data will greatly facilitate the study of the bioavailability of new APIs.

In the previous articles [1,2], we presented a comparison of statistical methods in the study of drug excretion into breast milk with the use of the M/P descriptor. It was shown that the multiple linear regression (MLR) and random forest (RF) analyses were most effective in describing this pharmacokinetic phenomenon, with the use of chromatographic data and physicochemical properties of the tested compounds. These analyses did not deviate from the known principles of bioavailability to breast milk and showed a close relationship between M/P and the level of drug–protein binding (PB) as well as the state of ionization of the API in the bloodstream.

The papers also describe the most effective conditions for thin layer chromatography (TLC) as an analytical model for predicting the penetration of drugs into breast milk. According to these results, it can be assumed that the use of drug–protein binding indices, together with chromatographic data, will make it possible to predict the level of drug distribution into breast milk.

The main aim of this study is to provide supplementary analyses, which include: determination of physicochemical parameters related to drug protein binding; searching for

a mathematical model of PB and/or M/P prediction; and the use of affinity chromatography data as an index of pharmacokinetic properties.

The goal of developing such a model is its further utility in predicting the PB of newly developed active pharmaceutical ingredients. Only easily available API properties are needed to use the model. It can facilitate the process of introducing a new drug to use and reduce expensive in vivo testing.

In this study the following statistical methods were used: cluster analysis (CA), discriminant function analysis (DFA) and principal component analysis (PCA) random forest regression (RF). All molecular descriptors used in this study are listed and described in Table 1.

Table 1. List of molecular and chromatographic descriptors used in statistical analyses.

Descriptor	Description	Reference/Database/Software
a/b/n code	acidic, basic or neutral character of the compound; describes the division into groups: a, b and n	CHEMBL database [3]
B1	calculation parameter B2, describes the bioavailability in the CNS and determines penetration through the blood-brain barrier: log bb = 0.139 + 0.152 log P	reference [4]
B2	calculation parameter B2, describes the bioavailability in the CNS and determines penetration through the blood-brain barrier: log bb = 0.547 − 0.016 PSA	reference [5]
B3	calculation parameter related to protein binding: log (bound fraction/unbound fraction) = 0.5 log P–0.665	reference [6]
CNS+/−	ability to penetrate into the central nervous system (+ or −)	DrugBank database [7]
DM	dipole moment	HyperChem, Hypercube, Inc.
eH	energy of the highest occupied molecular orbital (HOMO)	HyperChem, Hypercube, Inc.
eH-eL	ionization capacity	HyperChem, Hypercube, Inc.
eL	energy of the lowest unoccupied molecular orbital (LUMO)	HyperChem, Hypercube, Inc.
HA	number of hydrogen bond acceptors	ACD/Labs
HD	number of hydrogen bond donors	ACD/Labs
log D	distribution coefficient	ACD/Labs
log M/P	logarithm of M/P	
log MW	logarithm of MW	
log P	partition coefficient	HyperChem, Hypercube, Inc.
log U/D	the ratio of neutral to ionized form; determines the degree of ionization	Calculated using: pK_a-pH for acids; pH-pK_a for bases
M/P	milk/plasma drug concentration ratio	references [8–13]
MW	molecular weight	HyperChem, Hypercube, Inc.
PB	percentage of plasma protein binding	DrugBank
PhCharge	the charge of the API under physiological conditions	DrugBank
pK_a	negative logarithm of the acid dissociation constant (K_a)	ACD/Labs
PSA	polar surface area	ACD/Labs
Sa	the surface area of the molecule	HyperChem, Hypercube, Inc.
V	the volume of the molecule	HyperChem, Hypercube, Inc.
NP; RP	R_f (retention factor) obtained from TLC using impregnated with bovine serum albumin (BSA) plates in normal and reversed phase	TLC experiment
NP/C; RP/C	R_f from impregnated NP or RP plate/control R_f	TLC experiment
k_{HSA}	retention factor from HPLC using column with immobilized human serum albumin (HSA)	HPLC experiment
log k_{HSA}	logarithm of the retention coefficient obtained from $HPLC_{HSA}$	HPLC experiment
log k_{IAM}	logarithm of the retention coefficient obtained from $HPLC_{IAM}$ (column with immobilized artificial membrane)	HPLC experiment

2. Results

2.1. Correlation Analyses

The experiment investigated the results of using data from several chromatographic analysis experiments (HPLC$_{HSA}$, NP TLC, RP TLC and, additionally, HPLC$_{IAM}$) in predicting drug binding to protein, and thus bioavailability to breast milk. A group of 165 APIs was analyzed, in which acidic, basic and neutral drugs were observed. The best correlation with PB values was shown in the results of the HPLC$_{HSA}$ and NP TLC experiments, in the form of log k and R$_f$ values, (HPLC$_{HSA}$: $n = 165$, $R = 0.39$); (NP TLC: $n = 162$, $R = 0.31$). The relationship is directly proportional. This is the result for all kinds of relationships. Much better results were obtained for acidic drugs ($R = 0.50$), even considering the smaller number of cases ($n = 34$) (Table A1, Appendix A).

Then the effect of the most frequently mentioned molecular descriptors, related to drug distribution into breast milk and protein binding, was investigated. In all groups of APIs, molecular descriptors related to the hydro-lipophilic nature of drugs play a dominant role. The most important parameters are the partition coefficient and the distribution coefficient (log P and log D). The ability to form hydrogen bonds (HD, HA) is visible here and the correlation with PB is significant. The ratio of neutral to dissociated form (log U/D), dissociation constant (pKa), ionization capacity of compounds (eH-eL) and other electron descriptors: eL and eH, show no significance. The influence of hydrophobic parameters (Sa, V, MW) is visible only in the form of the surface area to volume ratio (Sa/V). As can be seen above, this factor correlates inversely with all types of cases (Table A2, Appendix A).

2.2. Discriminant Function Analysis

All of the descriptors most strongly related to the variability of the PB, which at the same time did not limit the number of cases studied, were introduced into the discriminant function analysis (DFA). All cases were tested using the a/b/n code.

In the stepwise DFA, the discriminant variables included 9 out of 16 entered variables: PhCharge, B2, pKa, M/P, log k$_{HSA}$, log k$_{IAM}$, NP, eL and log U/D (Table 2).

Table 2. Classification matrix for the model using discriminant variables: PhCharge, B2, pKa, M/P, log k$_{HSA}$, log k$_{IAM}$, NP, eL, log U/D.

API Group	Correctly Classified Cases (%)	a $p = 0.17895$	n $p = 0.52632$	b $p = 0.29474$
a	100,00	17	0	0
n	96,00	0	48	2
b	92,86	0	2	26
all	95,80	17	50	28

The PC1 factor discriminates the groups of APIs the most (PC 1 eigenvalue = 3.61). The variables PhCharge and pKa have the most important share in its value. The PC2 factor (PC2 eigenvalue = 0.81) was shaped by the chromatographic descriptors and the ability to ionize (log U/D). The means of the canonical variables (PC1) for group a = -3.52, for group n = 0.03 and for group b = 2.08, therefore PC1 most strongly discriminates between groups a and b. The means of the canonical variables (PC2) for group a = -0.93, for group n = 0.86 and for group b = -0.97. In this case, the centroids of groups a and b are almost equal, and the group of neutral compounds (n) is the most discriminated against (Figure 1).

2.3. Principal Component Analysis

PCA was performed to determine the effect of the primary descriptors on the characteristics of the drug's ability to pass into breast milk. In order to better visualize the obtained results from the analysis, the M/P values were converted into the scale of the drug penetration into milk—M/P$_{code}$. The values of this indicator are in the range 1–4. Code 1 corresponds to drugs with an M/P value <0.40—completely safe; 2 corresponds to the range of 0.40–0.80—at the safety limit; 3 range 0.81–1.20—possibly over the safety limit; and 4 is M/P >1.20—dangerous.

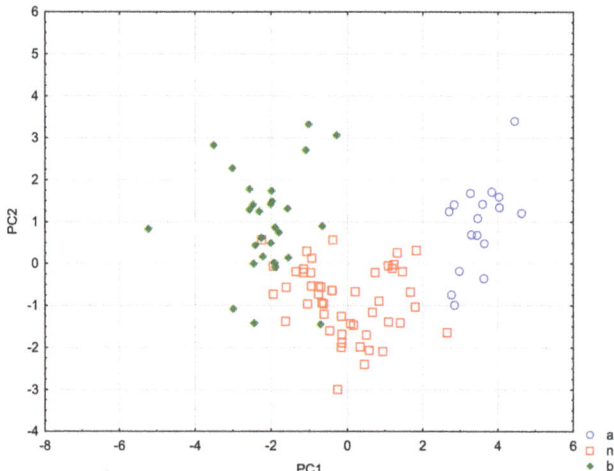

Figure 1. Discrimination against acidic (**a**), basic (**b**) and neutral drugs (*n*). The scatter plot of canonical values for root 1 relative to root 2. Discriminating variables: PhCharge, B2, pKa, M/P, log k_{HSA}, log k_{IAM}, NP, eL, log U/D.

In the course of the analysis, the smallest number of principal components explaining the maximum range of the total variance in the group was initially established. Five factors explain 100% of the variability in the levels of drug excretion into breast milk. The first two factors, PC1 and PC2 (principal components), are described by all used descriptors. As a result, two main components explaining a total of 72% of the variability were obtained. The $HPLC_{HSA}$, $HPLC_{IAM}$, NP TLC and RP TLC chromatographic data is responsible for the first component, PC1 (43.26%), the second component, PC2 (28.66%), is determined by the PB value.

The projection of cases on the PC1 × PC2 plane is presented below (Figure 2):

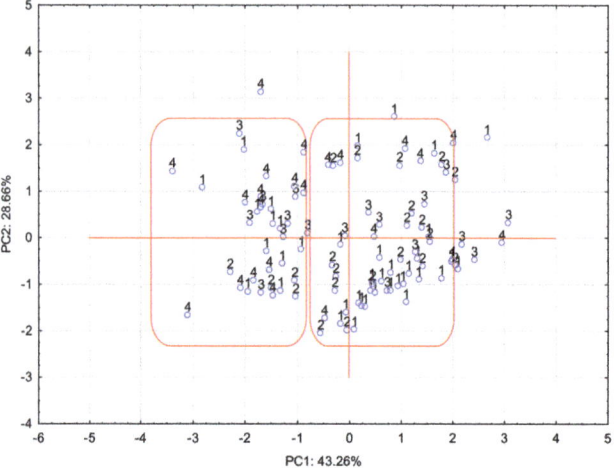

Figure 2. Projection of cases onto the PC1 × PC2 plane.

In the graph of the projection of cases onto the PC plane, where the grouping variable is the scale of drug penetration into breast milk (M/P$_{code}$), it can be seen that the tested APIs can be divided into two groups (surrounded by a box in the graph). One group included drugs with a lower level of M/P (1–2) penetration—safe, and the other group, M/P 3–4—dangerous. This division is not entirely obvious. It was created on the basis of factors explaining 75% of the variability. Few examples of misclassification are visible. The distinction between these groups is related to PC1. Derivatives with a low M/P are located on the right side of the plot and are clearly related to the positive values of PC1. APIs easily excreted into milk are on the left side of the chart and have negative PC1 values. The share of variables in this component, determined by the PC1-variable correlation (factor loadings), reveals the parameters of the greatest importance for the investigated pharmacokinetic feature of drugs. They are: log k$_{HSA}$, log k$_{IAM}$, NP and RP. Thus, affinity chromatography, based on protein binding, can predict the bioavailability of an API into breast milk.

The graph of the projection of variables onto the PC plane shows graphically the relationship between the component and the variable. The graph shows the so-called unit circle, i.e., the maximum correlation of 1 between the variable and the factor. The closer a given variable is to the unit circle line, the greater its correlation with the observed phenomenon (Figure 3).

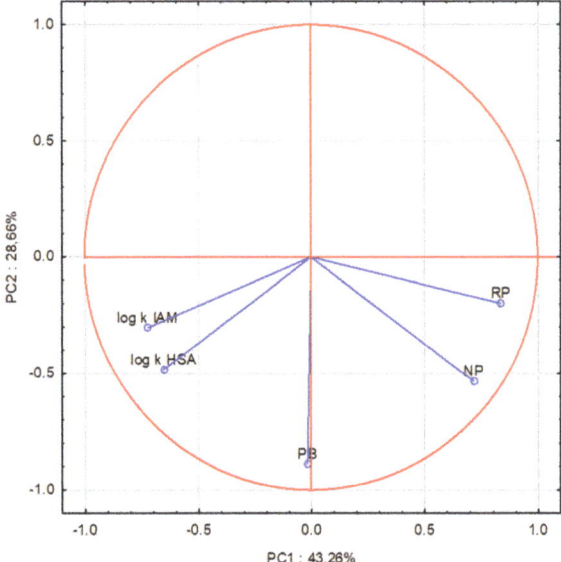

Figure 3. Projection of variables on the plane of factors PC1 × PC2.

2.4. Cluster Analysis

In order to emphasize the diagnostic value of the experiment and to determine the difference in the values of the parameters determining the ability of drugs to penetrate into breast milk, cluster analysis (CA) was also performed. CA was conducted in the proposed M/P$_{code}$ scale, using the k-means method. The means of the most important biological descriptors (CNS +/−, B1, PhCharge, acid/base, NP, RP, log k$_{HSA}$, log k$_{IAM}$ and PBcode) were compared for groups M/P$_{code}$ 1–4. As shown, all drug biological parameters showed a group variability (see Figure 4). The M/P code values range from 1 to 4 with a clear distinction between relatively safe and unsafe groups. Physicochemical parameters: PB, acid/base, HD, log P, eL, log D also show differentiation, but not in all cases. Unfortunately, M/P$_{code}$ is too clustered here, which indicates a smaller influence of the tested properties on the observed feature (Figure 5). The descriptors: log D and eL show the highest differentiation.

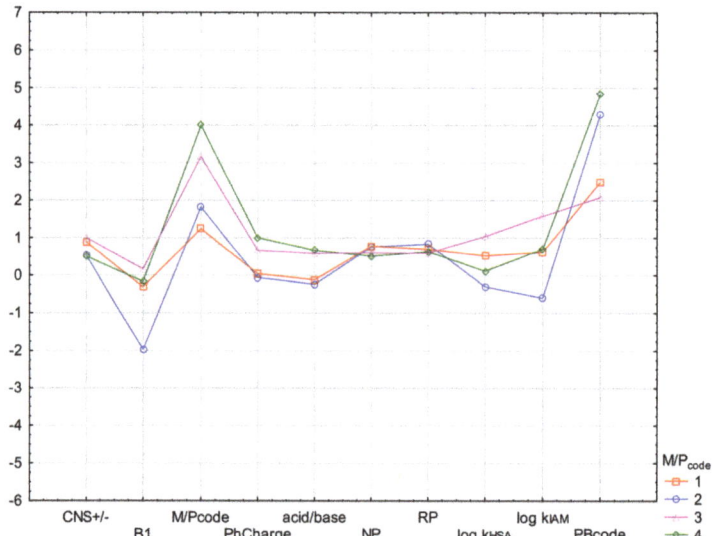

Figure 4. Mean descriptor values in M/P$_{code}$ cluster analysis (k-means method) using biological and chromatographic descriptors.

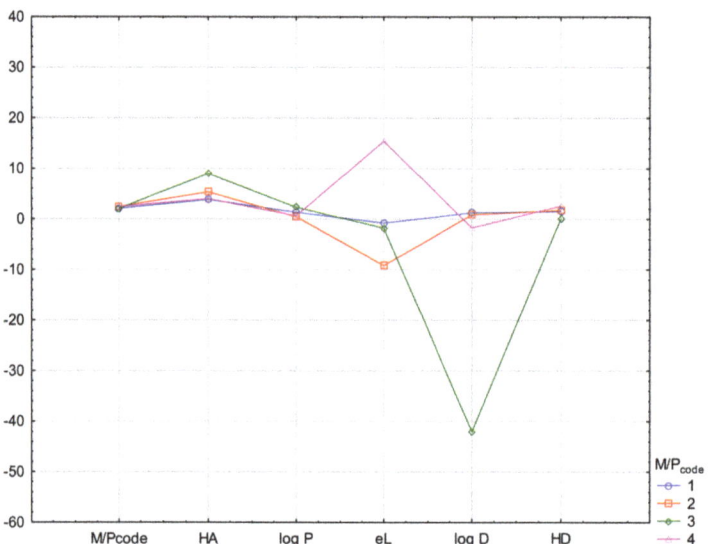

Figure 5. Mean descriptor values in M/P$_{code}$ cluster analysis (k-means method) using physicochemical descriptors.

The above analyses confirmed the values of the parameters HA, log P, log D and eL. The parameters of log D, HA and eL show the greatest differentiation. Unfortunately, the M/P$_{code}$ values are poorly differentiated and their values do not correspond to the variability of other descriptors.

2.5. Regression Methods

MLR failed to create a reliable PB prediction model, therefore an attempt was made to analyze protein binding by other regression methods. A total of 165 test compounds and 22–23 independent variables were used to perform partial least squares (PLS) and random forest regression (RF). The variables used are listed for each model (Tables 3 and 4). During the analyses, 165 compounds were randomly divided into a training set, 70% of the total (TRAIN, n = 115 compounds,) and a test set for external validation, 30% of the total (TEST, n = 50).

Table 3. Twenty-three independent variables with NP TLC data used to create the RF and PLS model for PB.

No.	Independent Variable	No.	Independent Variable	No.	Independent Variable
1.	B3	9.	NP/B2	17.	eH
2.	PhCharge	10.	NP/log P	18.	eL
3.	acid/base	11.	MW	19.	eH-eL
4.	pKa	12.	log MW	20.	logD
5.	log U/D	13.	PSA	21.	Sa
6.	C	14.	HD	22.	V
7.	NP	15.	HA	23.	logP
8.	NP/C	16.	DM		

Table 4. Twenty-two independent variables with HPLC$_{HSA}$ data used to create the RF and PLS model for PB.

No.	Independent Variable	No.	Independent Variable	No.	Independent Variable
1.	B3	9.	log k$_{HSA}$/log P	17.	eL
2.	PhCharge	10.	MW	18.	eH-eL
3.	acid/base	11.	log MW	19.	log D
4.	pKa	12.	PSA	20.	Sa
5.	log U/D	13.	HD	21.	V
6.	k$_{HSA}$	14.	HA	22.	log P
7.	log k$_{HSA}$	15.	DM		
8.	log k$_{HSA}$/B2	16.	eH		

2.5.1. Partial Least Squares Regression

The PLS model using 23 independent variables, including NP TLC data (Table 3) showed low values of R^2 and Q^2, approximately 0.40, and even lower results of external validation, approximately 0.22–0.24 (Figure A1, Appendix A). Even lower values are achieved with the HPLC$_{HSA}$ chromatographic data. This indicates that, as in the case of breast milk prediction models, the PLS method is again not widely applicable here and is not an appropriate method to analyze this type of data.

2.5.2. Random Forest Regression

RF regression was performed with the use of 150 generated random trees. NP TLC data was used first. The independent variables used for the analysis of all 165 cases (independent variable, PB$_{abn}$) are listed in Table 3.

The obtained model (Figure 6) showed satisfactory results, especially for the training set (n = 115): R^2_{train} = 0.81; Q^2_{train} = 0.73. The results of external validation using the test kit (n_{abn} = 50) were lower: R^2_{test} = 0.65; Q^2_{test} = 0.56. The Monte Carlo permutation test (MCPT) showed the average value of the Q^2_{test} parameter was equal to 0.56 (Appendix A, Figure A2), which is similar to that in the presented model. The influence of individual independent variables on the model is presented in the chart below (Appendix A, Figure A3). The order of the descriptors presented there is as shown in Table 3. The log D parameter shows the strongest influence on the model using NP TLC data.

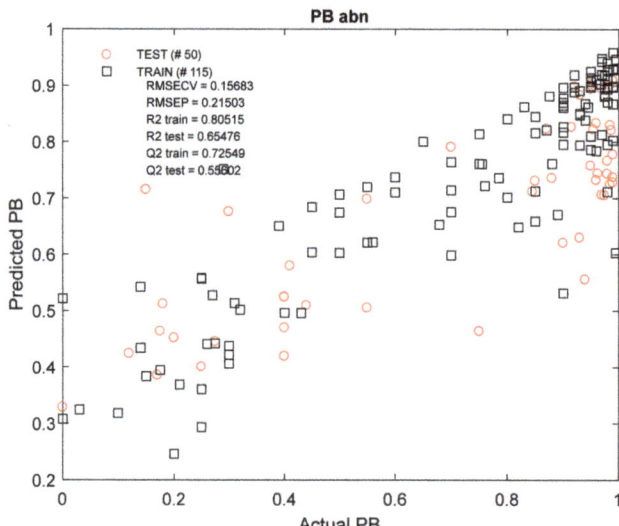

Figure 6. Actual versus predicted PB_{abn} values using RF regression modelling of molecular descriptor set containing 23 variables. $RMSE_{CV}$ = root-mean-square error of cross-validation, $RMSE_P$ = root-mean-square error of prediction, R^2 train/test = coefficient of determination for train/test set models, Q^2 train/test = coefficient of determination for the cross-validated models.

The data from the $HPLC_{HSA}$ experiment were then used for the RF regression (Table 4). The obtained model (Appendix A, Figure A4) again shows good results of the training set (n = 115): R^2_{train} = 0.81; Q^2_{train} = 0.78 but much lower parameters were obtained with external validation (n_{abn} = 50): R^2_{test} = 0.57; Q^2_{test} = 0.53. In the MCPT, the Q^2_{test} value was already at a low level and amounted to 0.35 (Appendix A, Figure A5).

Then, individual groups of compounds were dealt with, either separately, (a), (b) and (n), or combined, (an), (bn) and (ab). The results are shown in Table 5. Only the NP TLC data (Table 3) were used to construct the models, which gave the best results when tested for the complete set of compounds (n_{abn} = 165).

Table 5. Random forest regression results on individual drug combinations.

API Group	Train Set	Test Set
PB_a	n = 24 R^2 = 0.78; Q^2 = 0.62	n = 11 R^2 = 0.29; Q^2 = 0.11
PB_b	n = 35 R^2 = 0.88; Q^2 = 0.80	n = 15 R^2 = 0.33; Q^2 = 0.29
PB_n	n = 57 R^2 = 0.85; Q^2 = 0.81	n = 25 R^2 = 0.62; Q^2 = 0.59
PB_{an}	n = 82 R^2 = 0.82; Q^2 = 0.74	n = 35 R^2 = 0.60; Q^2 = 0.55
PB_{bn}	n = 92 R^2 = 0.85; Q^2 = 0.80	n = 40 R^2 = 0.44; Q^2 = 0.44
PB_{ab}	n = 59 R^2 = 0.80; Q^2 = 0.72	n = 26 R^2 = 0.38; Q^2 = 0.33

RF models for PB_a (n_a = 35) and PB_b (n_b = 50) gave poor results, especially in the external validation, similarly to their combined group (n_{ab} = 85), where the external validation results were in the range of Q^2 = 0.4–0.3.

The best results were obtained for the PB_n (n_n = 82) and PB_{an} (n_{an} = 117) groups. The R^2 and Q^2 values of the test kits ranged between 0.55 and 0.62 (Appendix A: Figures A6 and A7). In both models, the log D values are the most important in their creation (Appendix A: Figures A8 and A9).

3. Discussion

On the basis of the DFA analysis, it was possible to determine the influence of the acidic, basic and neutral properties of APIs on their protein binding capacity and to decide whether the analysis of the pharmacotherapy of nursing mothers (M/P predictions) should be divided into groups: a, n and b. The division into acidic, basic and neutral drugs is strongly related to the PB-related descriptors, so the use of groups a, b and n seems to bring value for further analysis. The low values of Wilks lambda for both roots, PC1 and PC2, confirm the value of the obtained results (0.11 and 0.54, respectively).

As the DFA analysis revealed a group of physicochemical and chromatographic parameters important for the bioavailability of drugs to milk, the use of CA emphasized the differentiation of their mean values in the M/P 1–4 groups. The above analyses confirmed the values of the parameters HA, log P, log D and eL. The parameters of log D, HA and eL show the greatest differentiation. Unfortunately, the M/P_{code} values are poorly differentiated and their values do not correspond to the variability of other descriptors. Based on the PCA, it can be concluded that the data of the drug–protein binding affinity chromatography, in the form of the proposed analytical models and the protein binding itself as the basis for the experimental design, are the most important parameters in predicting drug excretion into breast milk.

The final step in this study was to construct a model capable of predicting PB value, used as a trait strongly correlated with the bioavailability of breast milk. Unfortunately, it was not possible to obtain an MLR or PLS algorithm for protein binding prediction, that was reproducible for different groups. Models created by regression using the random forest method show a significant relationship, visible in the scatter plots (Figures 6, A4, A6 and A7). The influence of the determination coefficient (log D) and chromatographic parameters from the NP TLC and $HPLC_{HSA}$ experiments in each model are also noticeable. Unfortunately, they do not show the best predictive ability (external validation at the level of Q^2_{test} = 0.56 and 0.35 in MCPT tests).

The best results using random forest regression were obtained for the entire set of compounds, PB_{abn}, and for the PB_n and PB_{an} groups. It is the acidic and neutral compounds that bind primarily to albumin, which constitutes the majority of plasma proteins, so the literature values of protein binding (PB) refer mainly to the binding of drugs to HSA.

4. Materials and Methods

4.1. Molecular Descriptors

All tested drugs are listed in Supplementary Materials, along with molecular descriptors. Active pharmaceutical ingredients were extracted from pharmaceutical formulations, purchased in a generally accessible pharmacy. The main criterion used in composing the drug set was the availability of protein binding values (PB) along with milk-to-plasma ratios for each API, as these were the main pharmacokinetic phenomena studied.

The molecular descriptors selected for statistical analyses, which should have a significant effect on the penetration into breast milk and protein binding, are listed in Table 1. Some were taken from the literature, including M/P ratio obtained in vivo [8–13] or from online databases DrugBank [7] and CHEMBL [3]. Most of the physicochemical data were calculated in the following programs: HyperChem (HyperChem for Windows version 7.02, HyperCube Inc, Gainesville, FL, USA, 2002) and ACD/Labs (ACD/LabsTM Log D Suite 8.0, pKa dB 7.0, Advanced Chemistry Development Inc., Toronto, Canada, 2004).

Chromatographic descriptors were obtained in experiments, thin layer chromatography in normal (NP TLC) and reversed mode (RP TLC). The stationary phase was modified with bovine serum albumin (BSA). TLC was the source of retention factor (R_f) values, denoted in statistical models as NP and RP. High performance liquid chromatography was performed using immobilized human serum albumin column ($HPLC_{HSA}$) and immobilized artificial membrane ($HPLC_{IAM}$). HPLC was the source of the log k values (logarithm of retention factor), log k_{HSA} and log k_{IAM}. The TLC and HPLC experiments are detailed in Appendix B.

4.2. Statistical Analyses

DFA, PCA and CA were performed in STATISTICA 13.1 (TIBCO Software Inc., Palo Alto, CA, USA). DFA is a classification analysis determining which descriptors best define the assignment of individual cases to each of the predetermined groups. Wilks' lambda is a parameter used to evaluate the discriminant power of the entire model, i.e., all the independent variables used, and takes values from 0 to 1; the closer these values are to zero, the more discriminatory the model becomes.

PCA is used to combine highly correlated variables with one another into one new variable called the principal component (PC). The calculation of new factors consists in diagonalizing the correlation or covariance matrix. The choice of matrix depends on whether the original variables require standardization or centering to mean values. In this way, a reduced number of new variables is generated, but explaining the original variance as much as possible.

The purpose of cluster analysis (CA) is to combine cases into groups so that the association within the same group is as large as possible, and with cases from other groups as small as possible. The method of grouping the data used in the presented studies was the k-means method, in which the means for each cluster and in each dimension are examined, which allows assessment of to what extent the created clusters are different from each other. In the analysis of variance, the size of the F statistic performed in each of them shows how well a given dimension separates individual clusters. In the best situation, very different means are obtained for most of the dimensions analyzed.

PLS and RF regression were performed with MATLAB ver. 2019a (The MathWorks, Natick, MA, USA). The performance of the models was assessed by a double cross-validation. The statistical significance was then evaluated using permutation testing.

In the PLS method, the matrix of independent variables is analyzed for latent variables (LVs) that best describe the covariance between X and Y. Then these transformed independent variables are used in regression to predict the Y response. The RF method uses many decision trees which, based on the entered X variables, repeatedly "make a decision" about the predicted value of Y for each case, from which the mean value is then taken.

In regression analyses, it is good practice to divide the set of cases into two sets: training and testing, in order to perform external validation, which will demonstrate the predictive capacity of the model. The training set accounts for approximately 70% of all collected cases and is used to build a regression equation (training model). The rest, i.e., about 30% of cases, are included in the test set on which the equation is validated. The training and test sets are distributed randomly. In order to check the stability of the model and exclude random effects, it is worth carrying out such a division into two subsets and the construction of the equation several times. The Monte Carlo permutation test (MCPT) is used for this. For the training and test sets, RF regression was performed and RMSECV, Q^2 and R^2 were calculated. Then this procedure was repeated 100 times, each time the training and test sets were drawn anew. Furthermore, the distribution of Q^2 in the original and permuted models was compared and a one-way ANOVA was performed. In the next step, 100 training (70%) and test sets (30%) were prepared by randomly splitting the original data matrix. A similar MCPT (100 perm.) was then performed on the training and test sets that were derived from the permuted data matrix. The results of the original and permuted models were obtained and their Q^2 values were compared.

5. Conclusions

Positive results were obtained on the expediency of using chromatographic data in the study of protein binding and the penetration of drugs into breast milk. The presented statistical analyses showed a close relationship between HPLC and TLC analytical data (under set conditions) with the bioavailability of the drug into breast milk. The correlation of the PB and M/P ratios with these chromatographic data is high, also in the group of all cases (acidic, basic and neutral drugs) together. The most effective application of NP TLC and HPLC$_{HSA}$ data was found. There is also a greater correlation between PB and the chromatographic data in the group of acidic drugs (a), i.e., for specific binding to albumin.

The PCA and DFA analyses identified a group of physicochemical and chromatographic parameters important for the bioavailability of drugs in breast milk. The use of CA emphasized the differentiation of their mean values in groups M/P$_{code}$ 1–4.

NP TLC was proved to be the most useful chromatographic method in statistical analyses. In the case of HPLC$_{HSA}$ data, the relatively large share of the results from the column in the creation of the RF model turned out to be interesting. The second factor that emerges in almost all analyses is the high proportion of the log D parameter, i.e., lipophilicity associated with ionization.

Supplementary Materials: The following supporting information can be downloaded at: https://www.mdpi.com/article/10.3390/molecules27113441/s1, Tables S1–S9 contain all data used in statistical analyses.

Author Contributions: Conceptualization, E.B. and K.W.; methodology, E.B.; software, E.B and K.W.; validation, E.B.; formal analysis, K.W.; investigation, E.B.; resources, K.W.; data curation, E.B and K.W.; writing—original draft preparation, K.W.; writing—review and editing, E.B.; visualization, K.W.; supervision, E.B.; project administration, E.B.; funding acquisition, E.B. All authors have read and agreed to the published version of the manuscript.

Funding: This research was funded by internal grant from Medical University of Lodz number 503/3-016-03/503-31-001.

Institutional Review Board Statement: Not applicable.

Informed Consent Statement: Not applicable.

Data Availability Statement: Not applicable.

Conflicts of Interest: The authors declare no conflict of interest.

Appendix A

Table A1. Chromatographic data from TLC and HPLC experiments and their derivatives used in the analysis of analytical models.

Descriptor	n_{abn}	n_b	n_n	n_a	PB$_{abn}$ *	PB$_b$ *	PB$_n$ *	PB$_a$ *
NP	162	49	79	34	**0.31**	**0.31**	0.15	**0.50**
NP/C	162	49	79	34	0.00	−0.11	−0.02	**0.50**
NP/PSA	162	49	79	34	0.19	**0.28**	0.17	**0.37**
NP/B2	162	49	79	34	−0.10	0.02	0.18	**−0.69**
NP/log P	162	49	79	34	0.12	0.02	−0.20	**−0.44**
RP	162	49	79	34	0.01	−0.05	−0.20	0.17
RP/C	162	49	79	34	0.12	0.17	0.19	−0.10
RP/PSA	162	49	79	34	0.11	**0.21**	0.11	−0.03
RP/B2	162	49	79	34	−0.08	0.02	0.07	**−0.44**
RP/log P	162	49	79	34	0.12	0.09	0.18	0.16
log k$_{HSA}$	165	49	80	34	**0.39**	**0.28**	**0.45**	**0.55**
log k$_{HSA}$/B2	165	49	80	36	0.01	0.05	−0.04	0.08

Table A1. Cont.

Descriptor	n_{abn}	n_b	n_n	n_a	PB_{abn} *	PB_b *	PB_n *	PB_a *
log k_{HSA}/log P	165	49	80	36	−0.11	−0.04	−0.16	0.09
log k_{HSA}/PSA	165	49	80	36	0.16	0.11	**0.25**	**0.51**
log k_{IAM}	159	49	74	36	**0.20**	0.17	**0.41**	**0.28**
log k_{IAM}/PSA	159	49	74	36	−0.05	−0.04	0.07	−0.07
log k_{IAM}/log P	159	49	74	36	0.04	0.11	−0.05	−0.04
log k_{IAM}/B2	159	49	74	36	−0.03	−0.06	−0.04	−0.06

* correlation with chromatographic data.

Table A2. Physicochemical parameters of APIs and their correlation with data on PB.

Descriptor	n_{abn}	n_b	n_n	n_a	PB_{abn} *	PB_b *	PB_n *	PB_a *
acid/base	166				−0.15			
B1	129	34	66	29	**0.28**	**0.36**	**0.48**	0.13
B2	166	50	81	35	0.12	0.13	**0.27**	0.05
B3	166	50	81	35	0.13	0.11	**0.21**	0.05
log U/D	160	50	75	35	0.05	0.16	0.02	**0.22**
DM	160	47	79	34	−0.02	0.04	−0.04	−0.16
Sa/V	160	47	79	34	**−0.29**	**−0.34**	**−0.32**	−0.04
eH	160	47	79	34	0.05	0.13	−0.02	0.17
MW	162	48	79	35	−0.17	0.14	0.24	0.00
HD	166	50	81	35	**−0.23**	−0.07	**−0.39**	**−0.23**
HA	166	50	81	35	−0.14	−0.16	**−0.23**	−0.13
eL	160	47	79	35	0.03	0.14	0.00	−0.015
eH-eL	160	50	79	35	0.01	−0.08	−0.01	0.12
log P	160	49	79	35	**0.31**	0.10	**0.34**	**0.41**
log D	160	50	81	35	**0.28**	0.19	**0.38**	**0.30**
MW/V	160	47	79	35	0.03	0.18	0.09	0.03
PhCharge	165	50	80	35	−0.13	−0.05	0.06	**−0.20**
pKa	160	50	75	35	−0.05	−0.15	0.08	**0.22**
M/P	104	30	55	19	**−0.29**	**−0.20**	**−0.35**	0.11
CNS+/−	154	49	72	33	−0.18	−0.05	0.16	**0.33**

* correlation with physicochemical data.

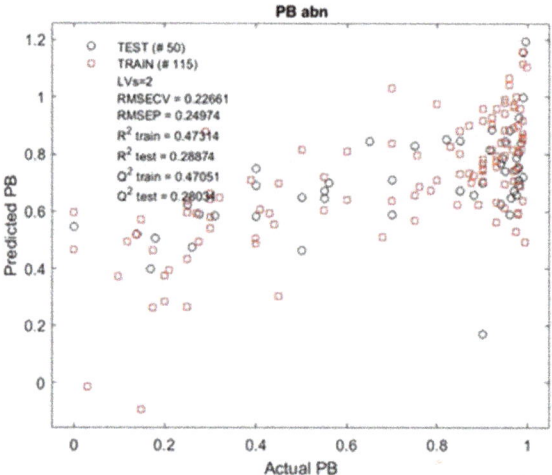

Figure A1. Actual versus predicted PB_{abn} values using PLS modelling and 23 molecular descriptors including NP TLC data. LVs = latent variables, $RMSE_{CV}$ = root-mean-square error of cross-validation, $RMSE_P$ = root-mean-square error of prediction, R^2 train/test = coefficient of determination for train/test set models, Q^2 train/test = coefficient of determination for the cross-validated models.

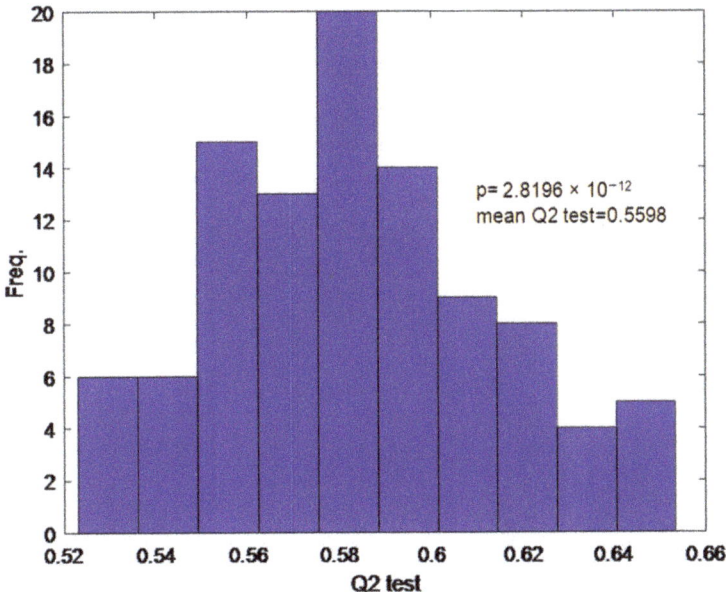

Figure A2. Monte Carlo permutation test (MCPT) showing Q^2 obtained from RF regression models developed on the test set, the number of repetitions was $n = 100$. The mean value of Q^2 was 0.5598 at the significance level $p = 2.8196 \times 10^{-12}$.

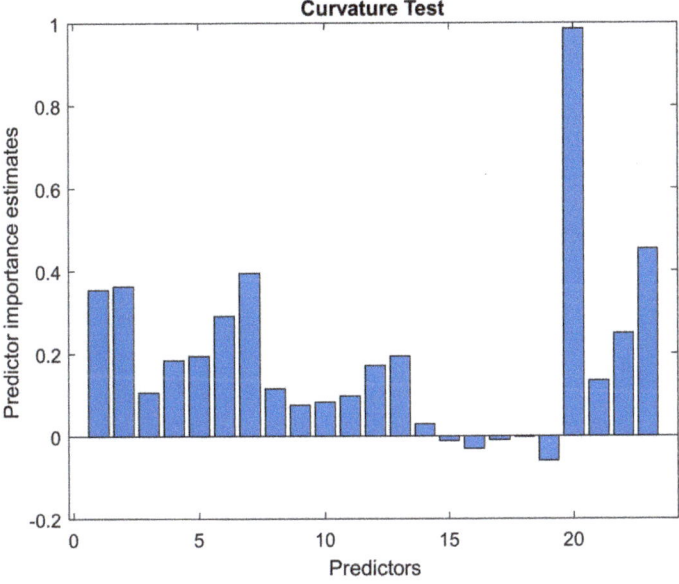

Figure A3. Contribution of individual descriptors to the generation of the RF regression model for PB_{abn}. The greatest influence is shown by the descriptor no. 20, i.e., log D.

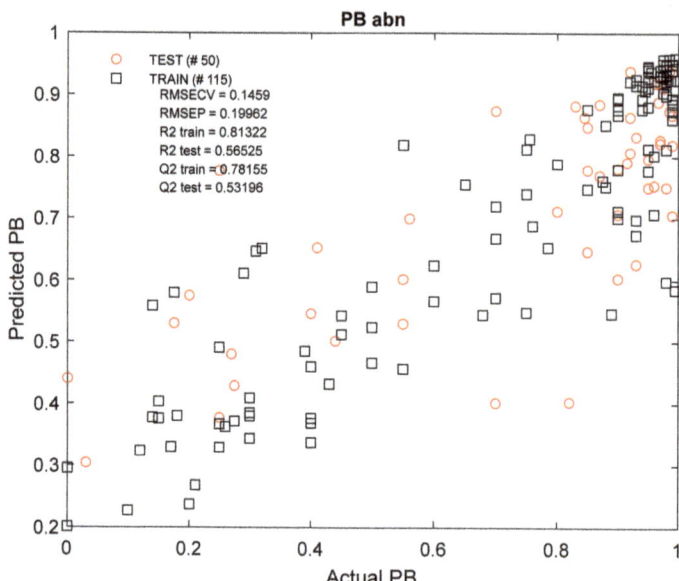

Figure A4. Actual versus predicted PB_{abn} values, using RF regression modelling of molecular descriptor set containing 22 variables along with $HPLC_{HSA}$ data. $RMSE_{CV}$ = root-mean-square error of cross-validation, $RMSE_P$ = root-mean-square error of prediction, R^2 train/test = coefficient of determination for train/test set models, Q^2 train/test = coefficient of determination for the cross-validated models.

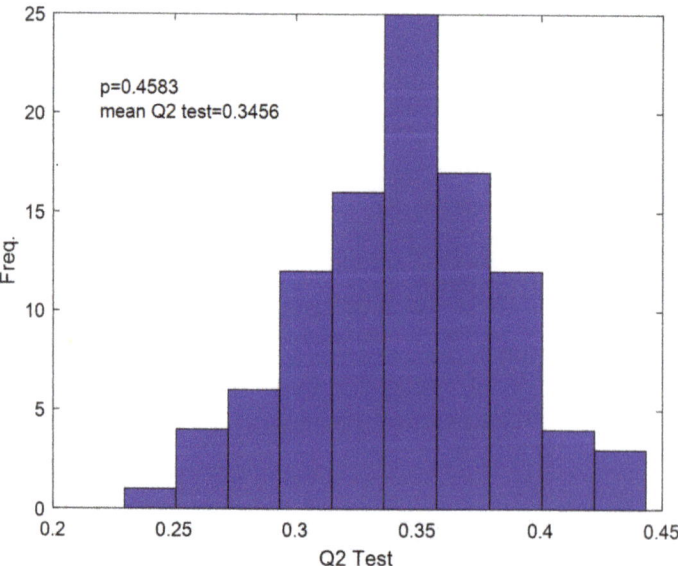

Figure A5. Monte Carlo permutation test (MCPT) showing Q^2 obtained from RF regression models developed on the test set, the number of repetitions was n = 100. The mean value of Q^2 was 0.3456 at the significance level p = 0.4583.

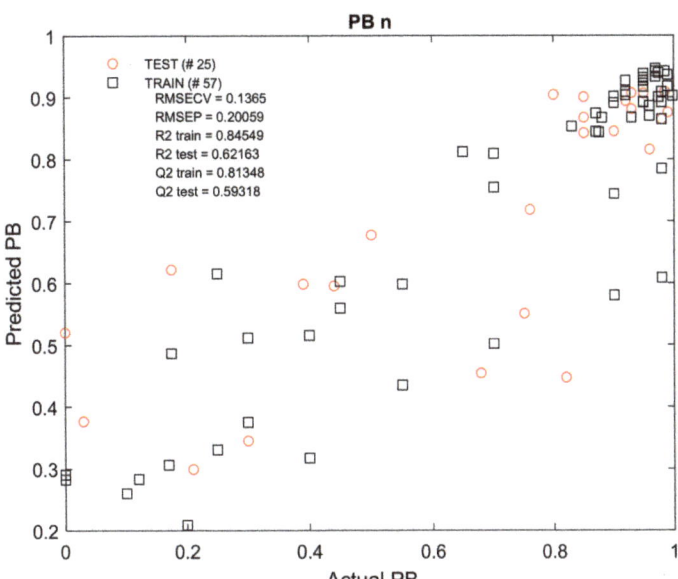

Figure A6. Actual versus predicted PB$_n$ values using RF regression modelling of molecular descriptor set containing 23 variables along with NP TLC data. RMSE$_{CV}$ = root-mean-square error of cross-validation, RMSE$_P$ = root-mean-square error of prediction, R^2 train/test = coefficient of determination for train/test set models, Q^2 train/test = coefficient of determination for the cross-validated models.

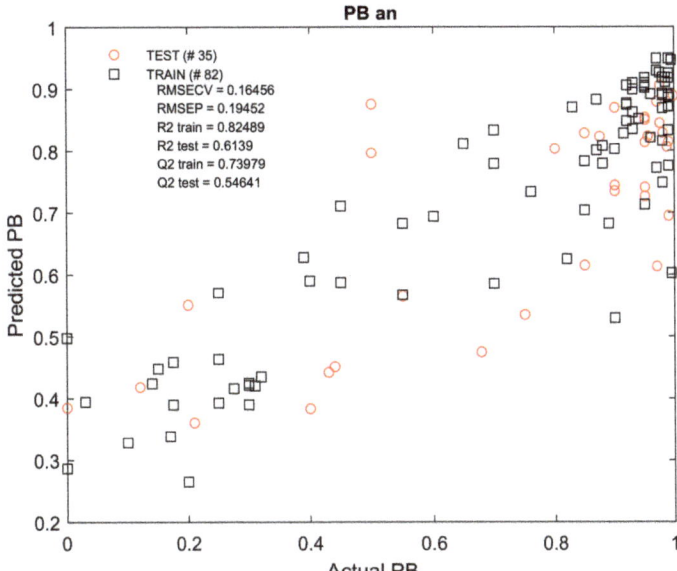

Figure A7. Actual versus predicted PB$_{an}$ values, using RF regression modelling of molecular descriptor set containing 23 variables along with NP TLC data.

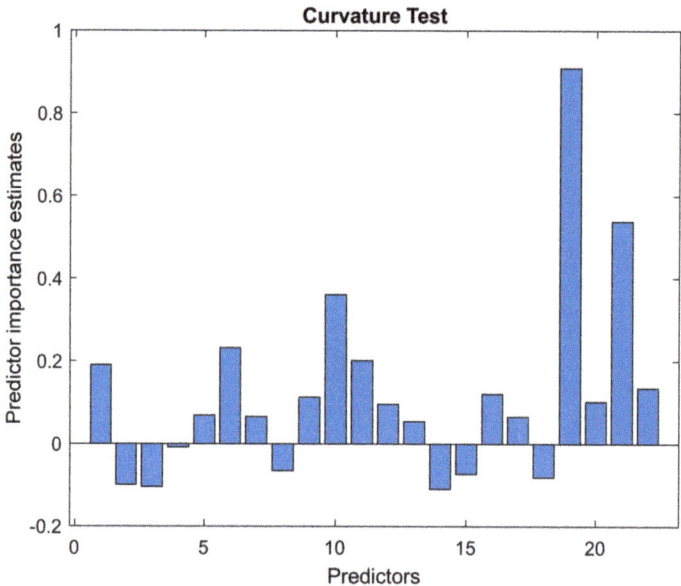

Figure A8. Contribution of individual descriptors to the development of the RF regression model for PB_n. The greatest influence is shown by the descriptor no. 19, i.e., log D, besides this, the molar weight (MW) and molar volume (V) are important.

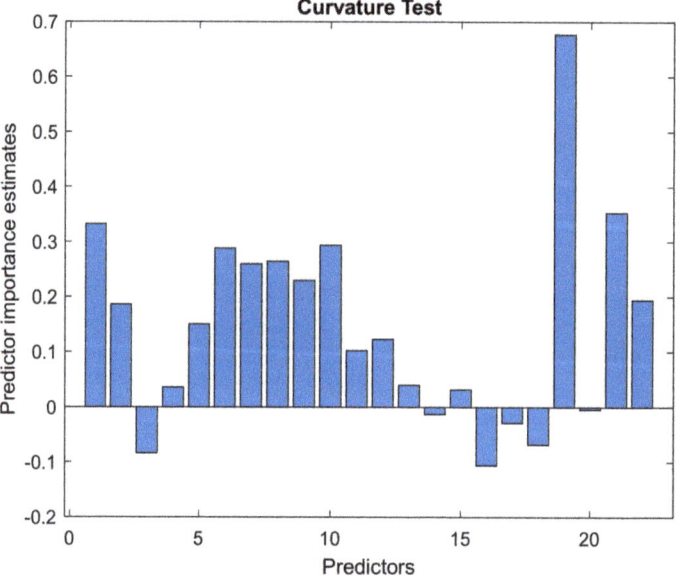

Figure A9. Contribution of individual descriptors to the development of the RF regression model for PB_{an}. The greatest influence is shown by the descriptor no. 19, i.e., log D, in this case a greater share of chromatographic parameters can be seen (descriptors nos. 6–9).

Appendix B Chromatographic Experiments

Appendix B.1 Materials and Reagents

For TLC chromatography, glass plates 20 × 20 cm from Merck, covered with silica gel with the addition of a fluorescent indicator, were used. Normal phase (NP) plates were used with standard Merck TLC Silica gel 60 F254 plates, while in reverse phase (RP) silanized plates RP-2: Merck TLC Silica gel 60 RP-2 F254 were used.

Solvents from J.T. Baker-Water, Methanol and Acetonitrile, with an HPLC gradient grade. Ammonium acetate p.a. was used to prepare an acetate buffer at pH 7.4.

The stationary phase of the plates, both NP and RP, was modified with an aqueous solution of bovine serum albumin purchased from Sigma Aldrich (bovine serum albumin, lyophilized powder).

Human serum albumin immobilized chromatography column was from Daicel: CHIRALPAK®HSA, 5 µm; 4 × 10 mm while column with IAM artificial membrane from Regis Technologies Inc.: IAM.PC.DD.2, 10 µm; 4.6 × 10 mm.

In HPLC chromatography, the organic solvents used (acetonitrile and methanol) and water were also obtained from J.T. Baker (HPLC gradient). LACH-NER ammonium acetate, ammonium acetate p.a. were used to prepare the acetate buffer (HPLCHSA), while to prepare the phosphate buffer (HPLCIAM) a ready-made reagent in the form of tablets (Sigma, Phosphate buffered saline, tablets)was used to be dissolved in a strictly defined amount of water for HPLC.

Appendix B.2 Isolation of Active Pharmaceutical Ingredients (APIs)

A total of 167 active pharmaceutical ingredients (APIs), isolated from pharmaceutical preparations, usually tablets or hard capsules, were used in the chromatographic experiments. Tablets (without coatings) or the contents of capsules crushed in a mortar were placed in 100 mL of 99.8% methanol, mixed with a magnetic stirrer for approximately 30 min and then passed to crystallization tanks through a funnel with a filter. The vessel with the filtrate was allowed to evaporate the solvent and the crystallized active substance was transferred to sealed vials, kept under refrigerated conditions.

The purity of the isolated substances was checked by TLC chromatography and densitometric scanning. All substances isolated gave single densitometric peaks and were used without further purification. The obtained API was dissolved in 99.8% methanol to give 1 mg/mL solutions which were then used in TLC and HPLC.

Appendix B.3 Impregnation of TLC Plates

The surface-modifying protein of the stationary phase of thin-layer chromatography plates was bovine serum albumin (BSA), which is a cheaper substitute for human albumin, with 76% homology and similar drug binding properties [14–18].

The impregnation of the plates was carried out with a 2 mg/mL solution applied to the surface using a Desaga SG 1 hand sprayer; the plates were then air dried. The best concentration was selected earlier—on NP plates impregnated with 1, 2 and 4 mg/mL BSA solutions, active substances were applied at a concentration of 1 mg/mL (solutions in 99.8% methanol), characterized by a different degree of protein binding described in the literature. Retention values differed significantly between plates coated with 1 and 2 mg/mL BSA, but no difference was found between 2 and 4 mg/mL. Therefore, it was decided to use a ratio of 1:2, drug concentration to BSA concentration on the plate.

Appendix B.4 TLC Chromatography

Normal and reversed phase thin layer chromatography (NP TLC and RP TLC respectively) was performed using silica-gel-coated glass plates. Half of them were covered with 2 mg/mL bovine serum albumin solution and half remained pure.

Solutions of the isolated APIs in 99.8% methanol (1 mg/mL) were applied to the plates using a Desaga HPTLC-Applicator AS 30 automatic applicator. The mobility of the compounds was also determined on the plates with no protein as a modifier. They have been marked as controls (C) and will allow evaluation of the influence of the modifier on API mobility. The plates were then developed in a mobile phase consisting of acetonitrile, acetate buffer pH 7.4 and methanol in the ratio 60:20:20 ($v/v/v$). The acetate buffer (20 mM) was prepared by dissolving 1.54 g of ammonium acetate in 1 L of distilled water. The pH was then adjusted with a concentrated ammonia solution using a pH meter. The plates were developed in standard, vertical chromatographic chambers, each time using 100 mL of the mobile phase, after the chamber was previously saturated with solvent vapors for approximately 1 h.

The unfolded-protein-impregnated plates and the control plates were scanned with a Desaga CD 60 densitometer. The values of the delay factor (R_f) were collected, i.e., the ratio of the distance traveled by the substance to be analyzed to the distance traveled through the front of the mobile phase. The analytical wavelengths were selected individually for each API using the multi-wavelength scanning option (values ranged from 200 to 300 nm). The experiment was repeated (for both BSA-coated and control plates) and the R_f values pooled are the mean of both series of experiments.

Appendix B.5 HPLCHSA Chromatography

High performance liquid chromatography was performed using a chromatography column with immobilized human serum albumin. The assay was performed on a Perkin Elmer Series 200 instrument connected to a UV-VIS spectrometer as detector. The analytical wavelength was the same for all compounds at 210 nm. The experiment was carried out with the 1 mg/mL methanolic solutions of active substances previously described. The mobile phase was a mixture of 10 mM acetate buffer pH 7.4, acetonitrile and methanol in the ratio 85:10:5 ($v/v/v$). The acetate buffer was prepared by dissolving 0.77 g of ammonium acetate in 1 L of distilled water. The pH was then adjusted with a concentrated ammonia solution using a pH meter.

The phase flow through the system was set to 0.9 mL/min as recommended by the column manufacturer. The solutions were delivered to the column using an autosampler syringe, the injection size was 10 µL. Since the column could not be thermostated, the room was kept at a constant temperature of 25 degrees Celsius.

Chromatographic data (retention coefficient, k, and derivative, log k) were obtained with TotalChrom software connected to an HPLC instrument. The k coefficient, which is the ratio between the amount of analyte in the stationary phase and its amount in the mobile phase, was obtained from the equation $k = (t_R - t_M)/t_M$, where t_R is the retention time of the analyzed substance and t_M is the dead time (the dead time marker was 99.8% methanol). The experiment was then repeated and the collected retention rates were the mean values of both series.

Appendix B.6 HPLCIAM Chromatography

The second experiment was performed using an immobilized artificial membrane (IAM) column. The assay was also performed on a Perkin Elmer Series 200 instrument connected to a UV-VIS spectrometer as the detector. The analytical wavelength was the same for all compounds, at 210 nm. The experiment was carried out with the 1 mg/mL methanolic solutions of active substances previously described. The mobile phase was a mixture of 10 mM phosphate buffer pH 7.4 and acetonitrile in the ratio 80:20 (v/v). The phosphate buffer was obtained by dissolving the finished tablet in the appropriate amount of distilled water (1 tablet per 200 mL). In this case, it was not necessary to adjust the pH of the buffer using a pH meter.

The phase flow through the system was set to 0.5 mL/min as recommended by the column manufacturer. The solutions were delivered to the column using an autosampler syringe, the injection size was 10 µL.

The collected chromatographic data, similar to the HSA column experiment, was the retention coefficient, k, and derivative, log k, which were obtained using TotalChrom software connected to the HPLC instrument. The experiment was then repeated and the collected retention rates were the mean values of both series.

References

1. Wanat, K.; Khakimov, B.; Brzezińska, E. Comparison of statistical methods for predicting penetration capacity of drugs into human breast milk using physicochemical, pharmacokinetic and chromatographic descriptors. *SAR QSAR Environ. Res.* **2020**, *31*, 457–475. [CrossRef] [PubMed]
2. Wanat, K.; Żydek, G.; Hekner, A.; Brzezińska, E. In silico plasma protein binding studies of selected group of drugs using TLC and HPLC retention data. *Pharmaceuticals* **2021**, *14*, 202. [CrossRef] [PubMed]
3. Chembl Database. Available online: https://www.ebi.ac.uk/chembl/ (accessed on 1 March 2022).
4. Norinder, U.; Haeberlein, M. Computational approaches to the prediction of the blood-brain distribution. *Adv. Drug Deliv. Rev.* **2002**, *54*, 291–313. [CrossRef]
5. Yang, F.; Zhang, Y.; Liang, H. Interactive association of drugs binding to human serum albumin. *Int. J. Mol. Sci.* **2014**, *15*, 3580–3595. [CrossRef] [PubMed]
6. Ozeki, S.; Tejima, K. Drug Interactions. II. Binding of Some Pyrazolone and Pyrazolidine Derivatives to Bovine Serum Albumin. *Chem. Pharm. Bull.* **1974**, *22*, 1297–1301. [CrossRef] [PubMed]
7. Drugbank. Available online: https://www.drugbank.ca/drugs/DB01174 (accessed on 29 March 2022).
8. Agatonovic-Kustrin, S.; Tucker, I.G.; Zecevic, M.; Zivanovic, L.J. Prediction of drug transfer into human milk from theoretically derived descriptors. *Anal. Chim. Acta* **2000**, *418*, 181–195. [CrossRef]
9. Hale, T.W. *Medications and Mother's Milk*, 15th ed; Pharmasoft Medical Publishing: Amarillo, TX, USA, 2012.
10. Katritzky, A.R.; Dobchev, D.A.; Hür, E.; Fara, D.C.; Karelson, M. QSAR treatment of drugs transfer into human breast milk. *Bioorg. Med. Chem.* **2005**, *13*, 1623–1632. [CrossRef] [PubMed]
11. Meskin, M.S.; Lien, E.J. QSAR analysis of drug excretion into human breast milk. *J. Clin. Pharm. Ther.* **1985**, *10*, 269–278. [CrossRef] [PubMed]
12. Wilson, J.T.; Brown, R.D.; Cherek, D.R.; Dailey, J.W.; Hilman, B.; Jobe, P.C.; Manno, B.R.; Manno, J.E.; Redetzki, H.M.; Stewart, J.J. Drug Excretion in Human Breast Milk: Principles, Pharmacokinetics and Projected Consequences. *Clin. Pharmacokinet.* **1980**, *5*, 1–66. [CrossRef] [PubMed]
13. Abraham, M.H.; Gil-Lostes, J.; Fatemi, M. Prediction of milk/plasma concentration ratios of drugs and environmental pollutants. *Eur. J. Med. Chem.* **2009**, *44*, 2452–2458. [CrossRef] [PubMed]
14. Tunç, S.; Çetinkaya, A.; Duman, O. Spectroscopic investigations of the interactions of tramadol hydrochloride and 5-azacytidine drugs with human serum albumin and human hemoglobin proteins. *J. Photochem. Photobiol. B Biol.* **2013**, *120*, 59–65. [CrossRef] [PubMed]
15. Carter, D.C.; Ho, J.X. Structure of serum albumin. *Adv. Protein Chem.* **1994**, *45*, 153–203. [CrossRef] [PubMed]
16. Ayranci, E.; Duman, O. Binding of fluoride, bromide and iodide to bovine serum albumin, studied with ion-selective electrodes. *Food Chem.* **2004**, *84*, 539–543. [CrossRef]
17. Ayranci, E.; Duman, O. Binding of Lead Ion to Bovine Serum Albumin Studied by Ion Selective Electrode. *Protein Pept. Lett.* **2004**, *11*, 331–337. [CrossRef] [PubMed]
18. Raoufinia, R.; Mota, A.; Keyhanvar, N.; Safari, F.; Shamekhi, S.; Abdolalizadeh, J. Overview of albumin and its purification methods. *Adv. Pharm. Bull.* **2016**, *6*, 495–507. [CrossRef] [PubMed]

Article

Anti-Inflammatory Potential of Fucoidan for Atherosclerosis: In Silico and In Vitro Studies in THP-1 Cells

Etimad Huwait [1,2,*], Dalal A. Al-Saedi [1,2] and Zeenat Mirza [3,4,*]

1 Department of Biochemistry, Faculty of Sciences, King Abdulaziz University, Jeddah 21589, Saudi Arabia; dalal.alsaadi@hotmail.com
2 Cell Culture Lab, Experimental Biochemistry Unit, King Fahd Medical Research Centre, King Abdulaziz University, Jeddah 21589, Saudi Arabia
3 King Fahd Medical Research Center, King Abdulaziz University, Jeddah 21589, Saudi Arabia
4 Department of Medical Laboratory Sciences, Faculty of Applied Medical Sciences, King Abdulaziz University, Jeddah 21589, Saudi Arabia
* Correspondence: ehuwait@kau.edu.sa (E.H.); zmirza1@kau.edu.sa (Z.M.); Tel.: +966-553-017-824 or +966-12-640-1000 (ext. 72074) (Z.M.); Fax: +966-12-6952076 (Z.M.)

Abstract: Several diseases, including atherosclerosis, are characterized by inflammation, which is initiated by leukocyte migration to the inflamed lesion. Hence, genes implicated in the early stages of inflammation are potential therapeutic targets to effectively reduce atherogenesis. Algal-derived polysaccharides are one of the most promising sources for pharmaceutical application, although their mechanism of action is still poorly understood. The present study uses a computational method to anticipate the effect of fucoidan and alginate on interactions with adhesion molecules and chemokine, followed by an assessment of the cytotoxicity of the best-predicted bioactive compound for human monocytic THP-1 macrophages by lactate dehydrogenase and crystal violet assay. Moreover, an in vitro pharmacodynamics evaluation was performed. Molecular docking results indicate that fucoidan has a greater affinity for L-and E-selectin, monocyte chemoattractant protein 1 (MCP-1), and intercellular adhesion molecule-1 (ICAM-1) as compared to alginate. Interestingly, there was no fucoidan cytotoxicity on THP-1 macrophages, even at 200 µg/mL for 24 h. The strong interaction between fucoidan and L-selectin in silico explained its ability to inhibit the THP-1 monocytes migration in vitro. MCP-1 and ICAM-1 expression levels in THP-1 macrophages treated with 50 µg/mL fucoidan for 24 h, followed by induction by IFN-γ, were shown to be significantly suppressed as eight- and four-fold changes, respectively, relative to cells treated only with IFN-γ. These results indicate that the electrostatic interaction of fucoidan improves its binding affinity to inflammatory markers in silico and reduces their expression in THP-1 cells in vitro, thus making fucoidan a good candidate to prevent inflammation.

Keywords: fucoidan; alginate; L-selectin; E-selectin; MCP-1; ICAM-1; molecular docking; THP-1 macrophage; monocyte migration

1. Introduction

Inflammation is the prime cause of cardiovascular diseases (CVDs), one of the most frequent reasons of death worldwide [1]. Atherosclerosis is one of the most frequent CVDs; it is an arterial hardening and subsequent narrowing caused by lipid deposition and gradual plaque buildup within an artery wall. This complex disease is initiated by inflammation and often leads to a stroke or heart attack [2]. The pathogenesis of atherosclerosis starts with the interaction between leukocytes and endothelium, followed by immune cell migration to the inflammatory lesion in a multi-step process called extravasation monocyte, involving adhesion and signaling molecules such as selectins and chemokines. This is characterized by tethering, the rolling of monocytes on vascular surfaces of endothelium, tight adhesion, and migration during inflammation [3].

Selectins (L- and E-selectin) are transmembrane receptors, which are expressed on leukocytes and activated endothelial cells, respectively. Their function is based on the extracellular lectin domain's calcium-dependent interaction with Lewisx sialyl (sLex) tetrasaccharide expressed on the glycoprotein [4,5], which mediates the initial stage of cell adhesion on the endothelial cell surface. Endothelial activation in response to proinflammatory cytokines secretes chemokines such as monocyte chemoattractant protein 1 (MCP-1 aka C-C motif chemokine 2, CLL2), which activates the C-C chemokine receptor type 2 (CCR2) on monocyte, followed by stimulating integrin β2 to a high-affinity state, enabling it to bind to intercellular adhesion molecule-1 (ICAM-1), which is overexpressed due to endothelial dysfunction [3,6]. The mechanism that hinders monocyte migration through blocking inflammatory biomarkers is critical to the early halting of inflammation.

Glycosaminoglycans (GAGs) are negatively charged linear polysaccharide chains that are covalently bound to proteins, for example, dermatan sulfate, heparan sulfate, and chondroitin sulfate, which mediate significant physiological functions, for instance, inflammation, through signaling and recognition [7–10]. Marine natural products are a promising therapeutic source of bioactive compounds. The polysaccharides derived from macroalgae have gained worldwide attention due to their myriad of structural, physicochemical, and biological activities. Fucoidan is a sulfated polysaccharide, which is mostly built from sulfated L-fucose molecules and other monomeric sugars, such as glucose, galactose, mannose, and uronic acid. Alginates are natural linear copolymers of α-L-guluronic acid and β-D-mannuronic acid, which widely exist in brown seaweeds [11,12]. The structure–activity relationship of fucoidan affects diverse biological activities; indicating promising pharmacological potential [13–16], although the molecular mechanism is still unknown. A recent study indicates that fucoidan extracts from different algal species, including *Fucus vesiculosus*, reduce the inflammatory cytokine levels in lipopolysaccharide-stimulated peripheral blood mononuclear cells and leukemia monocytic cell line (THP-1) in a dose-dependent fashion [17]. Moreover, fucoidan from *Fucus vesiculosus* inhibits lung cancer cell migration and invasion via phosphatidylinositol-3-kinase (PI3K)/Akt and the mammalian target of rapamycin (mTOR) signaling [18].

The computational prediction of the interaction between bioactive compounds and therapeutic target proteins rationally guides experimental methods and significantly reduces the cost of drug development [19]. Therefore, we predict and compare the ability of sulfated (fucoidan) and non-sulfated (alginate) polysaccharides to computationally interact with several targets that are implicated in inflammation, particularly endothelial dysfunction and monocyte migration, including L-selectin, E-selectin, MCP-1, and ICAM-1, by using the most frequently occurring monomer in polysaccharides. We also explored the pharmacodynamics of fucoidan on the above-mentioned inflammatory markers in THP-1 cells.

2. Results

This study aimed to understand the interaction between fucoidan and alginate with the potential inflammatory biomarkers (L-selectin, E-selectin, MCP-1, and ICAM-1, respectively) that are vital in monocyte migration. Firstly, the PPIs were predicted, followed by docking, to illustrate protein–ligand interactions. Finally, the best prediction was validated using experimental methods. Figure 1 summarizes our study approach.

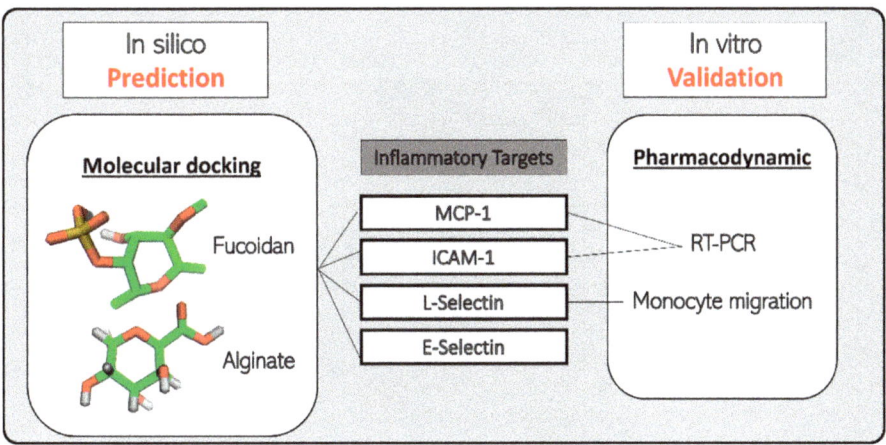

Figure 1. Study approach for prediction and validation of marine bioactive compounds (fucoidan and alginate) with inflammatory protein targets.

2.1. Protein–Protein Interaction

A summarized network of the predicted associations for inflammatory proteins is illustrated in Figure 2. Interestingly, the ICAM1 protein could be linked to seven predicted functional partners, namely, L-selectin (SELL), E-selectin (SELE), C-C chemokine receptor type 2 (CCR2), C-C motif chemokine 2 (CCL2), integrin subunit β2 (ITGB2), integrin α-M (ITGAM), and integrin αl (ITGAL). As such, all proteins were directly related to SELL, except for SELE, which was indirectly linked through GLG1. Integrin α-M acts as the network hub, while the evidence of co-expression showed the most associations between the above-mentioned protein entities.

2.2. Chemoinformatic Analysis

The ligand-based target prediction of classified targets fucoidan and alginate molecules by SwissTargetPrediction tool (Figure 2B,C) shows seven and four classes of human proteins, respectively. Alginate most likely interacts with G-protein coupled receptors (GPCRs) as (80%), whereas fucoidan can trigger GPCRs and secreted proteins at a similar rate of 13.13%. Physicochemical properties have great significance from the perspective of the medicinal chemistry of the drug development process. The SwissADME analysis for fucoidan and alginate monomer illustrates that both are hydrophilic (MLog p = −1.49 and −2.89, respectively). They have same number of hydrogen bond acceptors and a different number of hydrogen donors and rotatable bonds (Table 1). The 2D chemical structure of both the dietary ligands is shown in Figure 3.

Table 1. Analysis of swissADME for marine bioactive compounds (fucoidan and alginate).

Physicochemical Properties	Fucoidan	Alginate
MLogP	−1.49	−2.89
Molecular weight	256.27	193.13
Number of H-bond acceptors	7	7
Number of H-bond donors	2	4
Number Rotatable bonds	3	1

Note: LogP refers to the octanol–water partition coefficient.

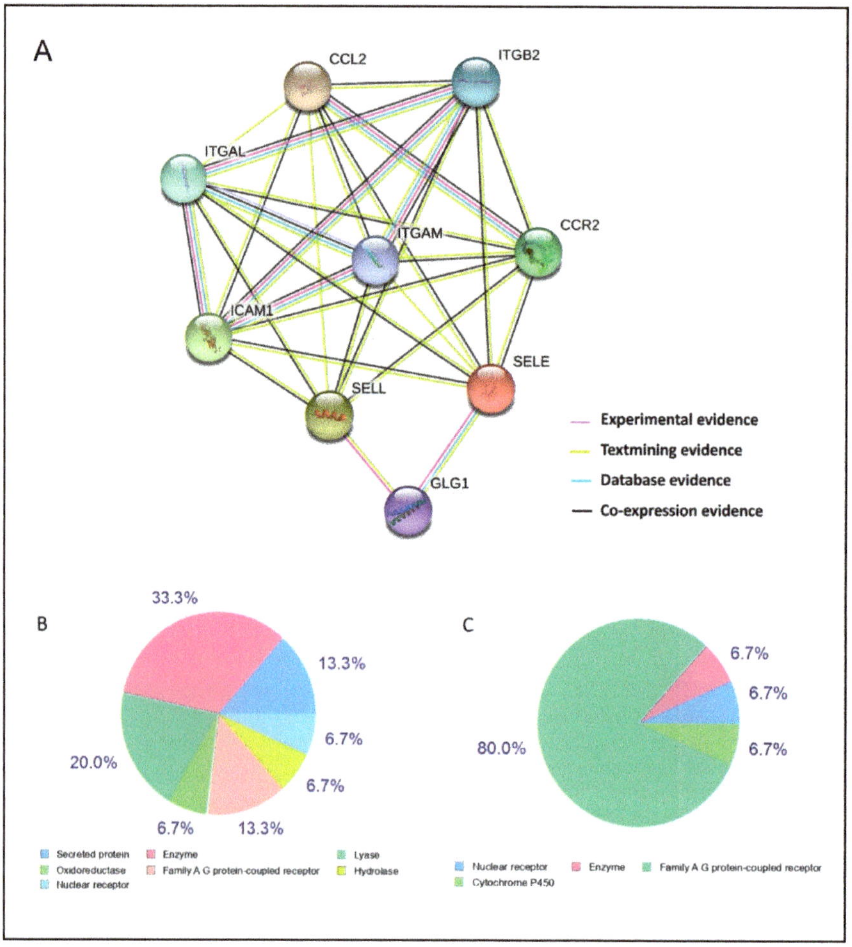

Figure 2. Prediction of biomolecular interactions. (**A**): STRING protein–protein interaction (PPI). Nodes in the network represent proteins, and different types of interaction evidence are indicated by interconnecting colored lines (co-occurrence: blue; purple; experimental: purple; text-mining: yellow; database: light blue; co-expression: black). (**B,C**): Swiss Target Prediction of the top 15 target categories for marine bioactive compounds: fucoidan and alginate, respectively.

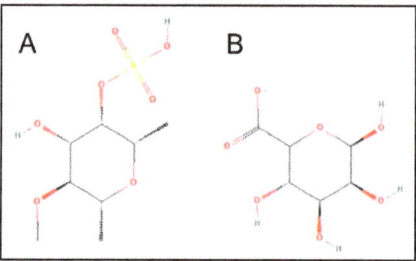

Figure 3. 2D chemical structure of fucoidan (**A**) and alginate (**B**).

2.3. Molecular Docking and Potential Binding Site Prediction

Molecular docking prediction was carried out to estimate binding affinity of fucoidan and alginate with target proteins, considering root mean square deviation <2Å. The lower binding energy corresponds to the higher affinity of the protein–ligand complex. As compared to alginate, fucoidan shows the higher binding affinity and lowest inhibition constant to the four target inflammatory proteins, as summarized in Table 2. The estimated free energy for the binding of L-selectin to fucoidan was −5.82 kcal/mol via the hydrogen bonds formed with Lys48 (1.9Å), Asn105 (2.0Å), Lys111 (2.0Å), and two bonds with Glu88 (1.9 and 2.0Å) in coordination with Ca^{2+} in the putative binding site (Figure 4A). In contrast, alginate binds via two hydrogen bonds with Glu88 (2.4 and 1.9Å) residues of L-selectin (Figure 4C) in a different position compared to fucoidan, without coordination with Ca^{2+}, resulting in a lower binding energy (−4.3 kcal/mol), and exhibit lesser electrostatic interactions, as shown in Figure 4B,D by the red region of the molecular surface. Based on energy and binding affinity values, the interaction of fucoidan with E-selectin is superior to that of alginate. The sulfate groups in fucoidan bind with the Asn83 (2.71Å) and Asp106 (2.06Å) residues at the carbohydrate recognition site in coordination with Ca^{2+} (Figure 4E). Stronger hydrogen bonds are formed by alginate without the assistance of Ca^{2+} (Figure 4G).

Table 2. AutoDock docking results of marine bioactive compounds (fucoidan and alginate) with inflammatory proteins.

Protein	PDB ID	Ligand	Binding Energy (Kcal/mol)	Inhibition Constant (Ki)	Interacting Residues
L-selectin	5VC1	Fucoidan	−5.82	54.41 μM	Lys84(1.9Å), Glu88(1.9Å), Tyr94(2.6Å), Asn105 (2.0Å), Lys111(2.0Å),
		Alginate	−4.3	704.72 μM	Lys55(2.2 Å), Trp60(2.6Å), Glu88(1.9 Å),
E-selectin	1G1T	Fucoidan	−5.69	67.62 μM	Lys55(2.0Å), Asn58(2.1Å), Asn83(2.7Å), Arg84(2.6Å), Asp106(2.1Å)
		Alginate	−4.09	997.16 μM	Asn58(1.9 Å), Trp60(2.0 Å), Lys74(1.9Å), Trp76(1.8 Å)
MCP-1	1DOK	Fucoidan	−5.67	69.96 μM	Cys11(1.8 Å), Tyr13(2.1 Å), Asn14(2.0Å), Cys52(1.9 Å)
		Alginate	−3.84	1.52 mM	Asn14(2.0 Å), Glu50(2.1Å), Cys52(1.7Å)
ICAM-1	1IAM	Fucoidan	−5.66	70.39 μM	Leu33(1.7Å), Lys39(2.4Å), Glu41(1.8Å), Lys50(1.8Å), Tyr52(1.9Å), Tyr66(2.1Å)
		Alginate	−4.98	224.33 μM	Leu33(1.9Å), Lys39(1.9Å), Glu41(2.3 Å), Lys50(2.5Å), Tyr52(2.1Å), Tyr66(2.2Å)

During molecular visualization, it was observed that fucoidan docked to MCP-1 within the N-loop (Tyr13 and Asn14) and β3-strand (Cys11 and Cys52) with an estimated free binding energy of −5.67 kcal/mol. Alginate binds with similar residues of MCP-1, except that the interactions between Cys 11 and Tyr13 are not seen; instead H-bonds with Glu50 are noticed, which decreases the binding energy to −3.84 kcal/mol (Figure 5A–D). ICAM-1 non-covalently interacts with fucoidan and alginate through similar residues (Lys50, Lys39, Tyr66, Tyr52, Glu41, and Leu33) but short-distance hydrogen bonds in the binding site. Fucoidan's binding interaction energy was observed to be −5.66 kcal/mol (Figure 5E–H). Based on in silico data, fucoidan was chosen to understand the nature of interactions with selected inflammatory markers (L-selectin, MCP-1, and ICAM-1) in further in vitro studies.

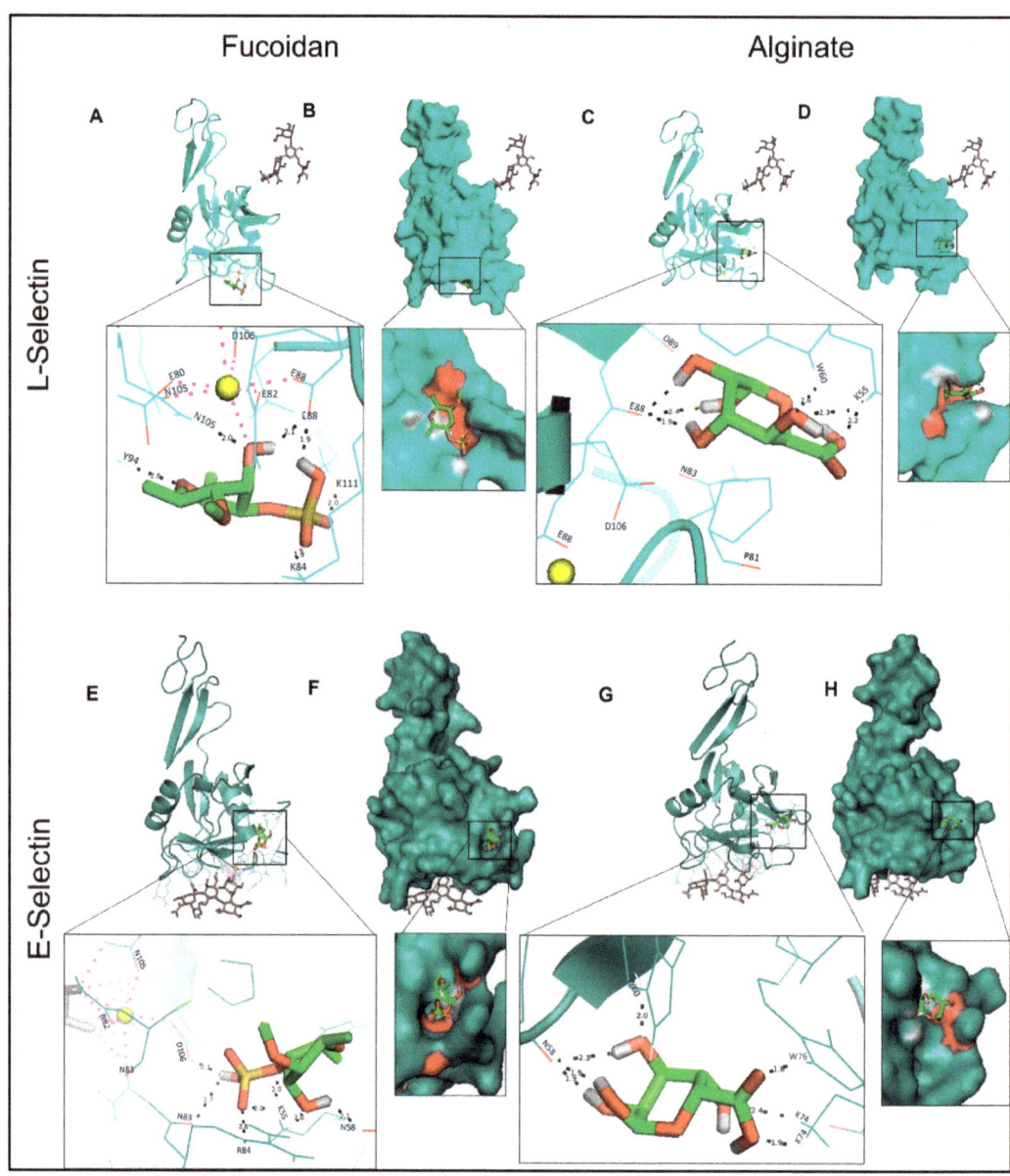

Figure 4. Molecular docking of L- and E-selectins with fucoidan and alginate. (**A,C,E,G**): 3D structures of proteins bound with N-linked glycan moieties (brown) and zoomed ligand-binding pocket. Black and purple dotted lines, respectively, describe the H-bonds and Ca^{2+} (yellow ball) coordination bonds. (**B,D,F,H**): molecular surface representation, and the red patches on the surface represent electrostatics of the binding cavity.

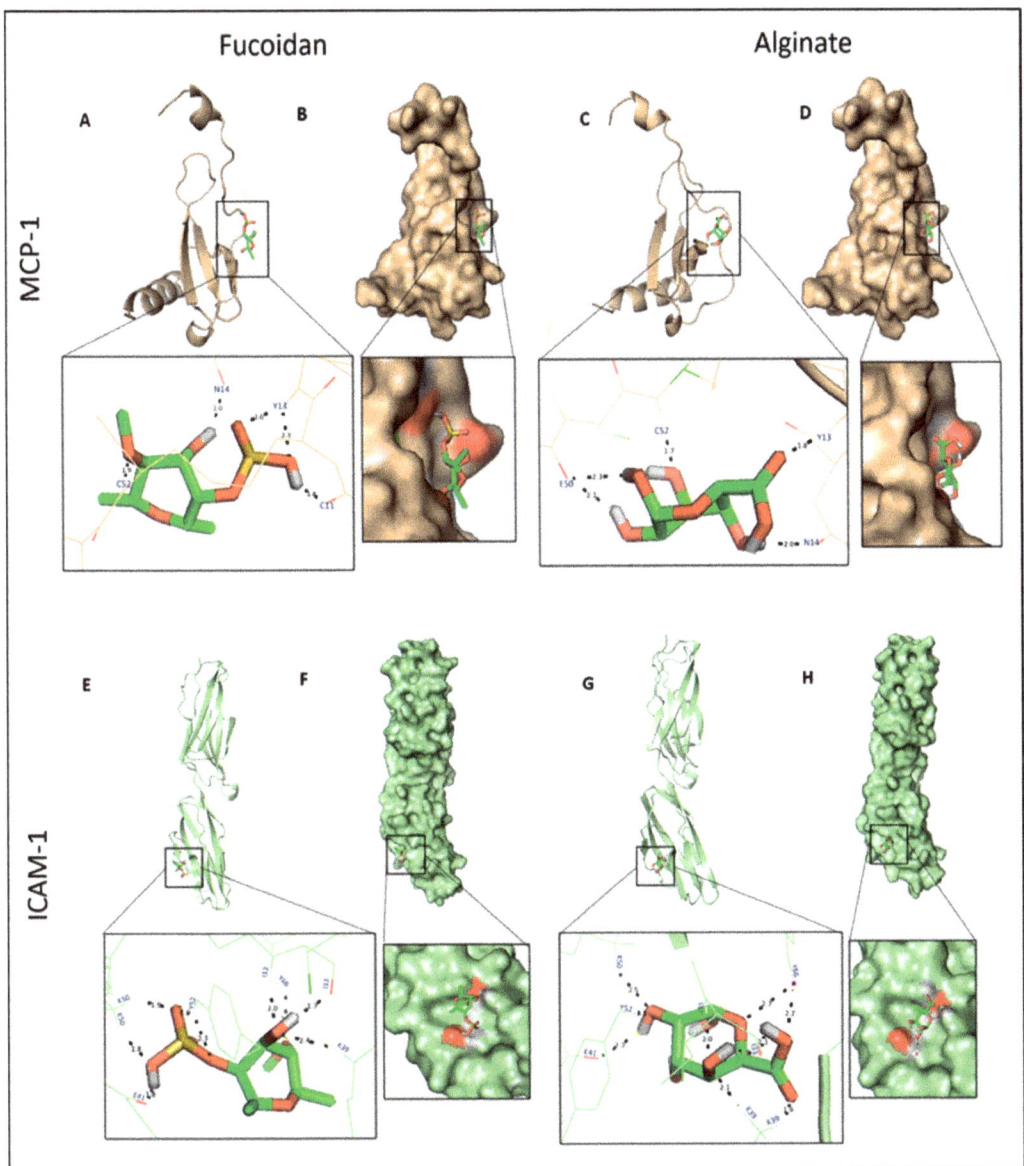

Figure 5. Molecular docking of MCP-1 and ICAM-1 to fucoidan and alginate. (**A,C,E,G**): 3D structures of proteins bound and ligand-binding pocket. Hydrogen bonds is represented by black dotted lines. (**B,D,F,H**): molecular surface representation, and the red patches on the surface represent electrostatic binding pocket.

2.4. Effect of Fucoidan on Viability and Proliferation of THP-1 Macrophages

To evaluate the effects of fucoidan on cell viability, an LDH assay was carried out, and the results were validated by assessing cell proliferation with crystal violet. Figure 6A demonstrates that fucoidan does not pose significant cytotoxicity to THP-1 macrophages when treated with increased doses compared to the vehicle. A total of 50 µg/mL fucoidan was chosen for further experiments in accordance with the published literature [20–22].

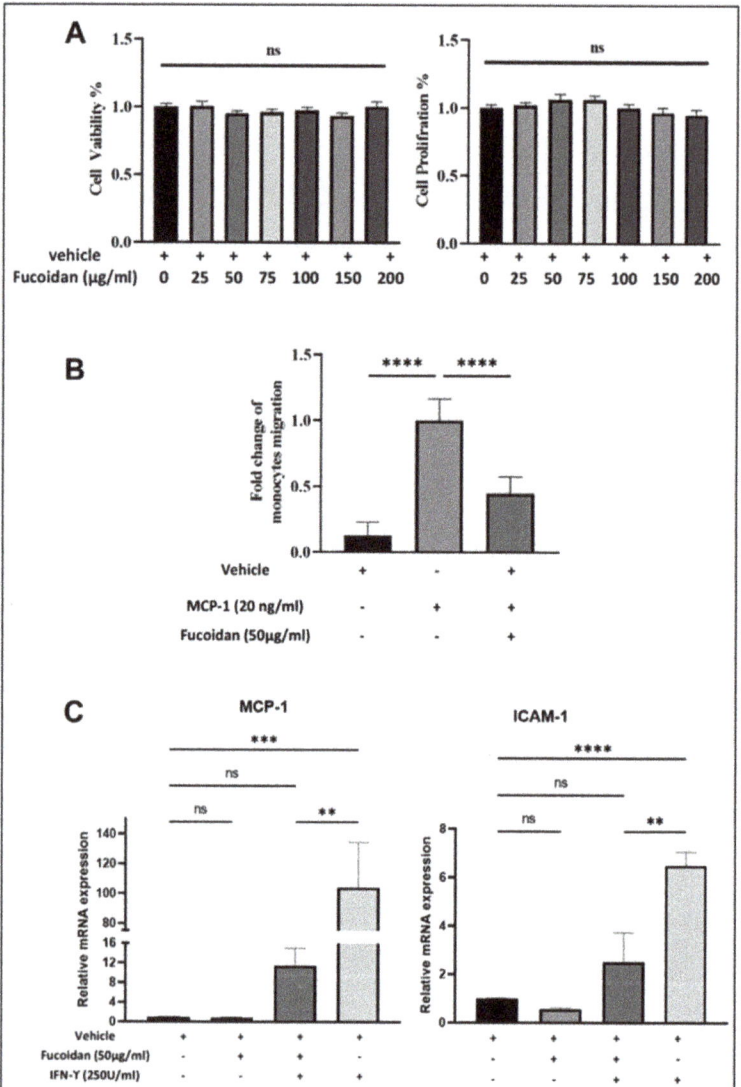

Figure 6. Biological activity of fucoidan on THP-1 cells. (**A**): Percentages of cell viability and proliferation on THP-1 macrophages subjected to various concentrations of fucoidan exposed for 24 h. (**B**): THP-1 monocyte migration assessed using a transwell chamber after 3 h stimulation with or without MCP-1 in the presence or absence of vehicle or fucoidan (50 µg/mL). (**C**): mRNA expression level of MCP-1 and ICAM-1 evaluated in THP-1 macrophages post treatment with fucoidan (50 µg/mL), vehicle, or alone for 24 h. Then cells were induced with or without IFN-γ for 3 h. Data were presented as mean ±SEM of triplicate three/two independent experiments ($n = 9$ for (**A**,**B**) and $n = 6$ for (**C**) and the p-values were non-significant (ns), ** $p < 0.001$, *** $p < 0.0005$ and **** $p < 0.0001$.

2.5. Fucoidan Inhibits Monocytes Migration to MCP-1

As shown in Figure 6B, the migration of THP-1 monocytes significantly increased in the presence of MCP-1 alone compared to vehicle, while the percentage migration of cells treated with fucoidan was significantly attenuated, by 50%, in response to chemokine MCP-1.

2.6. Fucoidan Modulates the Expression of Inflammatory Markers

MCP-1 and ICAM-1 are critical inflammatory genes for endothelium dysfunction. The transcriptomics of these genes in THP-1 macrophages, and their post-treatment with fucoidan and IFN-γ, is illustrated in Figure 6C. MCP-1 transcription was dramatically decreased by eight-fold in cells treated with IFN-γ in the presence of fucoidan compared to cells treated with IFN-γ alone. Interestingly, the effects of vehicle and fucoidan alone on the expression of both genes are not significantly different. Regarding the expression levels of ICAM-1 in THP-1 macrophages, fucoidan can attenuate the IFN-γ induced ICAM-1 expression in THP-1-derived macrophages by four-fold.

3. Discussion

Preventing leukocytes recruitment to inflammation sites can address the early stage of atherosclerosis, which is predominantly mediated by L-selectin [23,24]. L-selectin has a high affinity for binding sulfated carbohydrate moieties on *p*-selectin glycoprotein ligand-1 (PSGL-1), a glycoprotein located on leukocytes and endothelial cells that naturally binds to the selectin family. Upon the activation of endothelium, transcription-regulated E-selectin mediates the adhesion of neutrophils via PSGL-1 [25–27] or Golgi apparatus protein 1 (GLG1) within hours [28]. Endothelial activation triggers a chronic inflammatory response that involves the release of MCP-1, which subsequently binds and activates CCR2, the GPCRs embedded in the leukocytic cell membranes [29]. The signal transduction of the chemokine receptors initiates signaling to activate integrins, which are transmembrane heterodimeric proteins comprised of α and β subunits and responsible for firm adhesion to the extracellular matrix (ECM) and regulating the 'inside-out' cellular signaling. High-affinity integrins enable the tight adhesion of ICAM-1 to the transmigration of leukocytes through vascular endothelium [3,6]. Blocking these inflammatory biomarkers is crucial to stop or reduce atherosclerosis.

Polysaccharides are natural macromolecular polymers that can be found in a variety of dietary sources and have attracted a great deal of attention due to their important bioactivities [8]. The negative sulfate charges are known to play a role in the electrostatic interactions between GAGs and signaling proteins [10]. Fucoidan is a class of sulfated, fucose-rich polysaccharides present in diverse species of brown seaweed. Its unique features make it a promising candidate for nutraceuticals and pharmaceuticals for disease prevention [16,17,21]. Owing to the variety of chain structures, sulfation degrees, and positions, the structure–activity relationship between fucoidan and its mechanism of action is challenging to understand [9,17,21,22]. Therefore, we investigated the pharmacodynamics of fucoidan derived from *Fucus vesiculosus* as having anti-inflammatory potential for atherosclerosis on THP-1 cells.

Our results indicate that fucoidan has no significant cytotoxic effects on THP-1 macrophages, even at 200 μg/mL, which is consistent with several studies that examine cytotoxicity for 72 h [20–22]. Furthermore, molecular docking shows that, when fucoidan occupies the binding site in inflammatory proteins, it prevents the interaction between these proteins and other downstream regulatory partners and perturbs signaling. For instance, we found that it inhibits L-selectin, which is responsible for the adhesion of leukocytes, and suppresses MCP-1 and ICAM-1.

Hydrophilic drugs are desired for oral administration due to their bioavailability and easy formulation [30]. Lipinski's rule helps to estimate a compound's drug-likeness and includes molecular weight < 500 Da; LogP < 5; hydrogen-bond donors < 5 and hydrogen-bond acceptors < 10 [19]. The physiochemical features of both the ligands comply with these features. Moreover, the sulfated hydroxyl group imposes steric effect changes and electrostatic repulsion, causing flexion and extension of the polysaccharide chain and increased hydrophilicity, leading to improved affinity with proteins, and thereby altering biological activities [13,31]. Another indicator of a compound's flexibility is presence of rotatable bonds [32]. Our computational predictions refer to fucoidan's higher affinity for target proteins compared to alginate due to the flexibility that results from three rotatable bonds. Notably, the number of hydrogen bonds predicted via the Swiss tool was in accordance with the molecular docking of inflammatory proteins with selected ligands.

The electrostatic interactions of fucoidan most probably play a role in aiding the sulfate group's binding to Lys84 on L-selectin, similar to negatively charged Tyr51 of PSGL-1, which has 6-sulfo-sLex binding to L-selectin Lys85 [33]. The native binding site of selectins with Ca^{2+} in the lectin domain has identical residues, namely, Glu80, Glu88, Asn82, Asn105, and Asp106. This binding is explained by two conformations: it is extended with Asn83 coordinating Ca^{2+} and Glu88 away or bent with Glu88 coordinating Ca^{2+} and Asn83 away. This leads to a structural change that affects the re-orientation of the lectin and EGF-like domains, thereby stabilizing the high-affinity ligand-bonded state, which is vital to enduring the shearing force in the bloodstream and makes rolling less stable [34]. As shown in Figure 3A, fucoidan binds with Glu88, in coordination with Ca^{2+} molecule. It is worth mentioning that a comparative anti-inflammatory and anti-adhesive study investigated the origin and composition of fucoidans from diverse algal species, indicating that specific structural motifs of the fucoidans might mimic SLeX, resulting in suppressed L-selectin [35]. Our in silico results display that fucoidan binds strongly to L-selectin active sites. This supports the experimental findings of an inhibitory effect on the migration of THP-1 monocytes and suggests that fucoidan could be an antagonist for L-selectin, as previously mentioned [36]. A recent report also indicates that targeting L-selectin holds promise to control inflammation [37].

Residues in the alternative inflammatory target MCP-1's N-loop and B3 domain are necessary for binding interactions, while residues in the N-terminal area are important for receptor activation, according to structural–functional studies of chemokines [38,39]. To better understand the contribution of selective binding and activation by chemokine proteins to the chemokine receptor CCR2, Huma et al. assessed the binding of chemokine structure regions to CCR2 and observed that the N-terminal of chemokine is a major determinant of affinity and efficacy [29]. They postulated that chemokines attach to the receptor N-terminus via their N-loop and β3 residues (site1), and then the chemokine N-terminus (site2) activates the receptor by binding to its transmembrane helices, producing conformational changes and cellular signaling. Both bioactive compounds in this study bind between N-loop and β3 regions and could compete for CCR2 and obstruct binding. The results of other comparative study indicate that the hydroxyl groups of three types of flavanols (kaempferol, quercetin, and myricetin, respectively) bind with MCP-1 (-5.10, -5.28, and -6.39 kcal/mol, respectively) via common residues Cys11, Cys52, Asn14, Tyr13, and Lys16, which overlapped with that of the receptor-GAG-binding surface, hence indicating that chemokine-mediated leukocyte trafficking is likely reduced [40]. Although fucoidan and alginate both bind MCP-1 with same residues, alginate binds with a lower binding energy of -3.84 kcal/mol, while the sulfate group in fucoidan enhances this binding energy to -5.67 kcal/mol. The treatment of THP-1 macrophages with fucoidan for 24 h stimulates them to create inflammatory cytokines induced by IFN-γ, a macrophage-activating factor, as previously reported [41]. Fucoidan, hence, offers protective effect by drastically reducing MCP-1 expression.

The integrin's I domain-binding surface of ICAM-1 is relatively shallow, and Glu34 is present in the middle of the ICAM-1 coordination bond, with an Mg^{2+} ion in the I domain [42]. Furthermore, aromatic and hydrophobic residues on the ICAM-1 surround Glu34, Pro36, Tyr66, Met64, and the aliphatic portions of Gln62 and Gln73 contact the similar ring of hydrophobic residues on the I domain [43]. As a result, the electrostatic surface's contact regions have good charge complementarity. For ligand binding, a salt bridge between the I domain Glu241 and ICAM-1 Lys39 is required, allowing for ICAM-1 and the I domain to optimally interact [44,45]. Similar residues, including Lys39, were found in our study, participating in interactions with both ligands and ICAM-1. Polar interactions involving hydrogen bonds sustain this interaction, which is shorter in fucoidan, possibly due to its greater negative charges. Although MCP-1 and ICAM-1 have a similar binding affinity to fucoidan in terms of docking results, fucoidan suppresses ICAM-1 expression in THP-1 macrophages that undergo IFN-γ induction, with a lower fold change than MCP-1, which means fucoidan interacts with the non-specific protein [46]. Moreover, anionic polysaccharide can bind to distinct proteins with several levels of specificity to endothelial cells [47].

Even though this study lacks protein expression evaluation, an understanding of docking interactions with fucoidan and validated with gene expression experiments helps us gain knowledge of the effect at the protein level. It is worth mentioning that fucoidan can inhibit these proteins at 55–70 µM, according to the predicted inhibition values that are constant in molecular docking. That implies only a small amount is required to inhibit the protein's activity.

4. Materials and Methods

4.1. Protein–Protein Interaction Study

The significant protein–protein interactions (PPIs) existing between L-selectin, E-selectin, MCP-1, and ICAM-1 were explored using STRING protein database version 11.5. Network edges (evidence), and active interaction sources (text mining, databases, experiments, neighborhood, co-expression) were then employed as the primary settings, and limited to homo sapiens. Minimal required interaction score of >0.4 was applied to construct the PPIs networks [48].

4.2. Chemoinformatic Prediction

Chemoinformatic tools were employed to predict suitability of bioactive molecules (fucoidan and alginate) as a drug. SwissTargetPrediction predicts the most probable protein targets of biomolecules based on a blend of 2D and 3D structural and electrochemical complementarity [49]. SwissADME online tool evaluates physicochemical descriptors by computing the ADME features and drug likeliness of small molecules for consideration as an oral drug candidate [50].

4.3. Molecular Docking

Three-dimensional X-ray structures of inflammatory proteins, namely, L-selectin, E-selectin, MCP-1, and ICAM-1, were retrieved from RCSB's Protein Data Bank (PDB) (ID: 5VC1, 1G1T, 1DOK, and 1IAM, respectively) with a resolution of 1.85Å, 2.1Å, 1.94Å, and 1.5Å, respectively) [5,26,51,52]. The 3D structures of fucoidan and alginate (CID: 129532628 and 91666324, respectively) were downloaded from NCBI's PubChem database, and protein and ligand structures were prepared, followed by molecular docking to compute the binding energy in kcal/mol resulting from the interaction of fucoidan and alginate with proteins using Auto Dock 4.2.6 [53]. Docking was performed with monomeric unit of polysaccharides. Each protein structure was processed by selecting one chain and removing the water molecules and the existing co-crystallized ligand. The grid dimensions were generated according to the known binding sites of each protein. Docking was protein-rigid and ligand-flexible. Binding free energy of ligand-protein interaction was used to score various configurations. The best pose was chosen based on the lowest docking energy (kcal/mol) and lower RMSD [54]. Complex structures were visualized by PyMol 1. Level (DeLano Scientific LLC., Palo Alto, CA, USA).

4.4. Cell Culture

THP-1, a human monocytic leukemia cell line, was provided by Molecular Biomedicine Unit, King Faisal Specialist Hospital and Research Centre, Riyadh, KSA. THP-1 cells were maintained as an undifferentiated monocyte grown in suspension in RPMI medium 1640 (1×) supplemented with fetal bovine serum (FBS, 10% v/v), L-glutamine (200 mM, 1% v/v) and penicillin-streptomycin (100 U/mL) (Gibco™, ThermoFisherScientific, Waltham, MA, USA). Cell incubation was carried out in an atmosphere with 5% CO_2, 95% humidity and a 37 °C temperature.

4.4.1. Cell Viability and Proliferation Assays

A lactate dehydrogenase (LDH) cytotoxicity assay was carried out for cell viability measurement following the manufacturer's instructions (88953; ThermoFisherScientific, Waltham, MA, USA). Seeding of THP-1 monocytes was carried out with a density of 1×10^5 cells/cm^2 in 96-well plates and differentiation into macrophages was performed with 0.16 μL of phorbol myristate acetate (PMA, 1 mg/mL, ThermoFisher (Kandel) GmbH, Germany) overnight at 37 °C and 5% (v/v) CO_2. Fucoidan (≥95% HPLC, F8190; Sigma-Aldrich, St. Louis, MO, USA) was dissolved in pure distilled water (vehicle) at 10 mg/mL and then diluted in culture media at different concentrations to treat the macrophages for a further 24 h. Subsequently, 50 μL supernatants of treated THP-1 macrophages were transferred into new 96-well plates, along with a 50 μL assay buffer. Following incubation for 30 min at 25 °C, 50 μL of stopping solution was mixed. Absorbance was noted at 490 nm using a microplate reader (BioTek Synergy HT, Agilent Technologies, Santa Clara, CA, USA). Crystal violet dye was used to evaluate the proliferation of cells via binding to the DNA of viable cells [55]. Adherent macrophages remaining after the LDH test were employed for the cell proliferation assay. Cells were stained with 50 μL of 0.2% (w/v) crystal violet solution (dissolved in 10% ethanol) for 5 min at room temperature. THP-1 macrophages were washed 3–4 times with PBS prior to the addition of 50 μL of solubilization buffer (0.1 M NaH_2PO_4 ethanol solution). Treated plate was shaken for 5 min before measuring absorbance with a microplate reader at 570 nm. Results were tabulated as the percentage of viability related to control.

4.4.2. Migration Assay

Migration assay was used to estimate fucoidan's ability to inhibit monocyte migration in response to chemoattraction. A 1 mL culture media containing 20 ng/mL of monocyte chemoattractant protein (MCP-1/MCAF, Sigma-Aldrich, St. Louis, MO, USA, SRP3109) was added to the bottom of companion plates of the SPL Insert hanging (35224; SPL Life Sciences, Gyeonggi-do, Korea) in all wells except the control well. Undifferentiated THP-1 monocyte cells (5×10^5 cells/mL) were added to inserts with a 0.8 μm pore size. Then, immediately after being treated with either a control (vehicle) or 50 μg/mL of fucoidan, cells alone were used as a positive control for MCP-1. Plate chambers were incubated with 5% (v/v) CO_2 at 37 °C for 3 h. Cells that had migrated into the lower chambers were collected, and centrifuged at $250 \times g$ for 5 min. Cell pellets were resuspended in 1 mL of fresh media and cells were counted using a hemocytometer [56]. Monocyte migration was expressed as a fold-change relative to the fraction of cells that moved through the insert into the bottom wells in response to chemokine alone.

4.4.3. Quantitative Reverse Transcription-PCR

Two groups of THP-1 macrophages were taken (untreated and treated with 50 μg/mL fucoidan for 24 h). Inflammation was induced in both the groups with 0.13 μL of interferon-γ human (INF-γ, 13265; 1 mg/mL, Sigma-Aldrich) treated for 3 hrs. Total mRNA extraction was carried out for all (vehicle, fucoidan alone, fucoidan with IFN-γ and IFN-γ) using the RNeasy™ mini kit (74104; Qiagen, Germany) and transcribed into cDNA as per the instructions using the ImProm-II Reverse Transcription kit (A3800; Promega, Madison, WI, USA). A quantitative polymerase chain reaction (qPCR) was performed using the BioFACT™ 2X Real-Time PCR Master Mix (For SYBR® Green I) kit (DQ383–40h; Daejeon, Korea). Target genes (MCP-1 and ICAM-1) expression was analyzed by a StepOnePlus™ Real-time PCR system (Applied Biosystems, Waltham, MA, USA). Relative quantification of their expression with fold change and *p*-value was calculated using the comparative threshold method (Ct, 2−ΔΔCT) after normalization with glyceraldehyde-3-phosphate dehydrogenase (GAPDH) housekeeping gene. Table 3 enlists the primers that were used [56].

Table 3. Primer Sequences used for human MCP-1, ICAM-1 and GAPDH genes.

Gene	Primer Sequence
MCP-1	Forward: CGCTCAGCCAGATGCAATCAATG Reverse: CGCTCAGCCAGATGCAATCAATG
ICAM-1	Forward: GACCAGAGGTTGAACCCCAC Reverse: GCGCCGGAAAGCTGTAGAT
GAPDH	Forward: CTTTTGCGTCGCCAGCCGAG Reverse: GCCCAATACGACCAAATCCGTTGACT

4.4.4. Statistical Analysis

Statistical analysis was performed using one-way ANOVA to detect any statistically significant differences between the means of two or more independent groups, followed by a Sidak multiple comparison test. Excel Microsoft 365 and GraphPad Prism version 8 softwares were used for statistical analysis. The significance is represented using *p*-values as ns (non-significant), ** $p < 0.001$, *** $p < 0.0005$ and **** $p < 0.0001$.

5. Conclusions

Natural compounds can potentially alter or regulate cellular gene expression, aiding in the treatment and prevention of any diseases hallmarked by inflammation. The pharmacodynamically relevant ability of fucoidan to modulate key biomarker genes in the early stages of atherosclerosis was demonstrated. Fucoidan potentially blocks L-selectin and prevents monocyte migration, thereby modulating the expression level of MCP-1 and ICAM-1 in THP-1 macrophages. Our results support in silico molecular docking results, wherein fucoidan occupies the binding sites of inflammatory proteins. Future in vivo investigations will help us to better comprehend the underlying mechanisms at the molecular level, as well as the anti-inflammatory effects of natural substances and their use as dietary supplements. This emphasizes the benefits of a nutritionally orientated approach to prevent initial disease development. Pre-clinical trials are further needed to determine the efficacy of fucoidan and establish its role in the prevention and treatment of inflammatory disorders, including atherosclerosis.

Author Contributions: E.H. and D.A.A.-S. Conceived and designed the analysis; D.A.A.-S. Collected the data; E.H., D.A.A.-S. and Z.M. Contributed data or analysis tools; D.A.A.-S. Performed the analysis; D.A.A.-S. and Z.M. Wrote the paper. All authors have read and agreed to the published version of the manuscript.

Funding: This research received no external funding.

Institutional Review Board Statement: Not applicable.

Informed Consent Statement: Not applicable.

Data Availability Statement: Not applicable.

Acknowledgments: The authors really would like to express thanks and appreciation to Khalid Abu Khabar (Molecular Biomedicine Unit, KFSH&RC, Riyadh, KSA) for his kindly gift of the THP-1 cell line. We would also like to thank King Fahd Medical Research Center, KAU, Jeddah, KSA for their technical support.

Conflicts of Interest: The authors declare no conflict of interest.

Sample Availability: Samples of the compounds are not available from the authors.

References

1. WHO. Cardiovascular Diseases (CVDs). Available online: https://www.who.int/news-room/fact-sheets/detail/cardiovascular-diseases-(cvds) (accessed on 11 June 2021).
2. Flynn, M.C.; Pernes, G.; Lee, M.K.S.; Nagareddy, P.R.; Murphy, A.J. Monocytes, macrophages, and metabolic disease in atherosclerosis. *Front. Pharmacol.* **2019**, *10*, 1–13. [CrossRef] [PubMed]
3. Ley, K.; Laudanna, C.; Cybulsky, M.I.; Nourshargh, S. Getting to the site of inflammation: The leukocyte adhesion cascade updated. *Nat. Rev. Immunol.* **2007**, *7*, 678–689. [CrossRef] [PubMed]
4. Lasky, L.A. Selectin-carbohydrate interactions and the initiation of the inflammatory response. *Annu. Rev. Biochem.* **1995**, *64*, 113–140. [CrossRef]
5. Wedepohl, S.; Dernedde, J.; Vahedi-Faridi, A.; Tauber, R.; Saenger, W.; Bulut, H. Reducing Macro- and Microheterogeneity of N-Glycans Enables the Crystal Structure of the Lectin and EGF-Like Domains of Human L-Selectin To Be Solved at 1.9 Å Resolution. *ChemBioChem* **2017**, *18*, 1338–1345. [CrossRef] [PubMed]
6. Adrielle Lima Vieira, R.; Nascimento de Freitas, R.; Volp, A.C.P. Adhesion molecules and chemokines; relation to anthropometric, body composition, biochemical and dietary variables. *Nutr. Hosp.* **2014**, *30*, 223–236. [CrossRef] [PubMed]
7. Szabó-Fodor, J.; Bónai, A.; Bóta, B.; Szommerné Egyed, L.; Lakatos, F.; Pápai, G.; Zsolnai, A.; Glávits, R.; Horvatovich, K.; Kovács, M. Physiological Effects of Whey- and Milk-Based Probiotic Yogurt in Rats. *Polish J. Microbiol.* **2017**, *66*, 483–490. [CrossRef]
8. Zhong, Q.; Wei, B.; Wang, S.; Ke, S.; Chen, J.; Zhang, H.; Wang, H. The Antioxidant Activity of Polysaccharides Derived from Marine Organisms: An Overview. *Mar. Drugs* **2019**, *17*, 674. [CrossRef]
9. Takahashi, M.; Takahashi, K.; Abe, S.; Yamada, K.; Suzuki, M.; Masahisa, M.; Endo, M.; Abe, K.; Inoue, R.; Hoshi, H. Improvement of Psoriasis by Alteration of the Gut Environment by Oral Administration of Fucoidan from Cladosiphon Okamuranus. *Mar. Drugs* **2020**, *18*, 154. [CrossRef]
10. Huang, L.; Shen, M.; Morris, G.A.; Xie, J. Sulfated polysaccharides: Immunomodulation and signaling mechanisms. *Trends Food Sci. Technol.* **2019**, *92*, 1–11. [CrossRef]
11. Gacesa, P. Alginates. *Carbohydr. Polym.* **1988**, *8*, 161–182. [CrossRef]
12. Bouissil, S.; El Alaoui-Talibi, Z.; Pierre, G.; Michaud, P.; El Modafar, C.; Delattre, C. Use of Alginate Extracted from Moroccan Brown Algae to Stimulate Natural Defense in Date Palm Roots. *Molecules* **2020**, *25*, 720. [CrossRef]
13. Zayed, A.; El-Aasr, M.; Ibrahim, A.-R.S.; Ulber, R. Fucoidan Characterization: Determination of Purity and Physicochemical and Chemical Properties. *Mar. Drugs* **2020**, *18*, 571. [CrossRef] [PubMed]
14. Li, B.; Lu, F.; Wei, X.; Zhao, R. Fucoidan: Structure and bioactivity. *Molecules* **2008**, *13*, 1671–1695. [CrossRef] [PubMed]
15. Liu, J.; Wu, S.-Y.; Chen, L.; Li, Q.-J.; Shen, Y.-Z.; Jin, L.; Zhang, X.; Chen, P.-C.; Wu, M.-J.; Choi, J.; et al. Different extraction methods bring about distinct physicochemical properties and antioxidant activities of Sargassum fusiforme fucoidans. *Int. J. Biol. Macromol.* **2020**, *155*, 1385–1392. [CrossRef] [PubMed]
16. Chollet, L.; Saboural, P.; Chauvierre, C.; Villemin, J.-N.; Letourneur, D.; Chaubet, F. Fucoidans in Nanomedicine. *Mar. Drugs* **2016**, *14*, 145. [CrossRef]
17. Ahmad, T.; Eapen, M.S.; Ishaq, M.; Park, A.Y.; Karpiniec, S.S.; Stringer, D.N.; Sohal, S.S.; Fitton, J.H.; Guven, N.; Caruso, V.; et al. Anti-Inflammatory Activity of Fucoidan Extracts In Vitro. *Mar. Drugs* **2021**, *19*, 702. [CrossRef]
18. Lee, H.; Kim, J.-S.; Kim, E. Fucoidan from Seaweed Fucus vesiculosus Inhibits Migration and Invasion of Human Lung Cancer Cell via PI3K-Akt-mTOR Pathways. *PLoS ONE* **2012**, *7*, e50624. [CrossRef]
19. Moumbock, A.F.A.; Li, J.; Mishra, P.; Gao, M.; Günther, S. Current computational methods for predicting protein interactions of natural products. *Comput. Struct. Biotechnol. J.* **2019**, *17*, 1367–1376. [CrossRef]
20. Chen, L.-M.; Tseng, H.-Y.; Chen, Y.-A.; Tanzih, A.; Haq, A.; Hwang, P.-A.; Hsu, H.-L. Oligo-Fucoidan Prevents M2 Macrophage Differentiation and HCT116 Tumor Progression. *Cancers* **2020**, *12*, 421. [CrossRef]
21. Park, J.; Yeom, M.; Hahm, D.H. Fucoidan improves serum lipid levels and atherosclerosis through hepatic SREBP-2-mediated regulation. *J. Pharmacol. Sci.* **2016**, *131*, 84–92. [CrossRef]
22. Sun, J.; Sun, J.; Song, B.; Zhang, L.; Shao, Q.; Liu, Y.; Yuan, D.; Zhang, Y.; Qu, X. Fucoidan inhibits CCL22 production through NF-κB pathway in M2 macrophages: A potential therapeutic strategy for cancer. *Sci. Rep.* **2016**, *6*, 35855. [CrossRef] [PubMed]
23. Simon, S.I.; Chambers, J.D.; Butcher, E.; Sklar, L.A. Neutrophil aggregation is beta 2-integrin- and L-selectin-dependent in blood and isolated cells. *J. Immunol.* **1992**, *149*, 2765–2771. [PubMed]
24. Bargatze, R.F.; Kurk, S.; Butcher, E.C.; Jutila, M.A. Neutrophils roll on adherent neutrophils bound to cytokine-induced endothelial cells via L-selectin on the rolling cells. *J. Exp. Med.* **1994**, *180*, 1785–1792. [CrossRef] [PubMed]
25. Kansas, G.S. Selectins and their ligands: Current concepts and controversies. *Blood* **1996**, *88*, 3259–3287. [CrossRef]
26. Pouyani, T.; Seed, B. PSGL-1 recognition of P-selectin is controlled by a tyrosine sulfation consensus at the PSGL-1 amino terminus. *Cell* **1995**, *83*, 333–343. [CrossRef]
27. Sako, D.; Comess, K.M.; Barone, K.M.; Camphausen, R.T.; Cumming, D.A.; Shaw, G.D. A sulfated peptide segment at the amino terminus of PSGL-1 is critical for P-selectin binding. *Cell* **1995**, *83*, 323–331. [CrossRef]
28. Hidalgo, A.; Peired, A.J.; Wild, M.K.; Vestweber, D.; Frenette, P.S. Complete Identification of E-Selectin Ligands on Neutrophils Reveals Distinct Functions of PSGL-1, ESL-1, and CD44. *Immunity* **2007**, *26*, 477–489. [CrossRef]

29. Huma, Z.E.; Sanchez, J.; Lim, H.D.; Bridgford, J.L.; Huang, C.; Parker, B.J.; Pazhamalil, J.G.; Porebski, B.T.; Pfleger, K.D.G.; Lane, J.R.; et al. Key determinants of selective binding and activation by the monocyte chemoattractant proteins at the chemokine receptor CCR2. *Sci. Signal.* **2017**, *10*, eaai8403. [CrossRef]
30. Ritchie, T.J.; Macdonald, S.J.F.; Peace, S.; Pickett, S.D.; Luscombe, C.N. Increasing small molecule drug developability in sub-optimal chemical space. *Medchemcomm* **2013**, *4*, 673. [CrossRef]
31. Crijns, H.; Adyns, L.; Ganseman, E.; Cambier, S.; Vandekerckhove, E.; Pörtner, N.; Vanbrabant, L.; Struyf, S.; Gerlza, T.; Kungl, A.; et al. Affinity and Specificity for Binding to Glycosaminoglycans Can Be Tuned by Adapting Peptide Length and Sequence. *Int. J. Mol. Sci.* **2021**, *23*, 447. [CrossRef]
32. Lagorce, D.; Douguet, D.; Miteva, M.A.; Villoutreix, B.O. Computational analysis of calculated physicochemical and ADMET properties of protein-protein interaction inhibitors. *Sci. Rep.* **2017**, *7*, 46277. [CrossRef] [PubMed]
33. Bernimoulin, M.P.; Zeng, X.-L.; Abbal, C.; Giraud, S.; Martinez, M.; Michielin, O.; Schapira, M.; Spertini, O. Molecular Basis of Leukocyte Rolling on PSGL-1. *J. Biol. Chem.* **2003**, *278*, 37–47. [CrossRef] [PubMed]
34. Waldron, T.T.; Springer, T.A. Transmission of allostery through the lectin domain in selectin-mediated cell adhesion. *Proc. Natl. Acad. Sci. USA* **2009**, *106*, 85–90. [CrossRef] [PubMed]
35. Cumashi, A.; Ushakova, N.A.; Preobrazhenskaya, M.E.; D'Incecco, A.; Piccoli, A.; Totani, L.; Tinari, N.; Morozevich, G.E.; Berman, A.E.; Bilan, M.I.; et al. A comparative study of the anti-inflammatory, anticoagulant, antiangiogenic, and antiadhesive activities of nine different fucoidans from brown seaweeds. *Glycobiology* **2007**, *17*, 541–552. [CrossRef] [PubMed]
36. Thorlacius, H.; Vollmar, B.; Seyfert, U.T.; Vestweber, D.; Menger, M.D. The polysaccharide fucoidan inhibits microvascular thrombus formation independently from P- and l-selectin function in vivo. *Eur. J. Clin. Investig.* **2000**, *30*, 804–810. [CrossRef]
37. Smith, B.A.H.; Bertozzi, C.R. The clinical impact of glycobiology: Targeting selectins, Siglecs and mammalian glycans. *Nat. Rev. Drug Discov.* **2021**, *20*, 217–243. [CrossRef]
38. Jarnagin, K.; Grunberger, D.; Mulkins, M.; Wong, B.; Hemmerich, S.; Paavola, C.; Bloom, A.; Bhakta, S.; Diehl, F.; Freedman, R.; et al. Identification of Surface Residues of the Monocyte Chemotactic Protein 1 That Affect Signaling through the Receptor CCR2. *Biochemistry* **1999**, *38*, 16167–16177. [CrossRef]
39. Hemmerich, S.; Paavola, C.; Bloom, A.; Bhakta, S.; Freedman, R.; Grunberger, D.; Krstenansky, J.; Lee, S.; McCarley, D.; Mulkins, M.; et al. Identification of Residues in the Monocyte Chemotactic Protein-1 That Contact the MCP-1 Receptor, CCR2. *Biochemistry* **1999**, *38*, 13013–13025. [CrossRef]
40. Joshi, N.; Tripathi, D.K.; Nagar, N.; Poluri, K.M. Hydroxyl Groups on Annular Ring-B Dictate the Affinities of Flavonol–CCL2 Chemokine Binding Interactions. *ACS Omega* **2021**, *6*, 10306–10317. [CrossRef]
41. Yu, X.-H.; Zhang, J.; Zheng, X.-L.; Yang, Y.-H.; Tang, C.-K. Interferon-γ in foam cell formation and progression of atherosclerosis. *Clin. Chim. Acta* **2015**, *441*, 33–43. [CrossRef]
42. Lee, J.-O.; Bankston, L.A.; Robert, C.; Liddington, M.A.A. Two conformations of the integrin A-domain (I-domain): A pathway for activation? *Structure* **1995**, *3*, 1333–1340. [CrossRef]
43. Shimaoka, M.; Xiao, T.; Liu, J.-H.; Yang, Y.; Dong, Y.; Jun, C.-D.; McCormack, A.; Zhang, R.; Joachimiak, A.; Takagi, J.; et al. Structures of the alpha L I domain and its complex with ICAM-1 reveal a shape-shifting pathway for integrin regulation. *Cell* **2003**, *112*, 99–111. [CrossRef]
44. Edwards, C.P.; Fisher, K.L.; Presta, L.G.; Bodary, S.C. Mapping the Intercellular Adhesion Molecule-1 and -2 Binding Site on the Inserted Domain of Leukocyte Function-associated Antigen-1. *J. Biol. Chem.* **1998**, *273*, 28937–28944. [CrossRef] [PubMed]
45. Fisher, K.L.; Lu, J.; Riddle, L.; Kim, K.J.; Presta, L.G.; Bodary, S.C. Identification of the binding site in intercellular adhesion molecule 1 for its receptor, leukocyte function-associated antigen 1. *Mol. Biol. Cell* **1997**, *8*, 501–515. [CrossRef]
46. Rowe, A.; Berendt, A.R.; Marsh, K.; Newbold, C.I. Plasmodium falciparum: A Family of Sulfated Glycoconjugates Disrupts Erythrocyte Rosettes. *Exp. Parasitol.* **1994**, *79*, 506–516. [CrossRef]
47. Skidmore, M.A.; Mustaffa, K.M.F.; Cooper, L.C.; Guimond, S.E.; Yates, E.A.; Craig, A.G. A semi-synthetic glycosaminoglycan analogue inhibits and reverses Plasmodium falciparum cytoadherence. *PLoS ONE* **2017**, *12*, e0186276. [CrossRef]
48. Szklarczyk, D.; Gable, A.L.; Nastou, K.C.; Lyon, D.; Kirsch, R.; Pyysalo, S.; Doncheva, N.T.; Legeay, M.; Fang, T.; Bork, P.; et al. The STRING database in 2021: Customizable protein–protein networks, and functional characterization of user-uploaded gene/measurement sets. *Nucleic Acids Res.* **2021**, *49*, D605–D612. [CrossRef]
49. Daina, A.; Michielin, O.; Zoete, V. SwissTargetPrediction: Updated data and new features for efficient prediction of protein targets of small molecules. *Nucleic Acids Res.* **2019**, *47*, W357–W364. [CrossRef]
50. Daina, A.; Michielin, O.; Zoete, V. SwissADME: A free web tool to evaluate pharmacokinetics, drug-likeness and medicinal chemistry friendliness of small molecules. *Sci. Rep.* **2017**, *7*, 42717. [CrossRef]
51. Bella, J.; Kolatkar, P.R.; Marlor, C.W.; Greve, J.M.; Rossmann, M.G. The structure of the two amino-terminal domains of human ICAM-1 suggests how it functions as a rhinovirus receptor and as an LFA-1 integrin ligand. *Proc. Natl. Acad. Sci. USA* **1998**, *95*, 4140–4145. [CrossRef]
52. Lubkowski, J.; Bujacz, G.; Boqué, L.; Peter, J.D.; Tracy, M.H.; Alexander, W. The Structure of MC P-1 in Two Crystal Forms Provides a Rare Example of Variable Quaternary Interactions. *Nat. Struct. Biol.* **1997**, *4*, 64–69. [CrossRef] [PubMed]
53. Morris, G.M.; Huey, R.; Lindstrom, W.; Sanner, M.F.; Belew, R.K.; Goodsell, D.S.; Olson, A.J. AutoDock4 and AutoDockTools4: Automated docking with selective receptor flexibility. *J. Comput. Chem.* **2009**, *30*, 2785–2791. [CrossRef]

54. Ramírez, D.; Caballero, J. Is It Reliable to Take the Molecular Docking Top Scoring Position as the Best Solution without Considering Available Structural Data? *Molecules* **2018**, *23*, 1038. [CrossRef]
55. Yurdakok Dikmen, B.; Alpay, M.; Kismali, G.; Filazi, A.; Kuzukiran, O.; Sireli, U.T. In Vitro Effects of Phthalate Mixtures on Colorectal Adenocarcinoma Cell Lines. *J. Environ. Pathol. Toxicol. Oncol.* **2015**, *34*, 115–123. [CrossRef] [PubMed]
56. Moss, J.W.E.; Davies, T.S.; Garaiova, I.; Plummer, S.F.; Michael, D.R.; Ramji, D.P. A Unique Combination of Nutritionally Active Ingredients Can Prevent Several Key Processes Associated with Atherosclerosis In Vitro. *PLoS ONE* **2016**, *11*, e0151057. [CrossRef] [PubMed]

Article

Physicochemical and Microbiological Stability of Two Oral Solutions of Methadone Hydrochloride 10 mg/mL

Elena Alba Álvaro-Alonso [1,*], Ma Paz Lorenzo [2], Andrea Gonzalez-Prieto [3], Elsa Izquierdo-García [1], Ismael Escobar-Rodríguez [1] and Antonio Aguilar-Ros [4]

[1] Pharmacy Department, Infanta Leonor University Hospital, Av. Gran Vía del Este, 80, 28031 Madrid, Spain; elsa.izquierdo@salud.madrid.org (E.I.-G.); ismael.escobar@salud.madrid.org (I.E.-R.)
[2] Center for Metabolomics and Bioanalysis (CEMBIO), Faculty of Pharmacy, Universidad San Pablo-CEU, CEU Universities, Urbanización Montepríncipe, 28660 Boadilla del Monte, Spain; pazloga@ceu.es
[3] Central Laboratory of Madrid UR Salud, Infanta Sofía University Hospital, Paseo de Europa, 34, San Sebastián de los Reyes, 28703 Madrid, Spain; agonzalez@ursalud.com
[4] Faculty of Pharmacy, Universidad San Pablo-CEU, CEU Universities, Urbanización Montepríncipe, 28660 Boadilla del Monte, Spain; aguiros@ceu.es
* Correspondence: elenaalba.alvaro@salud.madrid.org; Tel: +34-911-918-404

Abstract: In this article, we studied physicochemical and microbiological stability and determined the beyond-use date of two oral solutions of methadone in three storage conditions. For this, two oral solutions of methadone (10 mg/mL) were prepared, with and without parabens, as preservatives. They were packed in amber glass vials kept unopened until the day of the test, and in a multi-dose umber glass bottle opened daily. They were stored at 5 ± 3 °C, 25 ± 2 °C and 40 ± 2 °C. pH, clarity, and organoleptic characteristics were obtained. A stability-indicating high-performance liquid chromatography method was used to determine methadone. Microbiological quality was studied and antimicrobial effectiveness testing was also determined following European Pharmacopoeia guidelines. Samples were analyzed at days 0, 7, 14, 21, 28, 42, 56, 70, and 91 in triplicate. After 91 days of storage, pH remained stable at about 6.5–7 in the two solutions, ensuring no risk of methadone precipitation. The organoleptic characteristics remained stable (colorless, odorless, and bitter taste). The absence of particles was confirmed. No differences were found with the use of preservatives. Methadone concentration remained within 95–105% in all samples. No microbial growth was observed. Hence, the two oral methadone solutions were physically and microbiologically stable at 5 ± 3 °C, 25 ± 2 °C, and 40 ± 2 °C for 91 days in closed and opened amber glass bottles.

Keywords: methadone hydrochloride; pharmaceutical solutions; drug compounding; high performance liquid chromatography; stability study; microbiology

1. Introduction

Methadone was synthesized in the 1940s. It is a pure synthetic opioid agonist with strong affinity and activity at the μ-opioid receptors. It is marginally more potent than morphine and has a longer duration of action, but a lower euphoric effect [1]. Thus, methadone is an alternative to morphine in that it has the same analgesic properties but has milder adverse effects [2].

In 1964, it was first used in clinical rehabilitation programs for opiate addictions, such as those associated with heroin. These programs were developed by a research team at the Rockefeller University of New York [3] and are known as methadone maintenance programs (MMP).

In Spain, heroin use peaked in the 1980s [4]. The first Spanish regulations on the prescription and dispensing of methadone for the treatment of opiate dependence appeared in 1983. However, it was not until 1990 that the prescription criteria were standardized and methadone treatment became widespread. Since then, a series of laws on the regulation and implementation of MMP have been passed and continue to be developed [5].

In the autonomous community of Madrid, resolution 189/2018 [6] was implemented in March 2018, which tasked the Hospital Pharmacy Service (HPS) of the Infanta Leonor University Hospital (ILUH) with supplying methadone to the 27 centers for the Comprehensive Care of Drug Addiction Patients (CDAP) of the Madrid Health Care Service, where MMP for opiate addictions are implemented. The aim of this resolution was to centralize the acquisition, preparation, distribution, and dispensing of methadone by the HPS. This initiative represented a first step in changing the pharmacotherapeutic health care model for the treatment of the patients in the program. To date, between 3000 and 5000 patients are prescribed methadone as an opiate substitute for the treatment of heroin-related addiction disorders.

The methadone solution prepared and supplied by the Pharmacy Service of the ILUH to MMP patients is described in the Spanish National Formulary [7] and is formulated with methadone hydrochloride in the raw material form and purified water at 10 mg/mL (M1). Methadone hydrochloride solution can precipitate at pH greater than 6.5 [8,9]. The solution should also be kept in waterproof and topaz glass containers and protected from light [10,11]. This formulation has assigned a beyond-use date (BUD) of 30 days in refrigerated storage. However, there is no mention or bibliography of any physicochemical or microbiological stability study.

This BUD causes several disadvantages, limiting the organizational capacity in the pharmacy service and in CDAP. A longer BUD would allow optimizing the daily workflow of the pharmacy service, to elaborate sufficiently in advance, as well as improve organizational aspects of the CDAP in the dosing tasks, by increasing the adaptability to the individual dispensing needs of methadone maintained patients [12]. It would even allow the dispensing of methadone for longer periods of time, something that nowadays would be a great advantage due to lockdowns imposed by COVID pandemic.

Furthermore, because the methadone daily prepared is an aqueous solution without preservatives in its formulation, a new compounding of methadone hydrochloride in oral solution was designed and validated in the Hospital Pharmacy Service of ILUH, given the possibility of microbial growth. Its composition included methylparaben and propylparaben as preservatives. The final methadone concentration was also 10 mg/mL (M2).

Due to the large volume of methadone solution to be dispensed (around 3500 L per year), the short BUD assigned, and the need to study the alternative methadone oral solution with preservatives, it became necessary to confirm the BUD of both formulations and even allow to increase them. This is only possible through physicochemical and microbiological stability studies.

In the literature, there are several works that describe stability studies of methadone, but they are either formulations with different concentrations, or with other vehicles in their compositions or even preparations for the intravenous route [10,11,13–17].

Finally, it is important to note that all of these stability studies have been made prior to the publication of the Spanish Good Practice guidelines for the preparation of drugs in hospital pharmacy services by the Ministry of Health in 2014 [18]. They affirm that the assignment of BUD longer than those indicated in the bibliography must be validated with physicochemical and microbiological stability of the non-sterile compounding. The same is described in the United States Pharmacopeia (USP) in its chapter 795 dedicated to the elaboration of non-sterile formulations [19]. These stability studies were also published prior to the resolution 189/2018 [6]. This led to the creation of a new compounding unit in the Pharmacy Service of the Infanta Leonor University Hospital, with different preparation material, raw materials, and packaging material [12] from those that had been used before in the literature and, therefore, this new preparation must be validated by means of a physicochemical and microbiological stability study, taking into account these new working characteristics. All of these aspects justified the need to carry out a stability study.

For all these reasons, and with the aim to confirm the BUD established in the Spanish National Formulary for the methadone hydrochloride oral solution and the new formulation with parabens, we designed and carried out a physicochemical and microbiological stability study in three storage conditions for 91 days, in opened and closed amber glass bottles.

2. Material and Methods

2.1. Reagents, Reference Standards, and Materials

Methadone hydrochloride was purchased from Esteve Pharmaceuticals S.A. (Barcelona, Spain). Methylparaben and propylparaben, acquired from Fagron Iberica (Terrassa, Spain), were used as preservatives. Purified water was obtained from Grifols laboratory (Barcelona, Spain). All were of Pharmacopoeia grade. In the mobile phase, we used acetonitrile HPLC gradient grade purchased from VWR Prolabo Chemicals (Fontenay-Sous-Bois, France). Phosphoric acid, sodium hydroxide (>99%), hydrochloric acid, and hydrogen peroxide were supplied from Panreac (Barcelona, Spain) and Milli-Q water. All reagents and solvents were of analytical grade.

2.2. Standard Operating Procedure (SOP) and Storage Conditions

Methadone oral solutions M1 and M2 were carefully prepared in the Pharmacy Service of the ILUH, one with parabens as preservatives and the other one without parabens, using the following SOP: (1) the required quantity of each ingredient for the total amount to be prepared was calculated; (2) each ingredient was accurately weighed; (3) the methadone hydrochloride was placed to the mortar and triturated until a fine powder was obtained; (4) purified water or preservative water (previously elaborated by dissolving and heating the preservatives in a water bath) was added to the powder and mixed to form a uniform solution; (5) an appropriate amount of water was used to make up the volume in a volumetric flask; (6) the final solution was then packaged in amber glass vials and bottles [10,11] that were previously sterilized.

Considering that each analysis must be done in triplicate, according to International Conference on Harmonization (ICH) guidelines [20], the storage environments and the analysis times, we elaborated 6 L per each oral solution (divided into three batches: 2 L for each storage condition). The composition of solutions M1 and M2 is shown in Table 1.

Table 1. Composition of methadone hydrochloride (10 mg/mL) oral solutions.

	M1 (g)	M2 (g)
Methadone hydrochloride	20	20
Methylparaben	-	1
Propylparaben	-	0.44
Purified water to	2 L	2 L

M1: methadone hydrochloride oral solution (10 mg/mL) without parabens; M2: methadone hydrochloride oral solution (10 mg/mL) with parabens.

Solutions were packaged in amber glass vials kept unopened until the day of the test (closed containers), and in a multi-dose umber glass bottle opened daily (opened containers) for carrying out the triplicated physicochemical and microbiological analysis, in three store conditions (refrigeration (5 ± 3 °C), room temperature (25 ± 2 °C), and 40 ± 2 °C) and 9 sample times for 91 days (0, 7, 14, 21, 28, 42, 56, 70, and 91). In total, we prepared 660 samples.

We carried out the physicochemical stability study at the Center for Metabolomics and Bioanalysis (CEMBIO) of the CEU San Pablo University and the microbiological stability study at the Central Laboratory of Madrid UR Health of the Infanta Sofia University Hospital (ISUH). For the conservation of the samples at 40 ± 2 °C the following chambers were used: chamber Vötsch Industrietechnik VC-0057 (Vötschtechnik, Balingen, Germany) and Binder CB Series 9040-0039 Model CB 210 (Binder, Tuttlingen, Germany) in CEMBIO and ISUH, respectively.

2.3. Physicochemical Stability Studies

2.3.1. Equipment and Chromatographic Conditions

HPLC analyses were performed on a qualified and calibrated chromatography system, Agilent-Technologies 1100 series (Madrid, Spain), comprising a quaternary gradient pump, an ultraviolet photodiode array detector (UV-DAD), a 100-vial programmable autosampler, a column oven compartment, an automatic injector, and a software controller.

We used a Waters-XTerra^{TM} RP18 (3.5 µm; 4.6 × 100 mm) column. The column temperature was maintained at 40 °C. The mobile phase consisted of acetonitrile as the organic phase (55%) and sodium phosphate 25 mM (phosphoric acid adjusted to pH = 10 with sodium hydroxide 6.0 M) as the aqueous phase (45%). The flow rate was 1.6 mL/min. The injection volume was 5 µL for each chromatographic analysis. The UV-DAD was set at λ = 254 nm.

2.3.2. Validation of the HPLC Method and Forced-Degradation Studies

The methods and their acceptance criteria were established on the basis of the ICH guidelines Q2 (R1) [21].

A new and improved HPLC method to determine methadone hydrochloride in the presence of preservatives was developed and validated prior to this work [22]. Linearity, precision and accuracy were calculated. The curve was constructed from methadone working concentrations of 75–125% (7.5, 9.0, 10.0, 11.0, and 12.5 mg/mL) to assess the linear relationship between the concentration of the analyte and the obtained areas. Detection limit (LOD) and quantification limit (LOQ) were also obtained following EURACHEM recommendations [23]. Forced-degradation studies were also performed in acid, in base, and in oxidation conditions. Peak purity was also calculated.

2.3.3. pH Measurements

Measurements were performed in triplicate using a 744 Metrohm® pH meter (Metrohm Ltd., Herisau, Switzerland) calibrated at pH 4.01 and 7.00 on days 0, 7, 14, 21, 28, 42, 56, 70, and 91.

2.3.4. Organoleptic Characteristics

We inspected the organoleptic characteristics, such as odor, color, flavor, presence of foreign particles or precipitates of the M1 and M2 samples stored at each storage environment. To perform the visual inspection of particles, 3 mL of each formulation were taken and transferred to a transparent test tube fitted with a stopper, inverted, and observed for 5 s against a black background, and 5 s against a white background under a very bright light, according to the Royal Spanish Pharmacopoeia [24].

Measurements were also performed on days 0, 7, 14, 21, 28, 42, 56, 70, and 91.

2.3.5. Data Analysis

Stability was defined as the 90–110% recovery of methadone hydrochloride in both solutions during the 91 days of the study following the guidelines on stability testing [18–20,25–27].

2.4. Microbiological Stability Study

The microbiological quality requirements for non-sterile pharmaceutical preparations, specifically for aqueous liquid oral formulas, according to Royal Spanish Pharmacopoeia and European Pharmacopoeia guidelines [28–32], are: <10^2 CFU/mL of aerobic bacteria (TAMC), <10^1 CFU/mL of yeast and mold (TYMC), and the absence of *Escherichia coli*. If the oral compounding includes preservatives, antimicrobial effectiveness testing must be previously demonstrated [33].

The following materials were used for the tests: 0.9% saline serum, ATCC reference strains of *Staphylococcus aureus* (ATCC 43300), Bacteroides thetaiotaomicrom (ATCC 29741), *Escherichia coli* (ATCC 25922), *Pseudomonas aeruginosa* (ATCC 9027), and *Candida parapsilopsis* (ATCC 2019). They were obtained from Thermo Fisher Scientific (Waltham, MA, USA).

The culture media used were MacConkey Agar, CNA agar (nalidixic acid and colistin) with 5% lamb blood, Schaedler agar with vitamin K, and 5% lamb blood and Sabouraud agar with chloramphenicol and gentamicin. They were supplied from Becton Dickinson (Franklin Lakes, NJ, USA). We had the MALDI-TOF technique for the identification of microorganisms [34,35].

Samples of opened bottles were opened every day for a few seconds in the HPS to simulate typical drug dosage. Then, the opened and closed triplicated M1 and M2 samples were sent from the ILUH Pharmacy Service to the ISUH on Tuesdays and were analyzed on days 0, 7, 14, 21, 28, 42, 56, 70, and 91.

Every day of the analysis, 1 mL of the samples was sown in MacConkey Agar and CNA for 48 h, in Schaedler agar in anaerobiosis for 72 h, and in Sabouraud agar for 5 days. TAMC and TYMC were calculated and detection of *Escherichia coli* was performed. For M2 samples, prior to sowing, it was diluted 1:10 with saline to neutralize the action of the preservatives.

For antimicrobial effectiveness testing, several containers of M2 were seeded with a suspension of one of the test organisms to give an inoculum of 10^5–10^6 microorganisms/mL. The volume of the suspension of inoculum did not exceed 1% of the volume of the product. They were incubated at 20–25 °C and followed for 28 days to observe the logarithmic reduction according to Royal Spanish and European Pharmacopoeia.

Finally, tests for specified microorganisms were also performed in accordance with Royal Spanish and European Pharmacopoeia [28,31,32]. The product would comply with the test if no colonies of the specified microorganisms were detected.

3. Results

3.1. Physicochemical Stability

3.1.1. HPLC Analysis and Forced-Degradation Studies

Retention times for methadone, methylparaben, and propylparaben were 4.34, 0.70, and 0.88 min, respectively. The chromatograms obtained are shown in Figures 1 and 2.

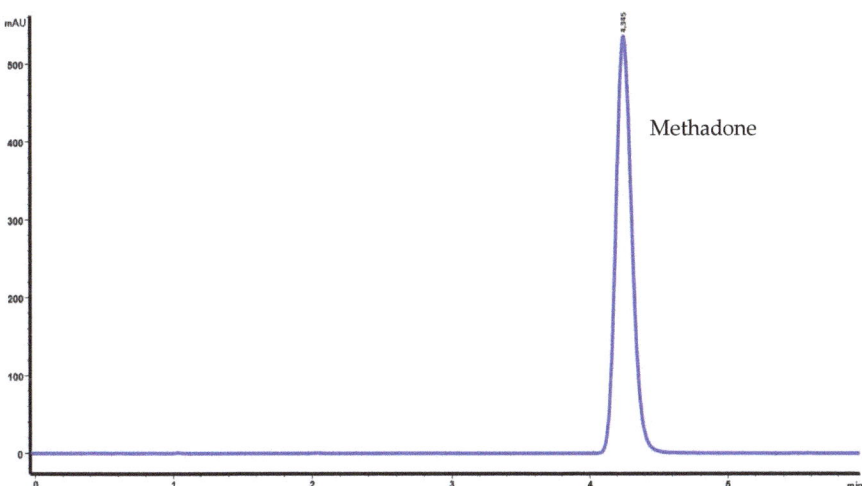

Figure 1. Chromatogram of methadone hydrochloride (10 mg/mL) without preservatives (M1). Retention time for methadone was 4.345 min.

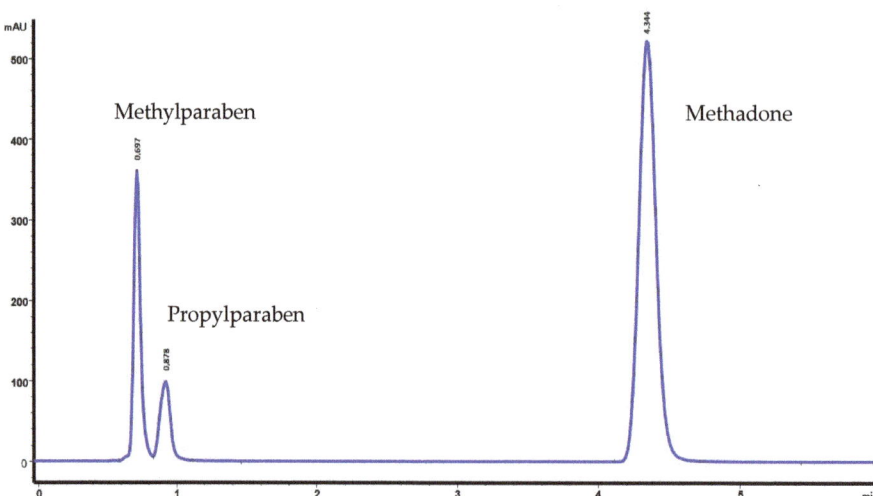

Figure 2. Chromatogram of methadone hydrochloride (10 mg/mL) with preservatives (M2). Retention times for methadone, methylparaben, and propylparaben were 4.344, 0.697, and 0.878 min.

The method was linear (y = 284.3x − 97.8, r = 0.996), with y being the obtained areas and x the concentration (mg/mL). Instrumental precision was 0.33% for standards (n = 10); intra-assay precision 0.53% (n = 6), and inter-assay precision 1.95% (n = 12). The relative standard deviation percentage for accuracy was 1.28%. The recovery percentage was 101.5 ± 1.5%.

We calculated that LOQ was 2.18 μg/mL. The LOD is the value capable of detecting the analyte. However, in the 0.0002 mg/mL concentration, four of the six samples were not detected, so the LOD was considered to be the previous concentration value in which methadone was detected. Therefore, LOD was 2.0 μg/mL.

The method was stability-indicating, with a complete separation of the degradation products from the peak of the methadone (in all situations, the methadone peak continued to be obtained at minute four). The most destabilizing conditions were oxidizing and alkaline. The methadone peak purity was 999.830 over 1000.

The percentage of recovery for all samples fulfilled the requirements of the compounding stability studies (90–110%) [26] because there was no significative degradation of methadone hydrochloride throughout the study in all batches in the three storage conditions (Table 2). This shows the stability of all samples throughout the 91 days of analysis. No differences were found between the samples of the closed and opened bottles.

Table 2. Percentage of recovery of both methadone hydrochloride oral solutions.

			Day 0	Day 7	Day 14	Day 21	Day 28	Day 42	Day 56	Day 70	Day 91
M1	Closed	Refrigeration	99.56 ± 0.26	99.92 ± 0.05	99.21 ± 0.14	99.86 ± 0.36	99.99 ± 0.19	99.61 ± 0.18	99.00 ± 0.20	100.89 ± 0.20	99.26 ± 0.08
		Room temperature	100.6 ± 0.41	101.17 ± 0.10	100.30 ± 0.14	100.09 ± 0.28	100.80 ± 0.21	100.25 ± 0.49	99.67 ± 0.51	101.88 ± 0.38	100.19 ± 0.29
		40 ± 2 °C	100.42 ± 0.20	100.28 ± 0.28	99.81 ± 0.12	100.16 ± 0.26	100.38 ± 0.32	99.99 ± 0.33	99.50 ± 0.09	101.64 ± 0.44	99.94 ± 0.63
	Opened	Refrigeration	100.28 ± 0.13	100.14 ± 0.52	99.26 ± 0.48	100.40 ± 0.44	99.89 ± 0.32	99.77 ± 0.39	99.51 ± 0.17	101.04 ± 0.29	99.27 ± 0.56
		Room temperature	104.29 ± 3.34	100.12 ± 0.05	99.36 ± 0.32	100.45 ± 0.32	99.71 ± 0.12	100.06 ± 0.30	99.84 ± 0.24	100.59 ± 0.40	99.76 ± 0.11
		40 ± 2 °C	99.72 ± 0.21	100.08 ± 0.17	99.19 ± 0.31	100.12 ± 0.05	99.69 ± 0.19	99.34 ± 0.31	99.43 ± 0.32	101.56 ± 0.24	99.62 ± 0.69
M2	Closed	Refrigeration	100.54 ± 0.10	100.43 ± 0.31	99.77 ± 0.09	100.39 ± 0.01	100.16 ± 0.06	99.82 ± 0.01	99.91 ± 0.02	100.70 ± 0.11	101.14 ± 0.02
		Room temperature	100.61 ± 0.37	101.01 ± 0.14	100.13 ± 0.02	100.69 ± 0.03	100.26 ± 0.04	100.63 ± 0.03	100.01 ± 0.05	101.65 ± 0.02	101.34 ± 0.04
		40 ± 2 °C	100.75 ± 0.10	100.44 ± 0.17	99.81 ± 0.01	100.37 ± 0.01	100.27 ± 0.05	100.30 ± 0.01	100.12 ± 0.03	101.94 ± 0.03	100.75 ± 0.07
	Opened	Refrigeration	99.79 ± 0.26	100.10 ± 0.23	99.29 ± 0.07	100.00 ± 0.01	99.48 ± 0.06	99.82 ± 0.01	99.71 ± 0.04	101.61 ± 0.03	99.31 ± 0.01
		Room temperature	100.07 ± 0.35	100.45 ± 0.17	99.15 ± 0.03	99.95 ± 0.02	99.83 ± 0.01	100.14 ± 0.02	99.79 ± 0.05	101.18 ± 0.04	99.62 ± 0.02
		40 ± 2 °C	99.76 ± 0.20	100.03 ± 0.20	99.19 ± 0.03	99.97 ± 0.06	99.79 ± 0.03	99.41 ± 0.02	99.94 ± 0.02	101.46 ± 0.05	99.76 ± 0.03

M1: methadone hydrochloride oral solution (10 mg/mL) without parabens; M2: methadone hydrochloride oral solution (10 mg/mL) with parabens.

3.1.2. pH Measurements

pH values increased throughout the 91 days of the study. We emphasize that lower values were obtained in the oral solution that contained the preservatives (M2). However, neither of the two solutions reached pH = 7 (Figures 3 and 4). This results ensured no risk of methadone precipitation.

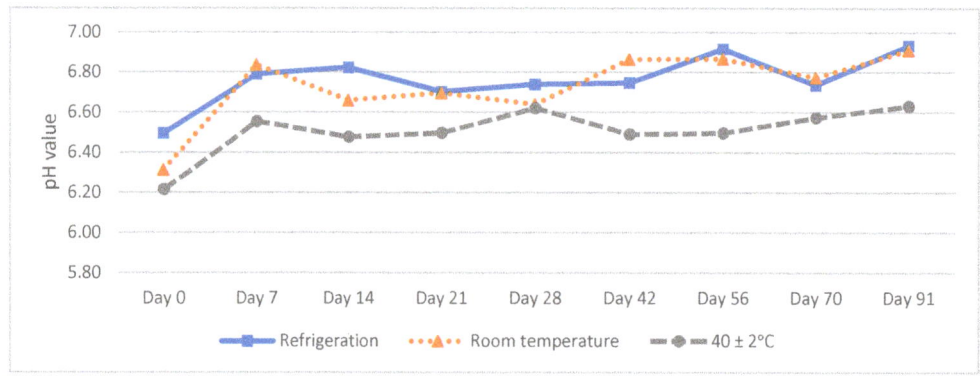

Figure 3. pH values of methadone hydrochloride (10 mg/mL) without the preservative (M1) oral solution.

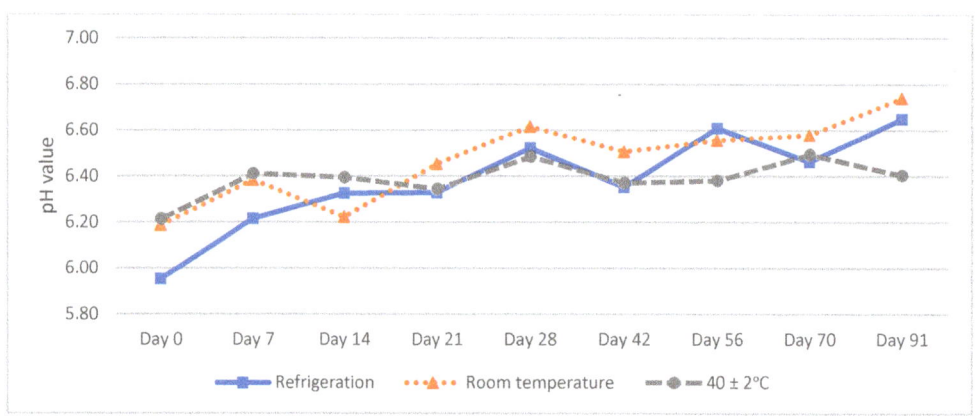

Figure 4. pH values of methadone hydrochloride (10 mg/mL) with the preservative (M2) oral solution.

3.1.3. Organoleptic Characteristics

Throughout the study, the samples of both solutions remained transparent, without coloration. No suspended particles or precipitates were observed. They showed no odor. However, the bitter taste (typical of methadone) on day 0 was maintained until the last day of the study. No differences were found in any of the three storage environments.

3.2. Microbiological Stability

The antimicrobial efficacy of the preservatives used in the M2 formulation was demonstrated prior to the microbiological stability study, as well as the fertility test of the culture media and the suitability of the counting method in the presence of the product.

Throughout the study, no microbial growth was observed in any sample analyzed (on day 70 it could not be cultured due to the SARS-CoV-2 lockdown). Despite this, the analysis on day 91 did not show any microbial growth either. Therefore, all samples from closed and opened bottles met the microbiological quality acceptance criteria for liquid oral formulas until the end time in the three storage environments.

4. Discussion

In this study, we carried out a physicochemical and microbiological stability study of two oral solutions of methadone hydrochloride (10 mg/mL) with and without preservatives, with an increased BUD until 91 days compared to the formulation described in the Spanish National Formulary.

The stability of methadone hydrochloride has already been studied in the literature. These previous studies have allowed us to avoid sampling in unstable conditions e.g., unprotected from light [11,16] or in plastic packaging material [10], where there was a greater loss of methadone. This fact led us to carry out our stability study in amber glass bottles.

Moreover, those stability studies were carried out on different pharmaceutical preparations of methadone compared to those used in this work. These preparations were elaborated at different concentrations, with other vehicles for oral administration different from water (such us Tang or simple syrup), for administration routes other than the oral route, or using methadone hydrochloride different from the raw material (tablets, injectables, etc.) [10,11,15–17]. We also find stability studies of formulations in which methadone was formulated with other drugs [13,36].

The use of preservatives with methadone was described in the literature for the first time by Beaseley and Ziegler [37] in 1977; they observed that the standard methadone solution was unstable and adding 1 mg of sodium benzoate per milliliter to the solutions can stabilize them for at least one month.

Due to the increased use of methylparaben as preservative in the 1980s, Ching et al. [14] carried out a stability study of methadone 5 mg/mL, demonstrating a stability of 4 months but without microbiological data. None of the previously described stability studies used in their composition the two parabens that we used in this work in the M2 solution. We used methylparaben and propylparaben since they are the most widely used preservatives in aqueous solutions formulated, such as preservative water, according to the Spanish National Formulary [38].

The HPLC method developed in our previous study was a rapid, simple, reliable, and economical technique analytically validated and has allowed for the efficient quantification of methadone hydrochloride in an oral solution without interference from preservatives and a concentration and composition not previously analyzed in the literature. This procedure was a new and an improved method in comparison to those described in the United States Pharmacopoeia (USP) [39] and the literature [37,40–43].

In our stability study, all samples comply with the recommendations. The two formulations (with and without preservatives) remained stable throughout the 91 days of study stored under refrigeration (5 ± 3 °C), room temperature (25 ± 2 °C) and 40 ± 2 °C, indicating that the presence of parabens does not alter the stability of methadone hydrochloride. With regard to pH, we want to highlight that, although the recommendations indicate that methadone hydrochloride should have a value lower than 6.5, in our study we demonstrated that this value can be extended (near to 7) because methadone hydrochloride remained stable at values close to that figure.

Regarding the small pH variations found in both methadone solutions over time, although they were relevant due to the results of our stability study, we can affirm that they were not due to microbial growth or the appearance of degradation products. Therefore, the variations found may be due to the small variations in temperature since the samples were removed from the chamber until their analyses with the pH meter, to the variations of the pH meter, due to the fact that they were not designed to operate in a solution that contained almost no ions.

Regarding the microbiology stability, it is important to highlight that aqueous solutions or suspensions are more unstable and much more vulnerable to microbiological contamination than solid forms. Most of the stability studies, being non-sterile pharmaceutical forms, do not include microbiological stability studies. Furthermore, contamination can reduce or inactivate the therapeutic action of the drug or even form toxic products affecting patient safety. Even the ingestion of pathogenic microorganisms can cause a serious infection in the pediatric population or in the immunosuppressed patient.

With these premises, it was necessary to carry out a microbiological study in which there was no microbial growth in the closed bottles or in those that were opened daily throughout the 91 days of the study, thus demonstrating that both solutions were microbiologically and physicochemically stable without toxic products.

After our results, unlike the results by Lauriault et al. [15], the addition of preservatives was not necessary to prevent microbial growth, since in our samples, both in the presence or absence of preservatives, no microbial growth was observed after 91 days of study, including those that were kept at room temperature and at 40 ± 2 °C. Furthermore, following the Good Practice Guidelines, standardizing procedures, cleaning, and evaluating the elaborating staff, the risk was minimized.

The evaluation of these two oral solutions has not been published before. Furthermore, a formulation of methadone hydrochloride solution with methylparaben and propylparaben in its composition has never been evaluated before. This fact allowed us to study the stability of an alternative methadone solution, in case the preservative-free solution daily prepared at the Hospital Pharmacy Service is not stable; we also analyzed whether the addition of preservatives, beyond increasing the microbiological quality, could alter the physicochemical stability of the solution.

Moreover, among other advantages of our study, we designed the stability study following literature recommendations and guidelines [20,25–27], choosing more sampling times to have more data in case we did not find stability throughout the study. We also chose three storage environments to obtain more information on the stability of methadone in a solution out of the fridge, both at room temperature or 40 ± 2 °C, situations that could arise in case of improper conservation or hot weather. These results offer efficient and necessary information that could be applied, day-to-day, at the Hospital Pharmacy Service, CDAP, and in a patient's home.

We also want to highlight the logistical and organizational complexity of this stability study, which involved a total of 660 samples (with difficulty in treating and identifying each sample) and three centers that had to be constantly coordinated.

After this study, the physicochemical and microbiological stability, the classical formulation (M1), and the new alternative solution (M2) were demonstrated. Thanks to our study, the main advantage and practical application is that we increased the beyond-use date of the methadone solution, which made it possible to improve the workflow organization in the daily preparations of the Hospital Pharmacy Service. More importantly, it also allows for better organization in centers, especially during holiday periods or due to a lack of staff. It even allows adapting the dispensing to the needs of patients receiving methadone treatment. We hope to publish these results (that we are measuring) in the future.

5. Conclusions

Two oral methadone hydrochloride solutions (10 mg/mL), with and without parabens as preservatives, were physically and microbiologically stable at 5 ± 3 °C, 25 ± 2 °C, and 40 ± 2 °C, for 91 days in previously sterilized closed and opened bottles.

The results obtained after this stability study allowed to increase the BUD of the methadone hydrochloride oral solution prepared daily at the Hospital Pharmacy Service of the Infanta Leonor University Hospital for the patients under methadone maintenance programs in the community of Madrid.

Author Contributions: Conceptualization, E.A.Á.-A.; methodology, E.A.Á.-A., M.P.L., A.G.-P. and E.I.-G.; software, E.A.Á.-A., M.P.L. and A.G.-P.; validation, E.A.Á.-A., M.P.L. and A.G.-P.; formal analysis, E.A.Á.-A.; investigation, E.A.Á.-A., M.P.L. and A.G.-P.; resources, E.A.Á.-A., M.P.L., A.G.-P. and E.I.-G.; data curation, E.A.Á.-A., M.P.L. and A.G.-P.; writing-original draft preparation, E.A.Á.-A.; writing-reviewing and editing, M.P.L., A.G.-P., E.I.-G., I.E.-R. and A.A.-R.; visualization, E.A.Á.-A.,

M.P.L., A.G.-P., E.I.-G., I.E.-R. and A.A.-R.; supervision: I.E.-R. and A.A.-R., project administration, E.A.Á.-A., I.E.-R. and A.A.-R.; funding acquisition, E.A.Á.-A., I.E.-R. and A.A.-R. All authors have read and agreed to the published version of the manuscript.

Funding: This research did not receive any specific grants from funding agencies in the public, commercial, or not-for-profit sectors.

Institutional Review Board Statement: Not applicable.

Informed Consent Statement: Not applicable.

Data Availability Statement: The datasets used and/or analyzed in the current study are available from the corresponding author upon reasonable request.

Acknowledgments: We would like to acknowledge the entire CEMBIO department of the San Pablo Ceu University Faculty of Pharmacy for allowing us to use their material, facilities, and equipment to carry out this work, and the Central Laboratory of Madrid UR Health of the Infanta Sofia University Hospital. We also thank the Subdirectorate of Pharmacy and Health Products and the Subdirection of Addictions of the Ministry of Health of the Autonomous Community of Madrid for the collaboration and support. These entities helped to develop this project and bring it to fruition. We extend special thanks to the Direction of the Infanta Leonor University Hospital, the Pharmacy Service, and the Central Services of the Infanta Leonor University Hospital, for their unqualified and total support. We would also like to thank María Chuecos Lozano, Javier Rupérez Pascualena, and Nélida Barrueco Fernández for the editorial help in the preparation of this paper.

Conflicts of Interest: The authors declare that they have no competing interests.

References

1. Spanish Agency of Medicines and Sanitary Products. Data Sheet Metasedin. 2010. Available online: https://cima.aemps.es/cima/pdfs/es/ft/62423/FT_62423.html.pdf (accessed on 16 January 2022).
2. Payte, J.T. A Brief History of Methadone in the Treatment of Opioid Dependence: A Personal Perspective. *J. Psychoact. Drugs* **1991**, *23*, 103–107. [CrossRef] [PubMed]
3. Dole, V.P.; Nyswander, M. A Medical Treatment for Diacetylmorphine (Heroin) Addiction. A Clinical Trial with Methadone Hydrocloride. *JAMA* **1965**, *193*, 646–650. [CrossRef] [PubMed]
4. Torrens, M.; Fonseca, F.; Castillo, C.; Domingo-Salvany, A. Methadone Maintenance Treatment in Spain: The Success of a Harm Reduction Approach. *Bull. World Health Organ.* **2013**, *91*, 136–141. [CrossRef] [PubMed]
5. Meneses Falcón, C.; Charro Baena, B. *Los Programas de Mantenimiento con Metadona en Madrid: Evolución y Perfil de los Usuarios*; Universidad Pontificia Comillas: Madrid, Spain, 2000; ISBN 978-84-89708-95-2.
6. Dirección General de Coordinación de La Asistencia Sanitaria. Resolución Del Director General de Coordinación de La Asistencia Sanitaria Por La Que Se Encomienda al Hospital Universitario Infanta Leonor Las Tareas de Suministro Diario de Metadona a Los Centros de Atención Integral a Drogodependientes de La Subdirección General de Asistencia En Adicciones Del Servicio Madrileño de Salud. N° 189/2018. 27 February 2018.
7. *Formulario Nacional*, 3rd ed.; Agencia Española de Medicamentos y Productos Sanitarios; Ministerio de Sanidad y Consumo: Madrid, Spain, 2020. Available online: https://www.aemps.gob.es/formulario-nacional/ (accessed on 15 January 2022).
8. *FN/2003/PA/022 Metadona, Hidrocloruro de. Formulario Nacional*, 3rd ed.; Agencia Española de Medicamentos y Productos Sanitarios; Ministerio de Sanidad y Consumo: Madrid, Spain, 2020. Available online: https://www.aemps.gob.es/formulario-nacional/ (accessed on 15 January 2022).
9. Martindale, D.; Brayfield, A. *Martindale: The Complete Drug Reference*, 39th ed.; Pharmaceutical Press: London, UK, 2017; ISBN 978-0-85711-309-2.
10. Provenza, N.; Calpena, A.C.; Mallandrich, M.; Pueyo, B.; Clares, B. Design of Pediatric Oral Formulations with a Low Proportion of Methadone or Phenobarbital for the Treatment of Neonatal Abstinence Syndrome. *Pharm. Dev.Technol.* **2016**, *21*, 755–762. [CrossRef] [PubMed]
11. Denson, D.D.; Crews, J.C.; Grummich, K.W.; Stirm, E.J.; Sue, C.A. Stability of Methadone Hydrochloride in 0.9% Sodium Chloride Injection in Single-Dose Plastic Containers. *Am. J. Health-Syst. Pharm.* **1991**, *48*, 515–517. [CrossRef]
12. Álvaro-Alonso, E.A.; Tejedor-Prado, P.; Aguilar-Ros, A.; Escobar-Rodríguez, I. Centralization of the Methadone Maintenance Plan in a Hospital Pharmacy Department in the Community of Madrid. *Farm. Hosp.* **2020**, *44*, 185–191. [CrossRef] [PubMed]
13. Little, T.L.; Tielke, V.M.; Carlson, R.K. Stability of Methadone Pain Cocktails. *Am. J. Hosp Pharm.* **1982**, *39*, 646–647. [CrossRef] [PubMed]
14. Ching, M.S.; Stead, C.K.; Shilson, A.D. Stability of Methadone Mixture with Methyl Hydroxybenzoate as a Preservative. *Aust. J. Hosp. Pharm.* **1989**, *19*, 159–161.
15. Lauriault, G.; Lebelle, M.; Lodge, B.; Savard, C. Stability of Methadone in 4 Vehicles for Oral-Administration. *Am. J. Hosp. Pharm.* **1991**, *48*, 1252–1256. [CrossRef] [PubMed]

16. Soy, D.; Roca, M.; Deulofeu, R.; Montes, E.; Codina, C.; Ribas, J. Stability of 0.1% and 0.5% Oral Methadone Clorhydrate Solutions in Saline. *Farm. Hosp.* **1998**, *22*, 249–251.
17. Friciu, M.M.; Alarie, H.; Beauchemin, M.; Forest, J.-M.; Leclair, G. Stability of Methadone Hydrochloride for Injection in Saline Solution. *Can. J. Hosp. Pharm.* **2020**, *73*, 141–144. [CrossRef] [PubMed]
18. Ministerio de Sanidad, Servicios Sociales e Igualdad. Guía de Buenas Prácticas de Preparación de Medicamentos en Servicios de Farmacia Hospitalaria. 2014. Available online: http://www.msssi.gob.es/profesionales/farmacia/pdf/GuiaBPP3.pdf (accessed on 15 January 2022).
19. The United States Pharmacopeia (USP). Chapter <795> Pharmaceutical Compounding—Nonsterile Preparations. In *The United States Pharmacopeia (USP)*; The United States Pharmacopeia: Rockville, MD, USA, 2013.
20. European Medicines Agency. *ICH Topic Q1A (R2) Stability Testing of New Drug Substances and Products*; European Medicines Agency: Amsterdam, The Netherlands, 2003.
21. European Medicines Agency. *ICH Topic Q2 (R1) Validation of Analytical Procedures: Text and Methodology*; European Medicines Agency: Amsterdam, The Netherlands, 2006.
22. Álvaro-Alonso, E.A.; Lorenzo-García, M.P.; Escobar-Rodríguez, I.; Aguilar-Ros, A. Development and Validation of a HPLC-UV Method for Methadone Hydrochloride Quantification in a New Oral Solution with Preservatives to Be Implemented in Physicochemical Stability Studies. *Rev. BMC Chem.* **2022**. Preprint. [CrossRef]
23. Eurachem. Guidance Document No. WGD 2, Accreditation for Chemical Laboratories: Guidance on the Interpretation of the EN45000 Series of Standards and ISO/IEC, Guide 25. 1993. Available online: https://www.eurachem.org/images/stories/Guides/pdf/Eurachem_CITAC_QAC_2016_EN.pdf (accessed on 15 January 2022).
24. Real Farmacopea Española. Capítulo 2.9.20. In *Contaminación Por Partículas: Partículas Visibles*, 5th ed.; Agencia Española de Medicamentos y Productos Sanitarios España; Ministerio de Sanidad y Consumo: Madrid, Spain, 2015.
25. European Medicines Agency. ICH Topic Q1C Stability Testing: Requirements for New Dosage Forms. 1998. Available online: https://www.ema.europa.eu/en/documents/scientific-guideline/ich-q-1-c-stability-testing-requirements-new-dosage-forms-step-5_en.pdf (accessed on 15 January 2022).
26. European Medicines Agency. Guideline on Stability Testing: Stability Testing of Existing Active Substances and Related Finished Products. 2007. Available online: https://www.ema.europa.eu/en/documents/scientific-guideline/guideline-stability-testing-stability-testing-existing-active-substances-related-finished-products_en.pdf (accessed on 16 January 2022).
27. Société Française de Pharmacie Clinique (SFPC); Groupe d'Evaluation et de Recherche sur la Protection en Atmosphère Contrôlée (GERPAC). *Methodological Guidelines for Stability Studies of Hospital Pharmaceutical Preparations*, 1st ed.; Print Conseil: Romagnat, France, 2013; ISBN 978-2-9526010-4-7; Available online: http://www.gerpac.eu/IMG/pdf/guide_stabilite_anglais.pdf (accessed on 16 January 2022).
28. European Pharmacopoeia. *Chapter Microbiological Quality of Non-Sterile Products for Pharmaceutical Use*, 7th ed.; Council of Europe: Strasbourg, France, 2010.
29. Real Farmacopea Española. Capítulo 5.1.4. In *Calidad Microbiológica de Las Preparaciones Farmacéuticas y de Las Sustancias Para Uso Farmacéutico No Estériles*, 5th ed.; Agencia Española de Medicamentos y Productos Sanitarios; Ministerio de Sanidad y Consumo: Madrid, Spain, 2015. Available online: https://extranet.boe.es/farmacopea/ (accessed on 15 January 2022).
30. Real Farmacopea Española. Capítulo 2.6.12. In *Control Microbiológico de Productos No Estériles; Ensayos de Recuento Microbiano*, 5th ed.; Agencia Española de Medicamentos y Productos Sanitarios; Ministerio de Sanidad y Consumo: Madrid, Spain, 2015. Available online: https://extranet.boe.es/farmacopea/ (accessed on 15 January 2022).
31. Real Farmacopea Española. Capítulo 2.6.13. In *Control Microbiológico de Productos No Estériles; Ensayo de Microorganismos Específicados*, 5th ed.; Agencia Española de Medicamentos y Productos Sanitarios; Ministerio de Sanidad y Consumo: Madrid, Spain, 2015. Available online: https://extranet.boe.es/farmacopea/ (accessed on 15 January 2022).
32. *European Pharmacopoeia (Ph. Eur.)*, 10th ed.; Council of Europe: Strasbourg, France, 2019; ISBN 978-92-871-8921-9.
33. Real Farmacopea Española. Capítulo 5.1.3. In *Eficacia de La Conservación Antimicrobiana*, 5th ed.; Agencia Española de Medicamentos y Productos Sanitarios; Ministerio de Sanidad y Consumo: Madrid, Spain, 2015. Available online: https://extranet.boe.es/farmacopea/ (accessed on 15 January 2022).
34. Dingle, T.C.; Butler-Wu, S.M. MALDI-TOF Mass Spectrometry for Microorganism Identification. *Clin. Lab. Med.* **2013**, *33*, 589–609. [CrossRef] [PubMed]
35. Mansilla, E.C.; Moreno, R.C.; García, M.O. *Aplicaciones de la Espectrometría de Masas MALDI-TOF en Microbiología Clínica*; Sociedad Española de Enfermedades Infecciosas y Microbiología Clínica: Madrid, Spain, 2019; ISBN 978-84-09-10307-2.
36. Lee, D.; Watson, N.; Whittem, T. Chemical Stability of Morphine and Methadone, and of Methadone in Combination with Acepromazine, Medetomidine or Xylazine, during Prolonged Storage in Syringes. *Aust. Vet. J.* **2017**, *95*, 289–293. [CrossRef] [PubMed]
37. Beasley, T.H.; Ziegler, H.W. High-Performance Liquid Chromatographic Analysis of Methadone Hydrochloride Oral Solution. *J. Pharm. Sci.* **1977**, *66*, 1749–1751. [CrossRef] [PubMed]
38. FN/2017/EX/028 Agua Conservante sin Propilenglicol. *Formulario Nacional*, 3rd ed.; Agencia Española de Medicamentos y Productos Sanitarios; Ministerio de Sanidad y Consumo: Madrid, Spain, 2020. Available online: https://www.aemps.gob.es/formulario-nacional/ (accessed on 16 January 2022).

39. United States Pharmacopeial Convention. *The United States Pharmacopeia: USP37: The National Formulary: NF32*; United States Pharmacopeial Convention Inc.: Rockville, MD, USA, 2014; ISBN 978-1-936424-25-2.
40. Hsieh, J.; Ma, J.; O'donell, J.; Choulis, N. High-Performance Liquid-Chromatographic Analysis of Methadone in Sustained-Release Formulations. *J. Chromatogr.* **1978**, *161*, 366–370. [CrossRef] [PubMed]
41. Derendorf, H.; Garrett, E.R. High-Performance Liquid Chromatographic Assay of Methadone, Phencyclidine, and Metabolites by Postcolumn Ion-Pair Extraction and on-Line Fluorescent Detection of the Counterion with Applications. *J. Pharm. Sci.* **1983**, *72*, 630–635. [CrossRef] [PubMed]
42. Adams, P.; Haines-nutt, R. High-Performance Liquid-Chromatographic Analysis of Methadone Hydrochloride in Pharmaceuticals. *J. Chromatogr.* **1985**, *329*, 438–440. [CrossRef]
43. Helmlin, H.J.; Bqurquin, D.; De Bernardini, M.; Brenneisen, R. Determination of Methadone in Pharmaceutical Preparations Using High-Performance Liquid Chromatography with Photodiode Array Detection. *Pharm. Acta Helv.* **1989**, *64*, 178–182.

Article

Lithium Ascorbate as a Promising Neuroprotector: Fundamental and Experimental Studies of an Organic Lithium Salt

Ivan Yu. Torshin [1], Olga A. Gromova [1,*], Konstantin S. Ostrenko [2], Marina V. Filimonova [3], Irina V. Gogoleva [4], Vladimir I. Demidov [4] and Alla G. Kalacheva [4]

1. Federal Research Center "Computer Science and Control" of Russian Academy of Sciences, 119333 Moscow, Russia; tiy135@yahoo.com
2. All-Russian Research Institute of Physiology, Biochemistry and Animal Nutrition, 249013 Borovsk, Russia; ostrenkoks@gmail.com
3. A. Tsyba Medical Radiological Research Center, 249036 Obninsk, Russia; vladimirovna.fil@gmail.com
4. Ivanovo State Medical Academy, 153012 Ivanovo, Russia; gogolev04@yandex.ru (I.V.G.); 13vid@mail.ru (V.I.D.); alla_kalacheva@mail.ru (A.G.K.)
* Correspondence: unesco.gromova@gmail.com

Abstract: Given the observable toxicity of lithium carbonate, neuropharmacology requires effective and non-toxic lithium salts. In particular, these salts can be employed as neuroprotective agents since lithium ions demonstrate neuroprotective properties through inhibition of glycogen synthetase kinase-3β and other target proteins, increasing concentrations of endogenous neurotrofic factors. The results of theoretical and experimental studies of organic lithium salts presented here indicate their potential as neuroprotectors. Chemoreactomic modeling of lithium salts made it possible to select lithium ascorbate as a suitable candidate for further research. A neurocytological study on cerebellar granular neurons in culture under conditions of moderate glutamate stress showed that lithium ascorbate was more effective in supporting neuronal survival than chloride or carbonate, i.e., inorganic lithium salts. Biodistribution studies indicated accumulation of lithium ions in a sort of "depot", potentially consisting of the brain, aorta, and femur. Lithium ascorbate is characterized by extremely low acute and chronic toxicity (LD50 > 5000 mg/kg) and also shows a moderate antitumor effect when used in doses studied (5 or 10 mg/kg). Studies on the model of alcohol intoxication in rats have shown that intake of lithium ascorbate in doses either 5, 10 or 30 mg/kg did not only reduced brain damage due to ischemia, but also improved the preservation of myelin sheaths of neurons.

Keywords: lithium therapy; neurocytology; toxicology; neuroprotection; chemoinformatics; big data

Citation: Torshin, I.Y.; Gromova, O.A.; Ostrenko, K.S.; Filimonova, M.V.; Gogoleva, I.V.; Demidov, V.I.; Kalacheva, A.G. Lithium Ascorbate as a Promising Neuroprotector: Fundamental and Experimental Studies of an Organic Lithium Salt. *Molecules* 2022, 27, 2253. https://doi.org/10.3390/molecules27072253

Academic Editors: Giovanni Ribaudo and Laura Orian

Received: 24 February 2022
Accepted: 25 March 2022
Published: 30 March 2022

Publisher's Note: MDPI stays neutral with regard to jurisdictional claims in published maps and institutional affiliations.

Copyright: © 2022 by the authors. Licensee MDPI, Basel, Switzerland. This article is an open access article distributed under the terms and conditions of the Creative Commons Attribution (CC BY) license (https://creativecommons.org/licenses/by/4.0/).

1. Introduction

Lithium is an essential trace element that exhibits pronounced neuroprotective and neurotrophic effects. Although lithium carbonate has been used in the pharmacotherapy of bipolar disorder for over 60 years [1], this is minimal use of lithium salts. Much more important is the fact that lithium is necessary for the normal functioning of all body systems and, above all, the nervous system [2]. Lithium is involved in the metabolism of simple sugars, lipids, regulation of blood pressure and hematopoiesis, regulation of inflammation, in the homeostasis of neurotransmitters, and, in general, exhibits neurotrophic and neuroprotective effects [3]. Organic lithium salts can enhance the effects of other neuroprotective agents [4].

Lithium ions exert their effects by activating neuroprotective and neurotrophic cell cascades. The mechanisms by which these effects of lithium are mediated include inhibition of glycogen synthetase kinase-3β (GSK-3β), induction of autophagy, inhibition of N-methyl-D-aspartate (NMDA) receptors, and anti-apoptotic effects. GSK-3β is a negative regulator of the Wnt cascade required for axonal growth. Inhibition of GSK-3β by lithium ions promotes activation of the neurotrophic cascade Wnt, accelerates the differentiation of neuronal progenitor cells, stimulates differentiation of astrocytes, myelin synthesis, supports neuronal survival, expression of brain-derived neurotrophic factor (BDNF), and of other neurotrophic factors [5]. Clinical studies show that lithium therapy induces an increase in gray matter volume [6]. It is advisable to use lithium preparations in patients with amyotrophic lateral sclerosis [7]. An interesting direction in developing neuroprotective drugs is related to the studies of organic lithium salts, characterized by low toxicity and high bioavailability [8].

Commonly used lithium carbonate has significant drawbacks, especially when taken for a long time (months) in high doses (grams). First, for the effective and safe use of these lithium carbonate drugs, it is necessary to regularly measure the concentration of lithium in the blood, which is an additional invasive procedure [9]. Secondly, lithium carbonate is characterized by a narrow therapeutic corridor of lithium concentrations in plasma (0.6–1.2 mmol/L). Li+ concentrations of 1.5–2.5 mmol/L are associated with mild toxicity, 2.5–3.5 mmol/L—with severe poisoning, and Li+ concentration in blood plasma exceeding 3.5 mmol/L can be life-threatening [10]. Thirdly, to achieve the desired therapeutic effect in psychiatry, it is necessary to use significant dosages of lithium carbonate (1–3 g/day, in the acute period—up to 9 g/day), which, due to the high toxicity of the salt, sharply reduces its spectrum of clinical applications [11].

Therefore, it is relevant to the search for such anions for the synthesis of lithium salts that (1) maximize the flow of lithium ions into neurons, (2) would be characterized by the most suitable spectrum of pharmacological activities, and (3) would not show toxic effects even with prolonged use.

This work presents the search results for organic anions most suitable for the synthesis of neuroprotective lithium salts. The actual screening and analysis of the pharmacological effects of organic anions were carried out in silico using the latest mathematical apparatus developed at the scientific school of Acad. RAS Yu.I. Zhuravlev (chemoreactomic modeling) and the proprietary software complex [12,13]. Then, a complex of in vitro and in vivo studies were carried out for the selected candidates. This paper presents the results of studies of lithium ascorbate, including neurocytological studies on cerebellar granular neurons in vitro (model of glutamate stress), analysis of biodistribution, acute and chronic toxicity, antitumor effects, and neuroprotective properties in vivo estimated on the model of alcohol intoxication.

2. Results

Chemoreactomic modeling made it possible to select lithium ascorbate as the most promising salt for further studies. After the synthesis of the lithium ascorbate its neurocytological study was carried out on the granular neurons of the cerebellum. Animal studies of the lithium ascorbate included assessments of lithium biodistribution, studies of acute and chronic toxicity, antitumor properties, adaptogenic and neuroprotective effects.

2.1. Results of Chemoreactomic Modeling of Lithium Salts

Chemoreactomic screening of lithium salts was performed in three stages: (A) assessment of acute toxicity (LD50) for 1245 water-soluble organic lithium salts, (B) assessment of bioavailability (%) and pharmacokinetic parameters (Cmax, tmax, Vd), (C) assessment of various pharmacodynamic effects (IC50, EC50). At the first stage, 38 minimally toxic lithium salts (LD50 > 1000 mg/kg) were selected. At the second stage, the 11 lithium salts with maximum bioavailability (>20%): ascorbate, nicotinate, oxybutyrate, orotate, citrate, gluconate, comenat, pyroglutamate, glycinate, asparaginate, lactate were identified. At the third stage, various biological and pharmacological effects of the selected lithium salts were assessed and an analysis of possible interactions of the lithium ion with human proteome proteins was carried out. In general, the chemoreactomic analysis showed that lithium ascorbate is a remarkable candidate for further studies. Furthermore, we present some of the estimates of various properties of lithium ascorbate in comparison with the "control" salts (lithium nicotinate, lithium oxybutyrate, and lithium carbonate). These salts were chosen as examples because they are used in normothymic (mood stabilizer) drugs.

In comparison with the control substances, lithium ascorbate was characterized by more prominent inhibition of the reuptake of serotonin and dopamine and by a greater affinity for the inhibition of glutamate and α-adrenergic receptors. Lithium ascorbate can also be characterized by anti-inflammatory action (due to modulation of prostaglandin metabolism), to exhibit moderate anticoagulant, antihyperlipidemic, antihyperglycemic and antitumor effects (Table 1).

Table 1. Results of chemoreactomic analysis of the lithium salts. "Const.", The common name for the type of investigated pharmacological constant ("Ki", "IC50", etc.); "Error"—the error of the estimated biological constant; "Unit", units of measurement.

Const.	Ascorbate	Nicotinate	Oxybutyrate	Carbonate	Error	Unit	Activity
Ki	648	2980	1503	8850	572	nM	Inhibition of serotonin reuptake
Ki	126	421	285	5450	559	nM	Inhibition of dopamine reuptake
Ki	622	1866	930	7402	320	nM	Inhibition of currents of N-methyl-D-aspartate glutamate receptors of the NR1a/NR2D type
Ki	173	438.7	862	6820	182	nM	Affinity for alpha-2 adrenergic receptors of rat brain homogenates, ligand
Ki	115	1838		5928	2539	nM	Displacement of N-methylscopolamine from M1 muscarinic receptors
IC50	79	2534β	1658	8240	191	nM	Inhibition of GSK-3β
EC50	60.4	111.1		-	82	mg/kg	Modeling anticonvulsant activity in a model of electroshock in mice
-	3.8	37.4	8.83	29.1	4.8	%	Percentage change in hematocrit at a dose of 10 mg/kg/day
-	17.2	12.9	eight	-	1.9	%	Fraction of unbound substance in human blood plasma
T1/2	8.4	54	25	-	4	Min	Metabolic stability in human liver microsomes
CL	29	19	21	-	16	mL/min	Characteristic clearance in human liver microsomes
Ki	5308	1868	2995	-	2118	nM	Affinity for the K-channel KCNH2
IC50	5069	2746	1545	-	1199	nM	Inhibition of the K-channel KCNH2
IC50	113.9	4967	495	9130	351	nM	Inhibition of MMP-2
Ki	165.3	347.5	344	10,239	54	nM	Inhibition of MMP-9
-	41.25	22.16	25.1	-	34	%	Anti-inflammatory activity as inhibition of carrageenan-induced paw edema (25 mg/kg orally)
IC50	125.7	383	598	>40,000	507	nM	Inhibition of clotting factor F10
EC50	126	2264	1987	82,303	295	nM	PPAR delta agonist
-	14.98	8.08		0	5.5	%	Antihyperglycemic activity in rats as a decrease in blood glucose at a dose of 100 mg/kg, orally
IC50	185.5	3842	1702	>20,000	929	nM	Antiproliferative Activity Against Human QG56 cells
IC50	110.8	3096	1100	>20,000	242	nM	Antiproliferative activity against human H460 cells

Lithium ascorbate is a highly effective antioxidant and, therefore, is metabolized and excreted from the body faster than the reference molecules. For lithium ascorbate, a lower T1/2 value in human liver microsomes (8.4 min, other molecules—more than 25 min) and a higher clearance value in liver microsomes (ascorbate—29 mL/min, nicotinate—19 mL/min, oxybutyrate—21 mL/min) were found.

It is important to note that lithium ascorbate may be more effective than nicotinate and oxybutyrate (see Table 1) in inhibiting the activity of the GSK-3β enzyme (ascorbate—47%, nicotinate—3.3%, oxybutyrate—6%)—one of the main target proteins of the lithium ions. The inhibition constant of GSK-3β was IC50 = 79 nM for lithium ascorbate and was much higher for the rest of the salts studied (IC50 = 1658–8240 nM). Let's remind the reader that the lower values of an inhibition constant correspond to a higher inhibition of the corresponding target protein.

2.2. Neurocytological Studies of Lithium Salts on Cultured Cerebellar Granular Neurons

A glutamate stress model evaluated the synthesized lithium ascorbate substance in neurocytological studies on cultured cerebellar granular neurons (CGN). Studies have shown that lithium ascorbate was more effective in supporting neuronal survival than lithium chloride or lithium carbonate in the 0.1–1.0 mM concentration range. At the same time, for lithium chloride and for lithium carbonate, a significant dispersion of the values of neuronal survival was observed, which did not allow us to register any statistically significant differences when compared to the control.

Following the analysis of the empirical distribution functions (e.d.f.) by the Kolmogorov–Smirnov method, the addition of lithium ascorbate in concentrations from 0.1 to 1 mM was non-toxic for CGN in the "blank" experiment (see Methods). Under conditions of glutamate stress, lithium ascorbate in concentrations of 0.2–1.0 mM significantly and dose-dependently increased the survival rate of CGN. The most pronounced neuroprotective effect was observed at lithium ascorbate concentration of 1 mM: neuronal survival increased, on average, by 11% (Kolmogorov's maximum deviation D = 0.45, $p < 0.001$ according to the Kolmogorov–Smirnov test, Figure 1). The use of lithium ascorbate even at the minimum concentration (0.1 mM) led to significant differences (maximum Kolmogorov's deviation D = 0.19; $p = 0.049$). With an increase in the concentration of lithium ascorbate, the significance of differences increased (the value of the maximum deviation D increased and the values of P decreased). An ANOVA test showed that neuronal survival under the cytotoxic effect of glutamate in the range of lithium ascorbate concentrations from 0.1 to 1 mM was significantly dose-dependently increased (on average, by 12%). At the same time, the use of a non-lithium salt of ascorbic acid (potassium ascorbate) was characterized by a much less pronounced neuroprotective effect: neuronal survival increased, on average, only by 5–6% (data not shown).

In accordance with the analysis of e.d.f. by the Kolmogorov–Smirnov method, under conditions of moderate glutamate stress, lithium carbonate at concentrations of 0.1–0.5 mM did not show a significant effect on the survival of CGN and at a concentration of 1 mM. On the contrary, Li_2CO_3 led to a significant decrease in neuronal survival (on average, by 25%, the maximum deviation D = 0.52, $p < 0.0001$). The latter result was confirmed by analyzing the survival rate of CGN using the Dunn test.

Under glutamate stress, lithium chloride at a concentration of 1 mM also led to a significant decrease in neuronal survival (on average, by 9%, maximum deviation D = 0.24, $p < 0.005$). Analysis of individual series of experiments using Dunn's test showed that the protective effect of lithium chloride (16%, $p < 0.05$ at a concentration of 0.1 mM) manifested itself only in the first series of experiments and was not reproduced in five other series of experiments. The differences in the effects of the studied lithium salts under glutamate stress are confirmed by the results of a comparative analysis using the Kolmogorov–Smirnov test (Figure 1E).

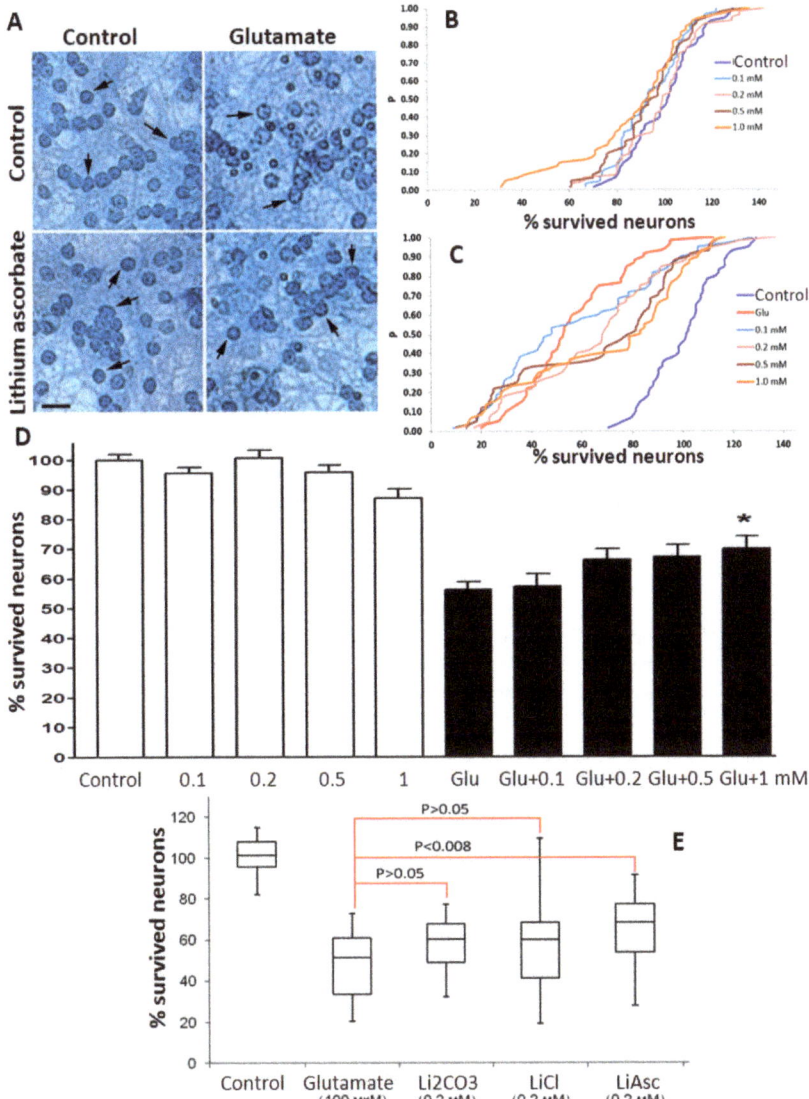

Figure 1. Results of neurocytological studies of lithium salts. (**A**) Representative photos of cultures of cerebellar granular neurons (indicated by arrows). Trypan blue staining of fixed cultures. Scale 1:0.000015. (**B**) Empirical distribution functions (e.d.f.) of the survival rate of neurons without the addition of glutamate (blank experiment); (**C**) E.d.f. of CGN survival under conditions of glutamate stress (100 µM glutamate). (**D**) Evaluation of the neuroprotective effect of lithium ascorbate according to the ANOVA test. White bars—lithium ascorbate (0.1–1.0 mM) in the absence of glutamate, black bars—at the same concentrations in the presence of 100 µM glutamate (Glu). K—control. For each column, 30 fields of view were calculated. *—$p < 0.01$ compared with glutamate without the addition of drugs. (**E**) Comparison of the neuroprotective efficacy of various lithium salts (at a concentration of 0.2 mM) under conditions of moderate glutamate stress (100 µM glutamate, 50% neurons survived). Li_2CO_3, lithium carbonate; LiCl, lithium chloride; LiAsc, lithium ascorbate. P, statistical significance in accordance with the Kolmogorov–Smirnov test.

2.3. Estimates of Lithium Biodistribution from Lithium Ascorbate Intake

A study of the compartmentalization of lithium in 11 biosubstrates of rats (brain, frontal lobe of the brain, heart, aorta, lungs, liver, kidneys, spleen, adrenal glands, femur, urine) was carried out after taking lithium ascorbate at a single dose of 1000 mg/kg. Within the mathematical framework of the non-compartmental analysis of the dynamics of concentrations in whole blood, we obtained the following values of the pharmacokinetic parameters of lithium ascorbate: C_{max} = 50.59 µg/L, T_{max} = 1.50 h, Clast = 33.7 µg/L, AUCt = 1750 µg/L·h, MRTt = 22.9 h, Lz = 0.005 1/h, $T_{1/2}$ = 141 h, CL = 0.029 l/h, Vd = 5.9 L.

The relevant pharmacokinetic curves (PK curves, i.e., the dependencies of concentrations on time) were obtained for homogenates of tissues of various organs (Figure 2A,B). Visual analysis of the PK curves showed that within 1–2 h after intake of lithium ascorbate, an intensive accumulation of lithium occurs in all tissues studied. The maximum values of the peak concentrations of lithium (C_{max}) were observed for homogenates of liver and heart tissues and the minimum values of C_{max}—in the lungs and aorta homogenates. It is important to note that the concentration of lithium in whole blood and in the frontal lobe of the brain remained stable for at least 40–45 h after passing the peak.

This observation indicates first that the predominant accumulation of lithium in whole blood and in the frontal lobes when lithium ascorbate is used. Secondly, this observation indicates that the maintenance of lithium concentrations in these organs occurs by means of a certain "depot" of lithium. Indeed, multi-compartmental analysis confirmed that the stabilization of lithium levels in the blood is supported by a special "depot" of lithium (which, most likely, consists of the aorta, femur, and brain, Figure 2C,D).

One-, two-, three- and four-compartment models in various configurations were investigated. As a result of modeling, it was found that the simplest model that most accurately described the studied PK curves was a three-chamber model, which included the gastrointestinal tract (first compartment), whole blood (central, second compartment) and lithium depot (third compartment). Moreover, in this model of the best quality possible (for the data collected), the elimination of lithium was carried out from the central compartment and not from the depot or the first compartment (Figure 2C). The quality of the investigated multi-compartmental models was characterized by the values of the standard deviation of concentrations between the theoretical and experimentally obtained PK curves (a = 3.4 µg/L with a correlation coefficient of 0.92.

Simulations have shown that the volume of the depot can be approximately half the volume of the central compartment (i.e., of the whole blood). In this model, the lithium ascorbate is quickly transferred from the gastrointestinal tract to the blood (k_{12} = 0.67 1/h) and is very slowly removed from whole blood (which corresponds to a small value of the constant ke_Central = 0.0068 1/h, Figure 2C). The rate of lithium exchange between the blood and the depot is comparable to the rate of transfer from the gastrointestinal tract to the blood, and the transfer of lithium from whole blood to the depot (K_{cd} = 0.41 1/h) is somewhat faster than the reverse process of (K_{dc} = 0.27 1/h).

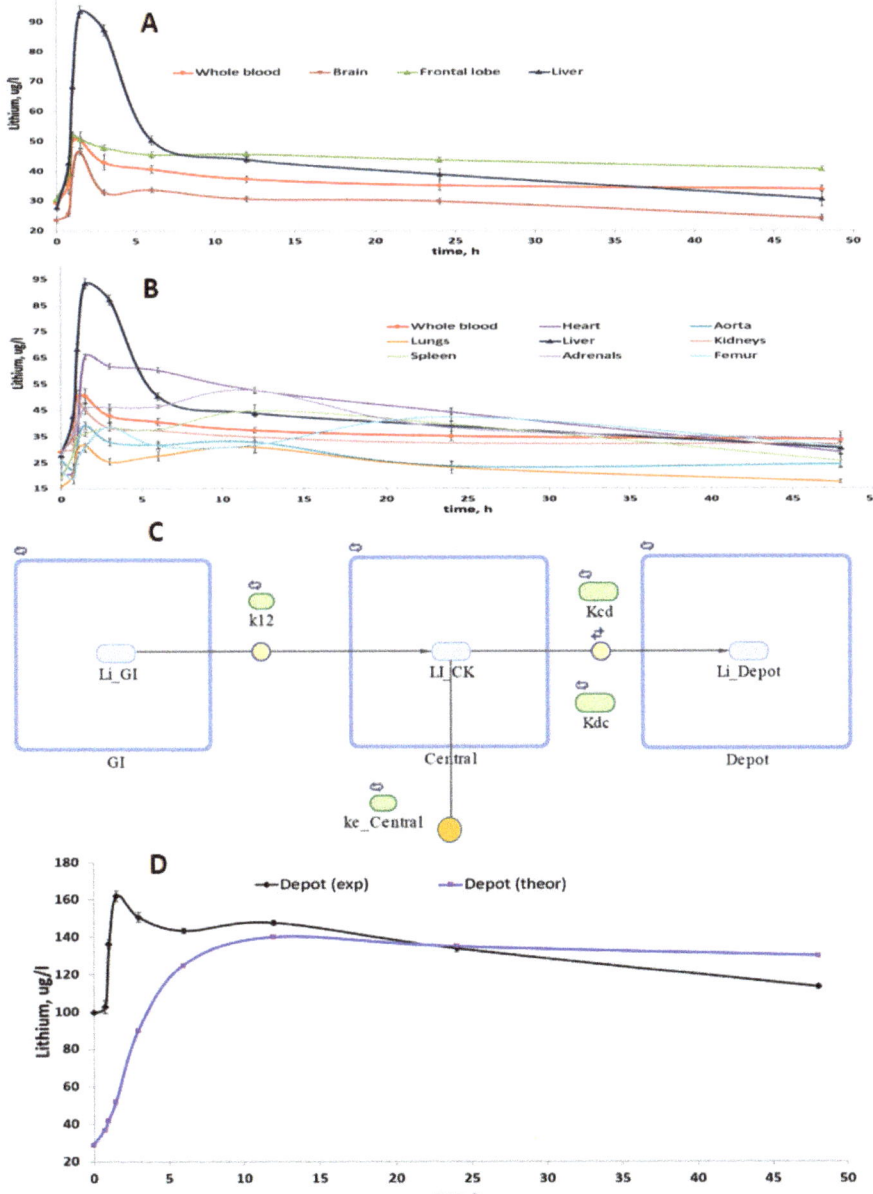

Figure 2. Analysis of lithium biodistribution after a single dose (1000 mg/kg) of lithium ascorbate. (**A**) Pharmacokinetic curves of lithium content in homogenates of brain tissue, whole blood and liver and, (**B**) other organs. (**C**) Three-chamber model of lithium ascorbate pharmacokinetics obtained as a result of multi-chamber pharmacokinetic modeling in MATLAB. The symbols for the corresponding constants are shown in the figure (k12, Kcd, Kdc, Ke_central). (**D**) Dynamics of concentrations in the "depot" of lithium are obtained due to multi-compartmental PK-modeling. The experimental data for the "depot" were obtained by summation of the lithium contents in the brain, aorta, adrenal glands, and femur.

2.4. Acute and Chronic Toxicity Studies of Lithium Ascorbate

Toxicity studies indicated that lithium ascorbate is characterized by low acute and chronic toxicity. In acute toxicity studies with a single dose of 3000 mg/kg of lithium ascorbate, the lethality was 0%, and any detectable pathological changes were absent (not even a localized irritation). At 4000 mg/kg, the (delayed) mortality was 20%, intoxication manifested itself as depression, diarrhea, tousled coat, bloody discharge from the nose and eyes in males, and diarrhea in females. Pathological changes included hemorrhages of the membranes of the brain, edema and hemorrhages in the lungs.

Finally, for white rats of the Wistar line, the LD50 of lithium ascorbate was estimated to be 6334 mg/kg of body weight, and the LD100 was 8000 mg/kg. Thus, lithium ascorbate belongs to the 5th class of "practically non-toxic compounds", LD50 $\geq$ 5000 mg/kg. Analysis of the results of determining the safety of lithium ascorbate by the Kerber and Pershin method for determining LD50 (Prozorovsky, 2007) showed that the differences do not exceed 0.1%. In comparison with lithium carbonate (LD50 = 531 mg/kg), lithium ascorbate was apparently 12 times less toxic.

In the study of chronic toxicity in Wistar rats, no teratogenic and embryotoxic effects of high doses of lithium ascorbate (1/10 LD50 per week) were observed in newborn rat pups. The cytogenetic effect of lithium ascorbate (according to the method of Zolotareva T.N.) led only to a decrease in mitosis though the difference between the control and the experimental groups was not significant.

The toxic effects of high doses of lithium ascorbate (1/10 LD50 daily during a month) on protein metabolism were absent. In particular, the intake of the lithium salts did not cause a decrease in the amount of essential amino acids. We also found that intake of lithium ascorbate did not cause a statistically significant effect on the decrease of the contents of free histidine, leucine, isoleucine, tyrosine, serine and methionine in the blood serum of rats. A trend (p = 0.06) was found for an increase in the content of linolenic, docosahexaenoic acids (increased by 15%), and dihomo-γ-linolenic acids (by 13%). The tendency to decrease the level of fatty acids was observed for oleic, pentadecanoic, palmitic acids (by 12–17%) and for myristic and arachidonic acids (by 5–7%). In brief, the results of toxicological studies indicate a good safety profile of the lithium ascorbate, even when used in high doses per os.

2.5. Assessment of the Antitumor Effects of Lithium Ascorbate

Studies of the dynamics of growth and metastasis of Lewis lung carcinoma transplanted mice F1 (CBAxC57 Bl/6j) have shown that lithium ascorbate exhibits moderate anti-tumor effects. Two series of experiments were carried out. In the first series, a comparative study of the effects of different doses of lithium ascorbate (5 and 10 mg/kg) was carried out. In the second series, a comparison of the effects of lithium ascorbate and lithium carbonate was carried out for the same dose (5 mg/kg). Analysis of the dynamics of LLC growth in different groups showed that both studied drugs, already after 3 days from the start of their use, caused moderate (by 10–15%) inhibition of LLC growth in tumor-bearing animals (Figure 3A,B). At the same time, the effect of lithium ascorbate was more pronounced and stable: a statistically significant effect of this drug was noted from day 10 and throughout the entire observation period, and the TPO index was at a fairly high level (30–40%). The antitumor effect of lithium carbonate in this experiment was less pronounced and stable (TPO = 20–30%). The studied lithium salts did not have any significant effect on the processes of metastasis of LLC and the growth of lung metastases (Figure 3C).

A

Group	Average relative tumor volume, rel. units (M ± SD)				
	7 days	10 days	13 days	17 days	21 days
Control	1.00 (n=13)	6.24 ± 1.88 (n=13)	9.21 ± 2.87 (n=13)	26.50 ± 7.77 (n=13)	35.75 ± 10.63 (n=13)
Lithium ascorbate	1.00 (n=14)	3.68 ± 0.82 # (n=14)	6.06 ± 1.79 * # (n=14)	17.56 ± 1.77 # (n=13)	25.69 ± 4.30 * (n=13)
Lithium carbonate	1.00 (n=15)	4.43 ± 1.39 # (n=15)	7.30 ± 3.09 (n=15)	19.96 ± 4.98 * (n=14)	28.02 ± 6.42 * (n=14)

* - $p < 0.05$ compared to control (Dunn's test), # - $p < 0.01$, Dunn's test

B

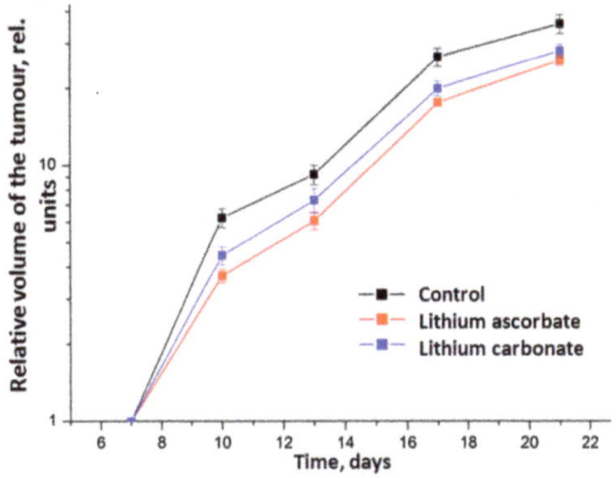

C

Group	Average number of lung metastases (M ± SD)		
	Large	Small	General
Control (n=13)	4.7 ± 4.2	17.3 ± 8.8	22.0 ± 12.5
Lithium ascorbate (n=13)	9.6 ± 9.3	24.6 ± 9.6	34.2 ± 15.7
Lithium carbonate (n=14)	4.3 ± 4.1	22.3 ± 12.7	26.6 ± 15.4

Figure 3. Assessment of the antitumor effects of lithium ascorbate and lithium carbonate. (**A**) Dynamics of the relative growth of Lewis lung carcinoma in the experimental groups. (**B**) Curves of relative LLC growth. (**C**) Calculation of the number of pulmonary metastases in the experimental groups on the 21st day of LLC growth.

2.6. Neuroprotective Properties of Lithium Ascorbate in the Model of Alcohol Intoxication

The effects of lithium ascorbate were investigated on a model of chronic alcohol intoxication, in which deviant behavior of animals is combined with irreversible degenerative changes in the liver and in the central nervous system (including demyelination of nerves). Lithium ascorbate at doses of 5, 10, and 30 mg/kg normalized behavioral responses in the open field and elevated plus-maze tests. An increase in the dose of lithium ascorbate (10 mg/kg, 30 mg/kg) did not lead to a significant improvement in the studied parameters of the state. Histological analysis showed that the use of lithium ascorbate minimized the level of ischemic damage to neurocytes to the level of a reversible state and promoted the preservation of the myelin sheaths of the nerves.

When the model of alcohol intoxication was reproduced, behavioral reactions were disturbed (chaotic movement without manifestations of exploratory and search behaviors). Histological analysis showed circulatory disorders, which were characterized by hemostasis in the capillaries and venules with the development of severe perivascular edema of the nervous tissue. Toxic damage to neurocytes in the cerebral cortex was characterized by acute swelling of the pyramidal cells with rounding of the cell body, swelling of the axon, homogenization of the cytoplasm with the disappearance of Nissl granulations, and disruption of the contours of the nucleus.

The "prophylactic" use of lithium ascorbate in alcohol intoxication improved vertical activity and other indicators of neurological tests (Figure 4A,B). Long-term alcohol intoxication caused an increase in the concentration of catecholamines in the blood, and the use of lithium ascorbate prevented an increase in the content of catecholamines in the blood of animals. The greater efficiency of lithium ascorbate was also manifested by the activation of ADH in the liver, which corresponds to the acceleration of ethanol elimination (Figure 4C,D). Alcohol caused a significant increase in serum MDA, while lithium ascorbate reduced MDA levels (which corresponds to inhibition of lipid peroxidation and activation of the body's antioxidant system). Higher doses did not lead to a significant increase in the effectiveness of lithium ascorbate (Figure 4B,D).

The "prophylactic use" of lithium ascorbate under conditions of chronic alcohol intoxication significantly alleviated the histological manifestations of cerebral ischemia. Compared with the placebo group, in all groups of animals receiving lithium ascorbate, circulatory disorders of the nervous tissue were characterized by focal hemostasis in the capillaries and moderately pronounced pericapillary edema of the nervous tissue, the cortex, the white matter of the cerebral hemispheres, and the brain stem.

When lithium ascorbate was taken, a significant part of the neurocytes of the cortex and subcortical nuclei were characterized by reversible changes, which were expressed by dispersion and indistinctness of the tigroid contours, the focal fusion of Nissl lumps in the cytoplasm, moderately pronounced swelling of the nucleus and axonal process. The macroglial reaction of the nervous tissue was minimal and was expressed by edema of the periarteriolar astrocytes. The impregnation of the pathways of the brain with silver showed the preservation of the myelin sheaths of the nerve fibers, which had clear contours (Figure 4E–H). Morphometric analysis showed that in the group of animals receiving lithium ascorbate at a dose of 5 mg/kg, the number of damaged nerve cells in the cortex was 18.5%, and in the placebo group, the number of damaged cells was 34.8% (with significant morphological signs of irreversible death of neurocytes). Higher doses of lithium ascorbate did not lead to a substantial improvement of the results of morphometric analysis.

In the "therapeutic" series of experiments, lithium ascorbate (5, 10, 30 mg/kg) was administered immediately after the alcohol solution was discontinued. After introducing the lithium ascorbate, the number of aggressive incidents was reduced. The animals showed a more profound and longer sleep, fell intoa stupor less often, and quickly adapted to the absence of alcohol in the diet. The histological analysis showed that lithium ascorbate as compared to placebo contributed to a reduction in circulatory disorders of the nervous tissue and focal hemostasis in the capillaries, to a less pronounced pericapillary edema of the nervous tissue in the cortex, to greater preservation of the white matter of the cerebral hemispheres and the brain stem.

In the groups on lithium ascorbate, only single neurocytes with slight injuries were noted. These injuries included the indistinctness of the tigroid contours, the focal fusion of Nissl lumps in the cytoplasm, and moderately pronounced swelling of the nucleus. The state of the conduction system of the brain of animals receiving placebo was characterized by a greater focal swelling of myelin fibers with an uneven distribution of myelin, which, when impregnated with silver, created a picture of fuzzy contours. Structural changes in the brain's nerve fibers in the group taking 5 mg/kg lithium ascorbate were limited to local edema of the nerve fibers with complete preservation of the myelin sheath. Higher doses of lithium salt did not improve the histological picture. In general, both "prophylactic" and "therapeutic" use of lithium ascorbate contributed to the relief of withdrawal symptoms and prevented formation of irreversible degenerative changes of the nervous tissue.

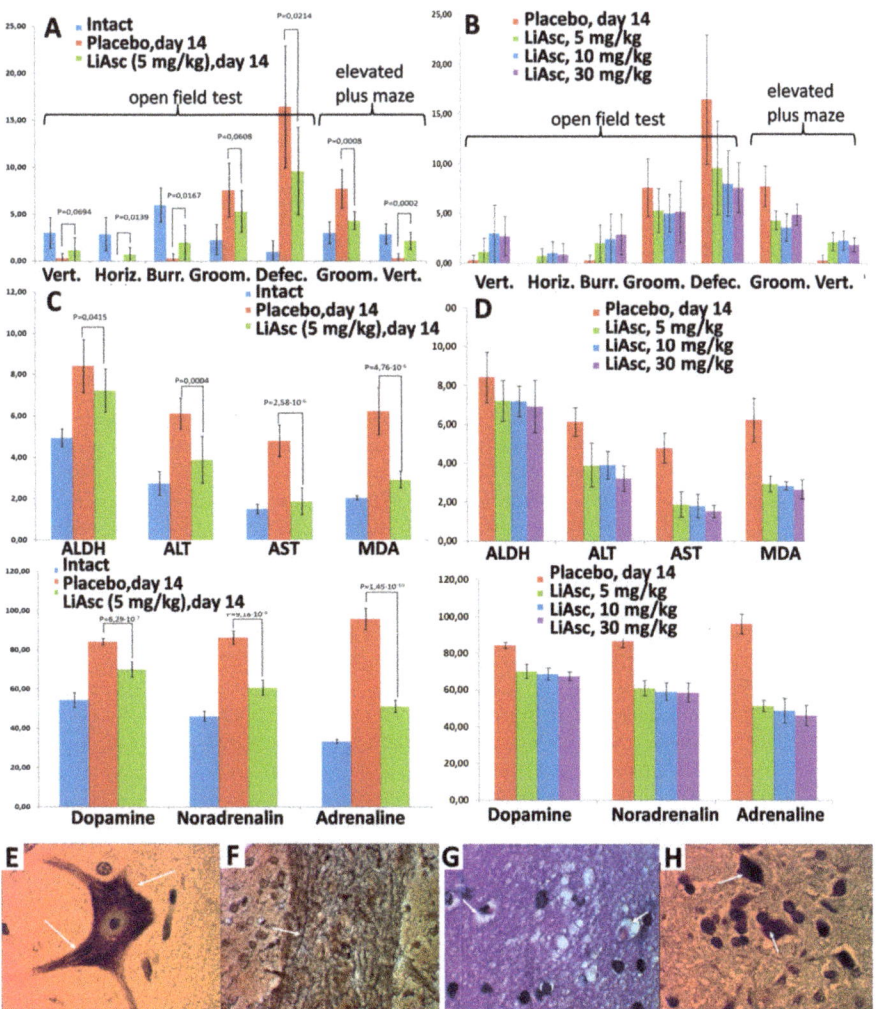

Figure 4. Effects of "prophylactic" use of lithium ascorbate in animals with the model of alcohol intoxication. (**A**) Indicators of neurological tests, 5 mg/kg. (**B**) Dose-dependence, neurological tests. (**C**) Biochemical indicators. (**D**) Dose-dependent effects on biochemistry. (**E**) Histological effects of lithium ascorbate. Focal fusion of Nissl lumps in the cytoplasm of the pyramidal cell. Nissl staining with toluidine blue. Magnification ×1200. (**F**) Clear contrasted contours of the commissural fibers of the brain with preservation of the integrity of the myelin sheath. Silver impregnation. Increase ×1200. (**G**) Histological picture. Pronounced pericapillary edema of the nervous tissue. Staining with hematoxylin and eosin. Magnification ×1200. (**H**) Hyperchromatosis, pycnosis of neurocytes. Astrogliocyte hypertrophy. Nissl staining with toluidine blue. Magnification ×1200.

3. Discussion

The present study combines the use of modern methods of data mining/artificial intelligence (chemoreactomic analysis) with experimental studies of the most promising candidates of lithium salts based on anions of organic acids. **Chemoreactomic modeling** showed that lithium ascorbate is a remarkable candidate for further studies. In comparison with the control substances, lithium ascorbate was characterized by more prominent inhibition of the reuptake of serotonin and of dopamine and by a greater affinity for the inhibition of glutamate and α-adrenergic receptors. It is important to note that lithium ascorbate may be more effective than nicotinate and oxybutyrate (see Table 1) in inhibiting the activity of the GSK-3β enzyme (ascorbate—47%, nicotinate—3.3%, oxybutyrate—6%)—one of the main target proteins of the lithium ions. The inhibition constant of GSK-3β was IC50 = 79 nM for lithium ascorbate and was much higher for the rest of the salts studied (IC50 = 1658–8240 nM).

Chemoreactomic modeling also indicated a lower affinity of lithium ascorbate for the potassium channel KCNH2 (Table 1). This potassium channel is an important "anti-targeting" protein interaction that should be avoided during drug development (since the disruption of KCNH2 activity can lead to the dangerous "long QT syndrome"). Prediction of the properties of lithium ascorbate showed that when taking doses of about 10 mg/kg/day, there will be no significant change in hematocrit (which is associated with changes in the water–salt balance of blood cells). Changes in hematocrit estimated for nicotinate and oxybutyrate lithium salts were apparently higher.

It is also important to notice that chemoreactomic assessments of the probability of interaction of the studied molecules with various transporter proteins suggested that lithium ascorbate can enter cells by means of SLC23A1, SLC23A2 transporters (vitamin C transporter proteins) [14,15]. The probabilities of the interaction of lithium ascorbate with these proteins were 0.9–0.95 a.u. and were much lower for the rest of the molecules (0.1–0.4 a.u.). Since these transporter proteins are present in neurons, the ascorbate anion in lithium ascorbate increases the bioavailability of the lithium ions. It promotes an increase in the accumulation of Li+ in the nervous tissue. Thus, lithium ascorbate represents a promising lithium compound for analyzing neuroprotective properties.

The results of **neurocytological studies** indicate a direct neuroprotective effect of lithium ascorbate on cerebellar granular neurons in culture. Treatment of neurons in culture with lithium ascorbate increased cell survival under conditions of glutamate stress, and this effect was almost indistinguishable for the inorganic lithium salts (carbonate, chloride) in the same range of concentrations (0.1–1.0 mM).

The studies of **lithium biodistribution** were performed using the standard math models of pharmacokinetic analysis. The biodistribution data collected herein (groups of animals taken at certain time periods) considerably differ from the classical pharmacokinetic study (samples of blood taken at certain time periods). Nevertheless, the usage of the pharmacokinetic models allowed us to describe quantitatively the peculiarities of the biodistribution of lithium in rats.

Pharmacokinetic-based math modeling has shown that the volume of the lithium depot can be approximately half the volume of the central compartment (i.e., of the whole blood, Figure 2). Lithium (ascorbate) is quickly transferred from the gastrointestinal tract to the blood (k12 = 0.67 1/h) and is very slowly removed from whole blood (which corresponds to a small value of the constant ke_Central = 0.0068 1/h, Figure 2C). The rate of lithium exchange between the blood and the depot is comparable to the rate of transfer from the gastrointestinal tract to the blood, and the transfer of lithium from whole blood to the depot (Kcd = 0.41 1/h) is somewhat faster than the reverse process of lithium transfer from the depot to the whole blood (Kdc = 0.27 1/h).

Comparison of the dynamics of the concentration of lithium in the depot, obtained as a result of the simulation, with the dynamics of the concentration of lithium in the "depot", consisting of the brain, aorta, adrenal glands, and femur, indicates certain similarities in changes in concentrations. Obviously, the depot, consisting of these organs, allows for the stabilization of lithium concentrations after the first 10–15 h of the experiment. Thus, lithium ascorbate contributes to the maintenance of stable concentrations of lithium ions in the blood and in the brain, which is important for the implementation of the preventive and therapeutic potential of the lithium ions.

The results of **toxicological studies** indicate a good safety profile of the lithium ascorbate, even when used in high doses. For white rats of the Wistar line, the LD50 of lithium ascorbate was estimated to be 6334 mg/kg of body weight, and the LD100 was 8000 mg/kg, indicating extremely low toxicity.

The mice model of lung Lewis carcinoma (LLC) studies indicated the onco-safety of lithium ascorbate and shown even a small but significant **antitumour effect**. Analysis of the dynamics of LLC growth in different groups showed that both lithium salts, ascorbate, and carbonate, after 3 days from the start of their use, caused moderate (by 10–15%) inhibition of LLC growth in tumor-bearing animals (Figure 3). At the same time, the effect of lithium ascorbate was more pronounced and stable: a statistically significant effect of this drug was noted from day 10 and throughout the entire observation period while the TPO index of tumor growth inhibition was at a fairly high level (30–40%). The antitumor effect of lithium carbonate in this experiment was less pronounced and stable (TPO = 20–30%). The studied lithium salts did not have any significant effect on the processes of metastasis of LLC and the growth of lung metastases.

The neurological effects of lithium ascorbate were investigated on a **model of chronic alcohol intoxication**. The deviant behavior of animals is combined with irreversible degenerative changes in the liver and central nervous system (including demyelination of nerves). Studies on the model of alcohol intoxication in rats have shown that intake of lithium ascorbate in doses either 5, 10 or 30 mg/kg reduced brain damage due to ischemia and also ramped up preservation of myelin sheaths of neurons.

In both "prophylactic" and "therapeutic" series of experiments **morphometric analysis** showed that in the group of animals receiving lithium ascorbate at a dose of 5 mg/kg, the number of damaged nerve cells in the cortex was 18.5%, and in the placebo group, the number of damaged cells was 34.8%, with significant morphological signs of irreversible death of neurocytes. The higher doses of lithium ascorbate did not result in any significant improvement in the results of morphometric analysis.

4. Materials and Methods

The complex of studies included in silico chemoreactomic modeling of lithium salts, synthesis of lithium salt(s), neurocytological studies, assessments of biodistribution, studies of acute and chronic toxicity, assessment of antitumor effects, adaptogenic, stress-protective and neuroprotective properties of lithium ascorbate on the model of alcohol intoxication.

During the animal studies, the animals were kept under standard conditions in accordance with Directive 2010/63/EU of the European Parliament and of the Council of the European Union of 22 September 2010 concerning the protection of animals used in scientific studies [16]. Indoor air control in compliance with environmental parameters (temperature 18–26 °C, humidity 46–65%). The rats were kept in standard plastic cages with bedding; the cages were covered with steel lattice covers with a stern recess. The floor area per animal met regulatory standards. The animals were fed in accordance with Directive 2010/63/EU. The animals were given water *ad libitum*. The water was purified and normalized for organoleptic properties in terms of pH, dry residue, reducing substances, carbon dioxide, nitrates and nitrites, ammonia, chlorides, sulfates, calcium and heavy metals in standard drinkers with steel spout lids.

Chemoreactomic modeling of lithium salts. The chemoreactomic approach to the analysis of the "structure-property" problem of molecules is the newest direction of the application of artificial intelligence systems in the field of post-genomic pharmacology. The methodology and the algorithms realized in the original proprietary software developed by the authors were described [12,13] and tested [15]. The training of the algorithms of metric analysis [15] was carried out on the basis of the data on structure and properties of the molecules presented in the PubChem, HMDB, STRING databases. The algorithms and software developed were realized within the framework of the topological theory of data analysis as described earlier [12]. The algorithms that predict molecular properties were validated using cross-validation estimates as described in [13]. Based on the algorithms developed, we estimated over 350 pharmacological properties of lithium salts relevant to their toxicity, pharmacokinetics, pharmacodynamics, etc. The list of the lithium salts studied (water-soluble organic lithium compounds with a molecular weight of less than 300 Da) along with the structural formulas was downloaded from the PubChem database (totally, 1245 compounds encoded in SMILES format). The algorithms for predicting molecular properties were extensively tested earlier on over 100,000 molecules [12,15]. Additional information about the algorithms is presented available online at www.chemoinformatics.ru and www.pharmacoinformatics.ru (both sites were accessed on 21 February 2022).

Synthesis of lithium salts. The synthesis of the lithium salts was based on the classical reaction of neutralization carried out with the organic acids and the lithium carbonate. A detailed description of the synthesis of lithium ascorbate is presented below. The synthesis was carried out in accordance with the equation of the reaction of neutralization $2C_6H_8O_6 + Li_2CO_3 = 2C_6H_7O_6Li + H_2O + CO_2\uparrow$. To a suspension of 17.6 g (0.1 M) of ascorbic acid (L-Ascorbic acid $C_6H_8O_6$, LLC Sigma-Aldrich Rus, Moscow, Russia, A5960, BioXtra, $\geq$99.0%) in 25 mL of distilled water we added the lithium carbonate (lithium carbonate Li_2CO_3, Merck KGaA, Moscow, Russia, 1.05671.1000, certificates Ph Eur, BP, USP) in the amount of 3.7 g (0.05 M). Li_2CO_3 was added in portions of 0.5 g with constant stirring and heating to 35–45 °C. Each successive portion of lithium carbonate was added only after termination of the copious production of the CO_2 gas. After adding the last portion of lithium carbonate, the suspension was heated to 70–80 °C until complete dissolution of the remaining mix of ascorbic acid and the lithium carbonate. As the result of neutralization, a clear, pale, straw-colored solution with pH$\approx$6–7 was obtained. When the solution was cooled to room temperature in 30–50 min, an abundant crystalline precipitate formed, which was filtered and washed with ethanol. The alcohol wash was combined with the filtrate in the 1:2 ratio and then placed into cold storage (+4 °C) overnight. The precipitate formed in cold storage was filtered, washed with a small amount of alcohol, and combined with the precipitate obtained earlier. The synthesized lithium ascorbate was dried in a vacuum oven at 30–50 °C. The total yield of lithium ascorbate was 80–90% (14–16 g).

Neurocytological studies of lithium salts on a culture of cerebellar granular neurons. We used seven-day cultures obtained by the method of enzymatic-mechanical dissociation of cerebellar cells of 7-day-old rats as described previously [17]. Cultivation was carried out in 96-well plastic plates for 7 days in a CO_2 incubator filled with gas mixture (95% air + 5% CO_2) at the temperature of 35.5 °C and the relative humidity of 98%. Solutions of lithium salts were prepared from dry anhydrous salts. The inorganic lithium salts were manufactured by Merck KGaA, Moscow, Russia (Lithium carbonate Li_2CO_3, 1.05671.1000, Ph Eur, BP, USP certifications; Lithium chloride LiCl, 1.05679.0100, Reag. Ph Eur certifications), the organic salts—as described above. Dry salts were dissolved in deionized water at a concentration of 10 mM, then sterilized by ultrafiltration and added to the cultivation medium on day 2 in vitro for the entire period of cultivation (up to 7 days).

The state of the neuronal cultures was monitored daily by visual inspection under microscope with phase contrast. The final concentrations of the lithium salts in the culture medium were 0.1, 0.2, 0.5 and 1 mM. The experimental design used was described earlier [4]. Briefly, monosodium glutamate (L-Glutamic acid monosodium salt 99–100%, LLC Sigma-Aldrich Rus, Moscow, Russia, NG-1626) was added to cultures at three different concentrations (50, 100 and 150 µM) to estimate the toxic effect. In this series of experiments the optimal glutamate concentration was 100 µM and it corresponded to the moderate-to-mild neurocytotoxic effect of glutamate (survival 30–70% of the neurons in control group which allowed us to assess the neuroprotective properties of the substances studied). The test substances were then introduced into the cultivation medium on the second day and left until the seventh day.

The study of the effects of each salt was carried out in two stages: (1) the "blank" experiment, in which the effects of different concentrations of lithium salt on the survival of neurons were studied without addition of glutamate, and (2) the "main" experiment, in which neurons were cultured under conditions of glutamate stress at different concentrations of lithium salts and the survival rates of the cultured neurons were measured.

The quantitative assessment of neuron survival was performed using direct counting of neurons with unchanged morphology [17]. For each salt, five experiments were performed; for each concentration of a salt at least three cultures were taken, each of which was photographed and counted using three consecutive visual fields. The number of neurons with unchanged morphology in control cultures was taken to be 100%. The ANOVA test with Bonferroni correction and the Dunn test were used for statistical analysis. Additionally, the analysis of empirical distribution functions (e.d.f.) of neuron survival values was carried out using the statistical test of Kolmogorov–Smirnov. The differences between groups were considered to be significant at $p < 0.05$ level, the results were expressed as mean ± standard deviation.

Estimates of the biodistribution of lithium after intake of lithium ascorbate. The experiments were carried out on male white Wistar rats (200–250 g, nine groups of six animals each). The animals were probed with solution of lithium ascorbate, 1 mL of which contained 250 µg of elemental lithium. The solutions were prepared using anhydrous powder of lithium ascorbate. The required volume of solution was calculated based on the weight of the animal in order to reach 1000 µg/kg dose of elemental lithium. After the probing with a solution of lithium ascorbate the biosubstrates studied (whole blood, brain, frontal lobe of the brain, heart, aorta, lungs, liver, kidneys, spleen, adrenal glands, femur) were taken successively from each of the 9 groups of animals at the nine time points (0 min, 45 min, 1 h, 1.5 h, 3 h, 6 h, 12 h, 24 h, and 48 h).

The lithium levels in homogenates of these 11 different biosubstrates were determined using mass spectrometry. The samples of homogenates were taken into plastic tubes and diluted 5 times with bidistilled and deionized water. The internal standard (indium at a concentration of 25 µg/L) was added to the solutions. Calibration solutions were prepared from VTRC standard solutions with a known content in the range from 5–1000 mgc/L. The resulting solutions were analyzed on inductively coupled plasma ionization mass spectrometer "Plasma Quad PQ2 Turbo" (VG Elemental, Winsford, Cheshire, UK). The power of the microwave generator was 1.3 kW, the flow rate of the plasma gas (argon) was 14 L/min, and the flow rate of the carrier gas was 0.89 mL/min. From 3 to 10 exposures of each sample were carried out; the signal integration time was 60 s.

The methods of multi-compartmental and non-compartmental pharmacokinetic analysis were used to quantify biodistribution of lithium. The multi-compartmental analysis was carried out using the SimBiology package as part of the MATLAB2016 software package [18], and the non-compartmental analysis was performed using Excel spreadsheets, supplemented by the modules of the PKSolver software package [19].

Study of acute and chronic toxicity of lithium ascorbate. The experiments were carried out on male and female Wistarrats 8 weeks of age at various concentrations of lithium ascorbate (0, 3000, 4000, 5000, 6000 and 7000 mg/kg). The total number of groups was 6, each group contained 10 rats (5 males 205.8 ± 1.89 g and 5 females 174.4 ± 1.66 g). Lithium ascorbate was administered intragastrically. In the study of acute toxicity, the tested and the control substances were taken once (fractionally, using a probe) with further observation of the animals for 14 days. Animals were deprived of food for 16 h before administration of the substances, before bodyweight recording and before euthanasia. Access to water was not restricted throughout the experiment. The cytogenetic effects were estimated after T.N. Zolotareva's methodology [20].

The animals were under continuous observation before administration of the lithium salt, within 30 min after the administration of the last portion of the salt, then hourly for 4 h, then after 24 h and then daily for 15 days. Behavior (depression/excitement), reactions to stimuli, condition of the skin and mucous membranes, secretions, muscle tone, coordination of movements were registered. Clinical examination of the animals was carried out before administration, then on the second, eighth and fourteenth days of the experiment. For histological examination, the material was fixed in a 10% solution of neutral formalin for 24 h, after which it was embedded in paraffin according to the generally accepted technique. Then sections with a thickness of 5–7 µm were made, which were stained with hematoxylin and eosin. The calculation of LD50 was carried out using the Bliss–Prozorovsky method [21]. Statistical analysis was performed using the Statistica 10.0 software (StatSoft, Tulsa, OK, USA).

Assessment of the antitumor effects of lithium ascorbate. The study was performed on 42 male mice hybrids F1 (CBA×C57 BL6j) at the age of 2.5–3 months, body weight 23–26 g. The animals had a veterinary certificate and passed the 20-day quarantine in vivarium. The studies were carried out on transplantable Lewis lung carcinoma (LLC). The LLC strain was obtained from the bank of tumor materials of the N.N. Blokhin's Institute and was maintained in male C57 BL6j mice. Transplantation of LLC to the male mice was carried out by subcutaneous administration of 1.9×10^6 tumor cells in 0.1 mL of the suspension in the region of the lateral surface of the right thigh after depilation of the area.

The animals were included in the experiment on the 7th day after LLC inoculation, when the tumor node had already formed and reached a measurable size in all mice. The animals were randomized into three groups—control (13 mice) and two experimental groups for each of the lithium salts (14 mice). The animals of the control group did not receive any further treatment. From 7 to 20 days after LLC transplantation, lithium salts carbonate in doses of 5 mg/kg (elemental lithium) in 1% starch gel, which was made *ex tempore*, were given through the probe to the animals of the experimental groups from 7 to 20 days after LLC transplantation.

Tolerability of the studied lithium preparations was assessed by daily observation of animals, in which the neurological status was studied according to the peculiarities of spontaneous locomotor activity, general excitability and response to tactile and sound stimuli, as well as by the food activity of animals and the dynamics of their body weight gain.

The effect of lithium preparations on the tumor process was studied by the dynamics of growth and the activity of metastasis of LLC. To do this, two diameters of tumor nodes were measured by a caliperevery 3–4 days: L—the maximum diameter of the node; W—the diameter orthogonal to L. The calculation of the volumes of the nodes was carried out using the approximation $V = (L \times W^2) \times (\pi/6)$ that was shown to correlate reliably with the MRI data [22]. The effect of drugs on the growth of LLC was assessed by comparing the volume of tumor nodes in the control and experimental groups at different periods of observation, and then by calculating the growth inhibition index TPO: $TPO = (V_K V_O)/V_K \times 100\%$, where V_K and V_O are the average volumes of nodes in the control and experimental groups, respectively. On the 21st day of carcinoma growth, the animals were removed from the experiment by cervical dislocation under ether anesthesia, the lungs were isolated, fixed for 24 h in Bouin's fluid, and then the numbers of large and small lung metastases was counted. The effect of lithium preparations on the processes of metastasis and the growth of metastases was assessed by comparing the numbers of metastases in the control and the experimental groups. Intergroup differences were assessed by Kruskal-Wallis rank analysis of variance using Dunn's test. Differences were considered significant at the 0.05 significance level. The calculations were performed using the Statistica 10.0 software package (StatSoft Inc., Tulsa, OK, USA).

Alcohol intoxication model of neural damage. The study was carried out on male white Wistar rats weighing 200–250 g ($n = 168$) as described in the guideline [23]. In brief, the selection criterion for rats, apart from absence of visible abnormalities in physical state and behavior, was the initial preference of 6% ethanol solution over the drinking water. To reveal this preference, a preliminary experiment was carried out for 3 days in individual cages with free access to both fluids. After selection, a 6% solution of ethyl alcohol was proposed as the only source of liquid; after a week, the concentration of alcohol was increased to 15%. After 2 weeks, the alcohol was replaced with drinking water.

The experiment was carried out in two series—"prevention" and "treatment", four groups in each series: (1) dose of 30 mg/kg, (2) a dose of 10 mg/kg, (3) 5 mg/kg dose, (4) placebo group. In a series of "prevention" experiments, the drug was administered in parallel with the initial intake of alcohol and in a series of "treatment" experiments—after the reproduction of the model of alcohol intoxication. Each group consisted of 21 animals.

At 14th day, the lithium salt was administered intragastrically through the probe, once a day in a volume of 0.5 mL for each animal.

A general assessment of the somatic and neurological status of animals included the estimates of the level of anxiety and other negative impacts of alcohol intake, the behavior of animals in the open field test and in the "elevated plus" maze test.

In the open field test (OFT), vertical locomotor activity, horizontal locomotor activity, the number of peeping into the holes ("burrow reflex"), the number of grooming acts, and the number of exits to the central zone were recorded. In the OFT test, the animal was placed in the same square located near the wall. The exposure time of each animal was 5 min. The round OFT involved an arena 1 m in diameter with a wall height of 0.4 m, the bottom of which was divided into sectors. In the open field, 3 zones were outlined: central, intermediate (6 sectors), peripheral (12 sectors). Lighting was provided by 2 lamps, 60 W each, which were located at a height of 1.5 m from the bottom of the chamber above the central segments of the field. The test recorded horizontal activity in the central and peripheral zones, vertical activity, the number of grooming acts, the number of defecation acts, etc. After each animal, the walls and bottom were treated with a wet and dry napkin.

In the "elevated plus" maze (EPM) test the setup used two arms, at the intersection of which there was an open area. One of the labyrinth arms had closed compartments. The labyrinth was installed at the height of 1 m from the floor. The stay of animals in the covered and closed arms and at the center of the labyrinth, the duration of grooming in the closed arms, number of grooming episodes in a closed arm were recorded. The time of testing animals in EPM was 5 min.

On days 0 and 14, the eight biochemical parameters (alanine transaminase ALT, aspartate aminotransferase AST, malondialdehyde MDA, dopamine, norepinephrine, adrenaline, serotonin in blood serum and alcohol dehydrogenase ADH in liver cells) were determined in all groups of animals. Blood sampling was performed from the sublingual vein by cutting the sublingual frenum. The blood contents were analyzed on an automatic biochemical analyzer "Konelab-20i" (Finland). Liver cells for determination of alcohol dehydrogenase were collected from animals after the experiment. ADH activity in hepatocytes was determined photometrically.

ALT and AST (μmol/mL/h) activities were determined on an automatic photometric analyzer ChemWell2910C (Palm City, FL, USA) using standard kits ALT-UTS and AST-UTS, respectively, produced by Eiliton LLC (Dubna, Russia). The metabolites MDA (nmol/mL) and AAD (nmol/mg protein/min) were measured on NanoDrop ™ 2000 microspectrophotometer. Serotonin was determined by the Michel's method. Catecholamines (nmol/mL) were determined on a Waters 590 liquid chromatograph with an amperometric detector (NPO Khimavtomatika, Moscow, Russia) (working electrode material—glassy carbon) using an Ascentic C18 column (5 μm, 4.6 × 250 mm). Electrophoretic determination was carried out on a capillary electrophoresis system, "Kapel-105" (OOO Lumex, Saint Petersburg, Russia), with a spectrophotometric detector, an unmodified quartz capillary with a total length of 60 cm, an effective length of 50 cm, and an inner diameter of 50 μm.

Histopathological analysis. After craniotomy, the brain was removed and fixed in a 10% solution of neutral formalin. After 1 day, the zone of the precentral gyrus of the forebrain was isolated using frontal incisions. The nerve tissue was processed according to the standard scheme (dehydration in ethyl alcohol, xylene), followed by the manufacture of the paraffin blocks. Histological sections 5–6 μm thick were prepared on a microtome "Microm" and were stained with hematoxylin and eosin. Duplicate sections were stained according to the Nissl method and impregnated with silver using Biovitrum (Saint Petersburg, Russia) reagent kit. The morphometric study of histological sections was carried out on image analyzer BioVision GmbH (Vienna, Austria) and consisted of counting the damaged neurocytes of the pyramidal layer of the cerebral cortex in 10 different fields of view with subsequent statistical processing of the results. Micrographs were obtained using a Micros research microscope and a DCM 900 digital eyepiece camera (Oplenic Optronics Equipment Co., Zhejiang, China). The results were processed using the Excel 2013 and

Statistica 10.0 (Tulsa, OK, USA) software packages. The significance of differences between groups was determined by a nonparametric U-test, the Wilcoxon–Mann–Whitney test.

5. Conclusions

Lithium ions have a significant effect on the homeostasis of acetylcholine, enkephalins, catecholamines, serotonin and exhibit neuroprotective properties. The toxic properties of the lithium carbonate drive the search for new salts to be used in lithium therapy. As shown by the results of this study, lithium ascorbate is a highly digestible and low toxicity lithium salt. A neurocytological study on cerebellar granular neurons in culture using a glutamate stress model showed that lithium ascorbate was more effective in supporting neuronal survival (+11%) than lithium chloride or carbonate. Estimates of the biodistribution of lithium have shown that when lithium ascorbate stimulates the accumulation of lithium ions mainly in the brain. Lithium ascorbate is characterized by extremely low acute and chronic toxicity and exhibits moderate antitumor effects. Studies in rats have shown adaptogenic and neuroprotective properties of relatively low doses of lithium ascorbate (5 mg/kg, which corresponds to ~1/1300 of LD50 of this substance).

6. Patents

A part of this work resulted in a patent: "Means with anti-stress, anxiolytic and antidepressant activity and composition based on it", RU2617512C1, https://patents.google.com/patent/RU2617512C1/ru (accessed on 21 February 2022).

Author Contributions: Conceptualization, I.Y.T., O.A.G., K.S.O.; methodology, all authors; software, I.Y.T.; validation, I.Y.T., O.A.G.; formal analysis, I.Y.T.; data collection and investigation, all authors (M.V.F., I.V.G., V.I.D., A.G.K., I.Y.T., K.S.O., O.A.G.); resources, O.A.G.; data curation, I.Y.T.; writing—original draft preparation, I.Y.T.; writing—review and editing, O.A.G.; visualization, I.Y.T.; supervision, I.Y.T., O.A.G.; project administration, I.Y.T. All authors have read and agreed to the published version of the manuscript.

Funding: The research was conducted under the contract No. 0063-2019-0003 "Mathematical methods of data analysis and forecasting" in the infrastructure of the Shared Research Facilities «High Performance Computing and Big Data» (CKP «Informatics») of FRC CSC RAS (Moscow).

Institutional Review Board Statement: The study was conducted according to the guidelines of the Declaration of Helsinki, and approved by the Institutional Review Board of Ivanovo State Medical Academy (protocol code 5, 24 March 2019).

Informed Consent Statement: Not applicable.

Data Availability Statement: The data are available upon reasonable request.

Conflicts of Interest: The authors declare no conflict of interest.

Sample Availability: Samples of the lithium ascorbate are available from the authors.

References

1. Ochoa, E.L.M. Lithium as a Neuroprotective Agent for Bipolar Disorder: An Overview. *Cell. Mol. Neurobiol.* **2021**, *42*, 85–97. [CrossRef] [PubMed]
2. Chouhan, A.; Abhyankar, A.; Basu, S. The feasibility of low-dose oral lithium therapy and its effect on thyroidal radioiodine uptake, retention, and hormonal parameters in various subcategories of hyperthyroid patients: A pilot study. *Nucl. Med. Commun.* **2016**, *37*, 74–78. [CrossRef] [PubMed]
3. Prokopieva, V.; Plotnikov, E.; Yarygina, E.; Bokhan, N. Protektornoe deĭstvie karnozina i organicheskikh soleĭ litiia pri étanol-indutsirovannom okislitel'nom povrezhdenii belkov i lipidov plazmy krovi u zdorovykh lits i bol'nykh alkogolizmom [Protective action of carnosine and organic lithium salts in case of ethanol-induced oxidative damage of proteins and lipids of blood plasma in healthy persons and alcoholic patients]. *Biomeditsinskaya Khimiya* **2019**, *65*, 28–32. [CrossRef]
4. Gromova, O.; Torshin, I.; Гоголева, И.В.; Pronin, A.V.; Stelmashuk, E.V.; Isaev, N.K.; Genrikhs, E.E.; Демидов, В.И.; Volkov, A.Y.; Khaspekov, G.L.; et al. Pharmacokinetic and pharmacodynamic synergism between neuropeptides and lithium in the neurotrophic and neuroprotective action of cerebrolysin. *Zhurnal Nevrol. I Psikhiatrii Im. SS Korsakova* **2015**, *115*, 65–72. [CrossRef] [PubMed]

5. Li, M.; Xia, M.; Chen, W.; Wang, J.; Yin, Y.; Guo, C.; Li, C.; Tang, X.; Zhao, H.; Tan, Q.; et al. Lithium treatment mitigates white matter injury after intracerebral hemorrhage through brain-derived neurotrophic factor signaling in mice. *Transl. Res.* **2019**, *217*, 61–74. [CrossRef] [PubMed]
6. Hozer, F.; Sarrazin, S.; Laidi, C.; Favre, P.; Pauling, M.; Cannon, D.; McDonald, C.; Emsell, L.; Mangin, J.-F.; Duchesnay, E.; et al. Lithium prevents grey matter atrophy in patients with bipolar disorder: An international multicenter study. *Psychol. Med.* **2021**, *51*, 1201–1210. [CrossRef] [PubMed]
7. Torshin, I.Y.; Gromova, O.A.; Kovrazhkina, E.A.; Razinskaya, O.D.; Prokopovich, O.A.; Stakhovskaya, L.V. Intelligent analysis of data on the relationship between blood trace elements and the condition of patients with amyotrophic lateral sclerosis indicated decreased levels of lithium and selenium. *Cons. Med.* **2017**, *19*, 88–96. [CrossRef]
8. Pronin, A.V.; Гоголева, И.В.; Torshin, I.; Gromova, O.A. Neurotrophic effects of lithium stimulate the reduction of ischemic and neurodegenerative brain damage. *Zhurnal Nevrol. I Psikhiatrii Im. SS Korsakova* **2016**, *116*, 99–108. [CrossRef] [PubMed]
9. Cousins, D.A.; Squarcina, L.; Boumezbeur, F.; Young, A.H.; Bellivier, F. Lithium: Past, present, and future. *Lancet Psychiatry* **2020**, *7*, 222–224. [CrossRef]
10. El-Balkhi, S.; Megarbane, B.; Poupon, J.; Baud, F.J.; Galliot-Guilley, M. Lithium poisoning: Is determination of the red blood cell lithium concentration useful? *Clin. Toxicol.* **2009**, *47*, 8–13. [CrossRef] [PubMed]
11. Van Deun, K.; Hatch, H.; Jacobi, S.; Köhl, W. Lithium carbonate: Updated reproductive and developmental toxicity assessment using scientific literature and guideline compliant studies. *Toxicology* **2021**, *461*, 152907. [CrossRef] [PubMed]
12. Torshin, I.Y.; Gromova, O.A.; Chuchalin, A.G.; Zhuravlev, Y.I. Chemoreactome screening of pharmaceutical effects on SARS-CoV-2 and human virome to help decide on drug-based COVID-19 therapy. *Farmakoekon. Mod. Pharm. Pharmacoepidemiol.* **2021**, *14*, 191–211. [CrossRef]
13. Torshin, I.Y.; Rudakov, K.V. On the Procedures of Generation of Numerical Features over Partitions of Sets of Objects in the Problem of Predicting Numerical Target Variables. *Pattern Recognit. Image Anal.* **2019**, *29*, 654–667. [CrossRef]
14. Torshin, I.Y.; Gromova, O.A.; Mayorova, L.A.; Volkov, A.Y. Targeted proteins involved in the neuroprotective effects of lithium citrate. *Neurol. Neuropsychiatry Psychosom.* **2017**, *9*, 78–83. (In Russian)
15. Torshin, I.Y.; Rudakov, K.V. Topological Chemograph Analysis Theory as a Promising Approach to Simulation Modeling of Quantum-Mechanical Properties of Molecules. Part II: Quantum-Chemical Interpretations of Chemograph Theory. *Pattern Recognit. Image Anal.* **2022**, *32*, 205–217. [CrossRef]
16. Directive 2010/63/EU of the European Parliament and of the Council of the European Union on the Protection of Animals Used for Scientific Purposes. Document 32010L0063. Available online: https://eur-lex.europa.eu/legal-content/EN/TXT/?uri=celex%3A32010L0063 (accessed on 21 February 2022).
17. Stelmashuk, E.V.; Novikova, S.V.; Isaev, N.K. Effect of glutamine on the death of cultured granular neurons induced by glucose deprivation and chemical hypoxia. *Biochemistry* **2010**, *75*, 1039–1044.
18. Ferreira, A.J.M.; Fantuzzi, N. *MATLAB Codes for Finite Element Analysis*; Springer: Berlin/Heidelberg, Germany, 2019; ISBN 978-1-4020-9199-5.
19. Zhang, Y.; Huo, M.; Zhou, J.; Xie, S. PKSolver: An add-in program for pharmacokinetic and pharmacodynamic data analysis in Microsoft Excel. *Comput. Methods Programs Biomed.* **2010**, *99*, 306–314. [CrossRef] [PubMed]
20. *Smirnov VG Cytogenetics*; M. Higher School: Moscow, Russia, 1991; 584p, ISBN 5-06-001024-4.
21. Prozorovsky, V.B. Statistical processing of the results of pharmacological studies. *Psychopharmacol Biol. Narcol* **2007**, *3*, 2090–2120.
22. Kawano, K.; Hattori, Y.; Iwakura, H.; Akamizu, T.; Maitani, Y. Adrenal tumor volume in genetically engineered mouse model of neuroblastoma determined by magnetic resonance imaging. *Exp. Ther. Med.* **2012**, *4*, 61–64. [CrossRef] [PubMed]
23. Khabriev, R.U. (Ed.) *Guidelines for Experimental (Preclinical) Study of New Pharmacological Substances*, 2nd ed.; JSC Publishing House "Medicine": Moscow, Russia, 2005; 832p, ISBN 5-225-04219-8.

Article

Protective Effect of Quercetin 3-*O*-Glucuronide against Cisplatin Cytotoxicity in Renal Tubular Cells

Daniel Muñoz-Reyes [1,2,3], Alfredo G. Casanova [1,2,3,4,5], Ana María González-Paramás [6], Ángel Martín [7], Celestino Santos-Buelga [6], Ana I. Morales [1,2,3,4,5], Francisco J. López-Hernández [2,3,4,5,*] and Marta Prieto [1,2,3,4,5]

[1] Toxicology Unit, Universidad de Salamanca, 37007 Salamanca, Spain; danimr@usal.es (D.M.-R.); alfredogcp@usal.es (A.G.C.); amorales@usal.es (A.I.M.); martapv@usal.es (M.P.)
[2] Department of Physiology and Pharmacology, Universidad de Salamanca, 37007 Salamanca, Spain
[3] Group of Translational Research on Renal and Cardiovascular Diseases (TRECARD), 37007 Salamanca, Spain
[4] Institute of Biomedical Research of Salamanca (IBSAL), 37007 Salamanca, Spain
[5] National Network for Kidney Research REDINREN, RD016/0009/0025, Instituto de Salud Carlos III, 28029 Madrid, Spain
[6] Polyphenols Research Group (GIP-USAL), Nutrition and Bromatology Unit, Faculty of Pharmacy, Universidad de Salamanca, 37007 Salamanca, Spain; paramas@usal.es (A.M.G.-P.); csb@usal.es (C.S.-B.)
[7] High Pressure Processes Group, BioEcoUva, Bioeconomy Research Institute, Department of Chemical Engineering and Environmental Technology, Universidad de Valladolid, 47011 Valladolid, Spain; mamaan@iq.uva.es
* Correspondence: flopezher@usal.es; Tel.: +34-923-294-400 (ext. 1444)

Abstract: Quercetin, a flavonoid with promising therapeutic potential, has been shown to protect from cisplatin nephrotoxicity in rats following intraperitoneal injection, but its low bioavailability curtails its prospective clinical utility in oral therapy. We recently developed a micellar formulation (P-quercetin) with enhanced solubility and bioavailability, and identical nephroprotective properties. As a first aim, we herein evaluated the oral treatment with P-quercetin in rats, which displayed no nephroprotection. In order to unravel this discrepancy, quercetin and its main metabolites were measured by HPLC in the blood and urine after intraperitoneal and oral administrations. Whilst quercetin was absorbed similarly, the profile of its metabolites was different, which led us to hypothesize that nephroprotection might be exerted in vivo by a metabolic derivate. Consequently, we then aimed to evaluate the cytoprotective capacity of quercetin and its main metabolites (quercetin 3-*O*-glucoside, rutin, tamarixetin, isorhamnetin and quercetin 3-*O*-glucuronide) against cisplatin toxicity, in HK-2 and NRK-52E tubular cell lines. Cells were incubated for 6 h with quercetin, its metabolites or vehicle (pretreatment), and subsequently 18 h in cotreatment with 10–300 μM cisplatin. Immediately after treatment, cell cultures were subject to the MTT technique as an index of cytotoxicity and photographed under light microscopy for phenotypic assessment. Quercetin afforded no direct cytoprotection and quercetin-3-*O*-glucuronide was the only metabolite partially preventing the effect of cisplatin in cultured tubule cells. Our results identify a metabolic derivative of quercetin contributing to its nephroprotection and prompt to further explore exogenous quercetin-3-*O*-glucuronide in the prophylaxis of tubular nephrotoxicity.

Keywords: quercetin; quercetin 3-*O*-glucuronide; cisplatin; nephrotoxicity; cytoprotection

Citation: Muñoz-Reyes, D.; Casanova, A.G.; González-Paramás, A.M.; Martín, Á.; Santos-Buelga, C.; Morales, A.I.; López-Hernández, F.J.; Prieto, M. Protective Effect of Quercetin 3-*O*-Glucuronide against Cisplatin Cytotoxicity in Renal Tubular Cells. *Molecules* **2022**, *27*, 1319. https://doi.org/10.3390/molecules27041319

Academic Editors: Giovanni Ribaudo and Laura Orian

Received: 19 January 2022
Accepted: 12 February 2022
Published: 15 February 2022

Publisher's Note: MDPI stays neutral with regard to jurisdictional claims in published maps and institutional affiliations.

Copyright: © 2022 by the authors. Licensee MDPI, Basel, Switzerland. This article is an open access article distributed under the terms and conditions of the Creative Commons Attribution (CC BY) license (https://creativecommons.org/licenses/by/4.0/).

1. Introduction

Cisplatin (cis-diaminnedichloroplatin (II)) is an inorganic compound with antineoplastic activity. Its pharmacological mechanism of action is based on the formation of a covalent bond with nitrogen 7 in purine bases, blocking cell division and inducing apoptosis [1]. As a chemotherapeutic drug, the use of cisplatin is curtailed by its toxicity, with nephrotoxicity posing the limiting effect [2]. The incidence of cisplatin nephrotoxicity is high, occurring in one in every three patients under treatment [3]. This nephrotoxicity is due to the drug accumulation in the kidney, since its elimination from the body is carried out through this organ by glomerular filtration and tubular secretion, without being metabolized. The copper transporter 1 (Ctr1) and the organic cation transporter 2 (OCT2) are mainly responsible for the entry of cisplatin into tubular renal cells, since they have a high affinity for the drug. In this way, its concentration in the epithelial cells of the proximal tubule is around 5 times higher than in plasma [4].

Several molecular mechanisms of cisplatin tubular cytotoxicity have been described, including: (a) direct DNA damage from adducts formations [5], (b) alteration of cellular transporters [6], (c) mitochondrial dysfunction and oxidative stress [5], (d) MAP kinase activation [7,8], (e) induction of apoptosis [3], and (f) inflammation [9]. These mechanisms constitute potential pharmacological targets for the prevention and alleviation of cisplatin nephrotoxicity [6].

Quercetin is a flavonoid found naturally in many fruits and vegetables. It has a wide variety of biological properties, among which its antioxidant, antiobesity, antiviral, antibacterial and anti-inflammatory effects stand out. Furthermore, quercetin is of special interest in the treatment of certain cancers, since it inhibits the proliferation of cancer cells and limits their growth [10].

In natural media, quercetin is usually found in glycosylated form. Quercetin glycosides can be partially hydrolyzed in saliva [11] and further in the small intestine by the enzyme lactase-phlorizin hydrolase in the intestinal epithelium releasing the aglycone that may be subsequently absorbed [12]. Efficient glucuronidation of quercetin can already occur in the small intestine by the action of UDP-glucuronyltransferases, as well as methylation by the action of catechol O-methyltransferases. Later, they are transported to the liver through the portal vein, where secondary metabolism occurs in the form of methylation, sulfation and conjugation with glucuronide. Conjugation of quercetin with sulfate is carried out by sulfotransferases. Thus, the circulating forms able to reach the biological targets are quercetin metabolites, which activity can differ from the original compound [13]. Once in the bloodstream, the metabolites bind to plasma proteins, such as albumin [10]. Finally, they accumulate in some organs, such as the kidney, which is involved in their excretion [14].

In the kidney, more than 90% of quercetin is in its conjugated form. Tubular cells have the enzymatic capacity to carry out a third biotransformation of quercetin. The metabolic conversions that occur include a complex combination of metabolite deconjugation followed by immediate sulfation, glucuronidation, methylation, and glucosylation [12].

We have previously demonstrated that quercetin protects against cisplatin nephrotoxicity in an in vivo model, without compromising the antineoplasic activity of the drug [3,15]. However, in our model, quercetin was administered intraperitoneally (i.p.), due to the low oral bioavailability of quercetin. The poor absorption and bioavailability of quercetin are mainly due to its limited solubility in aqueous fluids, which compromises the therapeutic application of the flavonoid. In order to improve its bioavailability, we have developed a micellar formulation (P-quercetin), which increases quercetin solubility approximately 10 times. Our results showed that this formulation, administered i.p., increased the plasma concentration of quercetin compared to the dose-equivalent administration of the unformulated flavonoid, and maintained the nephroprotective capacity when it was coadministered with cisplatin [16].

Since the i.p. via is not useful from a clinical point of view, and therefore the intended via of administration in humans is the oral one (p.o.), our first aim with this work was to verify whether P-quercetin protects against the nephrotoxicity of cisplatin by that via. However, the nephroprotection afforded by P-quercetin via i.p. was not observed following the administration of the same dosage through the oral via (see Section 2.1). In view of this result, a pilot study has been carried out to check whether the absorption of P-quercetin p.o. was lower than by the i.p., which might explain the absence of nephroprotection of the oral administration. With this objective, the presence of quercetin and its metabolites in biological samples (plasma and urine) was analyzed by HPLC, after the administration of the same dosage of P-quercetin by both routes. We found that P-quercetin was similarly absorbed via i.p. and p.o., whereas the profile of the metabolites found was slightly different between both vias (see Section 2.2). Our hypothesis in relation to these results is that quercetin in vivo protection may be related to the activity of some of its metabolite(s), and not to quercetin, at least partially. Thus, the pharmacokinetics after intraperitoneal administration would favor renal accumulation of the active metabolite/s to a greater extent than oral administration. So, the second aim of this study was to evaluate the cytoprotective capacity of quercetin and its main metabolites against cisplatin toxicity in renal tubular cells.

2. Results
2.1. Nephroprotection Study with Oral P-quercetin

In our in vivo model, renal function was heavily damaged by cisplatin. However, this effect was not ameliorated by oral P-quercetin, as was the case with intraperitoneal administration of the formulation in our previous experiments [16]. Rats in the cisplatin (CP) group experienced an acute kidney injury (AKI), as they showed a progressive increase in their plasma creatinine (Crpl) and urea levels compared to those of the controls (Figure 1a). AKI is defined and diagnosed according to elevations in Crpl concentration [17–19], an indirect marker of glomerular filtration rate (GFR). On the other hand, plasma urea concentration is a marker of azotemia [20]. These biomarkers also increased in the CP+PQor (cisplatin + oral P-quercetin) group, even slightly more extensive than in the CP group. In accordance with these data, cisplatin induced a severe drop in creatinine clearance (CrCl), a standard method for GFR measurement [21]. Oral P-quercetin also was unable of mitigate the drop in CrCl produced by cisplatin (Figure 1b). Finally, a significant increase in proteinuria was detected in both groups, CP and CP+PQor, on day 7. Although proteinuria may have glomerular origin, in the case of cisplatin nephrotoxicity it arise from defective tubular reabsorption due to tubular injury [22]. Oral P-quercetin also did not reverse in our model the excess urinary excretion of proteins produced by cisplatin (Figure 1b). The same pattern was found when twice the dose of P-quercetin was administered and the number of days of flavonoid administration prior to cisplatin was increased to 10 days (data not shown).

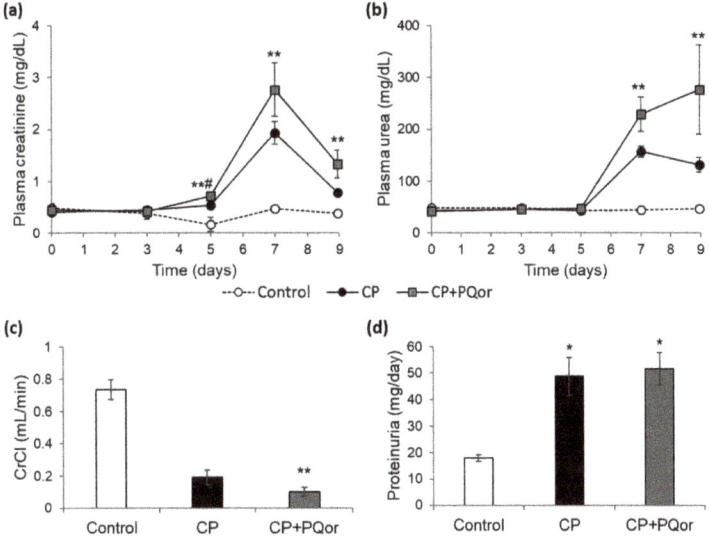

Figure 1. Evaluation of renal function in rats. (a) Evolution of plasma creatinine; (b) evolution of

plasma urea; (**c**) creatinine clearance on day 7; (**d**) proteinuria on day 7. Values are expressed as the mean ± SEM. * $p < 0.05$; ** $p < 0.01$ vs. Control group; # $p < 0.05$ vs. CP group. CP: cisplatin (6.5 mg/kg, i.p.) on day 3; CP+PQor: P-quercetin (100 mg/kg, p.o.) for 9 days and cisplatin (6.5 mg/kg, i.p.) on day 3. CrCl: creatinine clearance.

2.2. Metabolites Distribution in Plasma and Urine after P-quercetin Administration

The distribution of quercetin and its metabolites was analyzed in the plasma and urine of rats that had been administered 3 doses of P-quercetin, i.p. or p.o. In the case of plasma samples, a fairly similar profile in the chromatogram was obtained after flavonoid administration for each of the vias. This fact indicates that absorption occurred by both vias. However, differences were observed in the levels of some metabolites, namely methylquercetin glucuronide sulfate and quercetin sulfate, which appeared in higher concentrations after i.p. than after oral administration (Figure 2a). The metabolites profile found in urine samples was more complex than the one observed in plasma, with a greater number of quercetin metabolites detected. In this case, differences, mainly quantitative, were observed in some metabolites depending on the via of administration (Figure 2b), but also on the individuals.

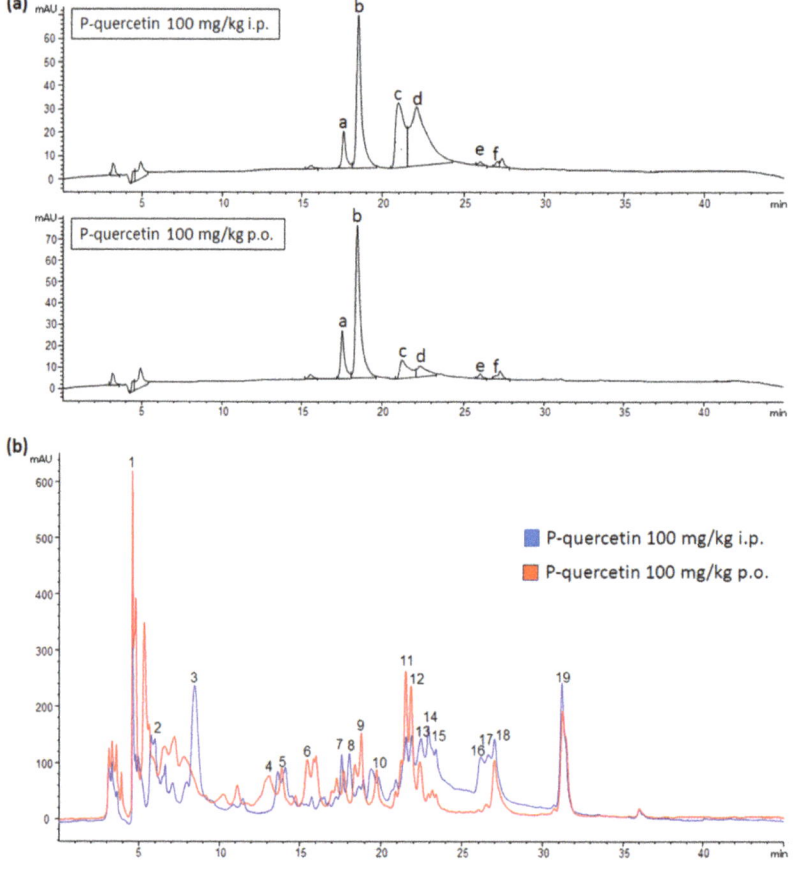

Figure 2. HPLC chromatograms recorded at 360 nm of plasma (**a**) and urine (**b**) samples from rats

treated with P-quercetin. The formulation (100 mg/kg) was administered for 3 days, p.o. or i.p. Tentative identification of peaks: **a** and **b**: quercetin glucuronide sulfate; **c**: methylquercetin glucuronide sulfate; **d** and **e**: quercetin sulfate; **f**: methylquercetin sulfate. **1**: Protocatechuic acid derivative; **2**: hydroxyphenylacetic sulfate; **3** and **8**: quercetin glucuronide sulfate; **4**: protocatechuic acid; **5**: methylquercetin diglucuronide; **6**: quercetin sulfate derivative; **7**: quercetin diglucuronide + quercetin glucose; **9**: quercetin glucuronide; **10**: methylquercetin glucuronide sulfate; **11**: methylquercetin glucuronide + quercetin glucuronide sulfate; **12** and **15**: methylquercetin glucuronide; **13**, **14** and **16**: quercetin sulfate; **17** and **18**: methylquercetin sulfate; **19**: methylquercetin. Information on MS data of the peaks is given in Table S1, Supplementary Material.

2.3. Evaluation of the Protective Capacity of Quercetin and Its Metabolites against Cisplatin Cytotoxicity in HK-2 Cells

2.3.1. Titration of Cisplatin in HK-2 Cells

Cisplatin cytotoxicity was determined in HK-2 cells, a cell line of human proximal tubule cells [23].

As it can be observed in Figure 3, a decrease in the cell viability of HK-2 cells was produced as the concentration of cisplatin increased. At 10 µM, a decrease in cell viability of approximately 30% was already observed compared to cells without treatment, but the difference was not statistically significant. From a concentration of 30 µM, a statistically significant decrease in cell viability was observed, indicating the presence of cell damage at such concentrations of cisplatin for HK-2 cells.

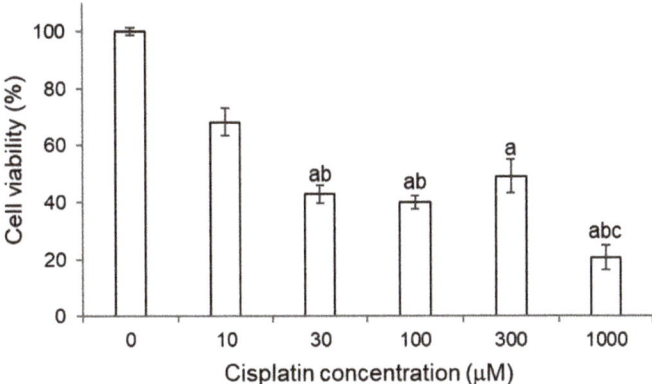

Figure 3. Titration of cisplatin cytotoxicity for 18 h of treatment in HK-2 cells. Cell viability was determined using the MTT assay. Values are expressed as the mean ± SEM. Significant differences ($p < 0.05$): a vs. 0 µM; b vs. 10 µM; c vs. 300 µM.

Microscopic images of HK-2 cells treated with different doses of cisplatin confirmed the presence of cell damage, as manifested in cell viability assays (MTT). Compared with the Control group, at a concentration of 10 µM of cisplatin, no cell damage was observed, so this concentration seems to exert an antiproliferative effect (according to MTT). At a concentration of 30 µM of cisplatin, apoptotic bodies and cisplatin-induced changes in cell morphology began to be detected. On the other hand, at 300 µM, necrosis was observed. The cellular phenotype after cisplatin treatment can be a key when evaluating the possible cytoprotection of quercetin metabolites. These results confirm those previously obtained by Sancho-Martínez et al. [5].

2.3.2. Titration of Quercetin and Its Metabolites in HK-2 Cells

In order to know the highest non-toxic concentrations for quercetin and its metabolites, the corresponding cell safety experiments were performed. Concentrations between

0.4–200 µM (Figure S1, Supplementary Material) were tested. The results are summarized in Table 1.

Table 1. Higher non-toxic concentrations of quercetin metabolites in HK-2 cells. The data were obtained by cell viability tests (MTT assay; Figure S1, Supplementary Material).

	Maximum Non-Toxic Concentration (µM) in HK-2 Cells
Isorhamnetin	1.56
Quercetin	12.5
Quercetin 3-O-glucoside	0.78
Quercetin 3-O-glucuronide	25
Rutin	25
Tamarixetin	12.5

Our results indicated that the highest metabolite concentration not toxic to cells is different among compounds of the same group, such as rutin and quercetin 3-O-glucoside, which are glycosylated compounds. Furthermore, quercetin 3-O-glucoside was the most toxic metabolite of those tested. On the other hand, rutin and quercetin 3-O-glucuronide were the compounds with a greater safety margin, being the concentration of 25 µM, in both cases, the highest non-toxic.

2.3.3. Protection Assays of Quercetin and Its Metabolites against Cisplatin Cytotoxicity in HK-2 Cells

Next, experiments were carried out to test the efficacy of quercetin and its metabolites against the damage of cisplatin. Cytotoxic concentrations of the drug were used according to the viability experiments and the images obtained in the light microscope (see Section 2.3.1): antiproliferative (10 µM), apoptotic (30 µM) and necrotic (300 µM) effect. On the other hand, the highest non-toxic concentration of each metabolite was used for cotreatment with cisplatin, according to Table 1.

HK-2 cells were subjected to 6-h pretreatments with quercetin/quercetin metabolites and then 18-h cotreatment with cisplatin and the corresponding metabolite. The results obtained for quercetin, quercetin 3-O-glucuronide and tamarixetin are shown in Figure 4. The results for the rest of the metabolites (quercetin 3-O-glucoside, isorhamnetin and rutin) can be consulted in Figure S2 (Supplementary Material).

Quercetin, at the dose tested (12.5 µM), did not affect the viability of HK-2 cells. However, its cotreatment with cisplatin was not effective in reversing the decrease in the number of viable cells caused by the drug. Therefore, according to our results, quercetin does not prevent the cytotoxicity of cisplatin in HK-2 cells (Figure 4).

Regarding the action of the glycosylated metabolites, quercetin 3-O-glucoside (0.8 µM) and rutin (25 µM), in cotreatment with cisplatin, no differences were observed with respect to the treatments only with cisplatin at any of the doses tested for the cytotoxic agent (Figure S2).

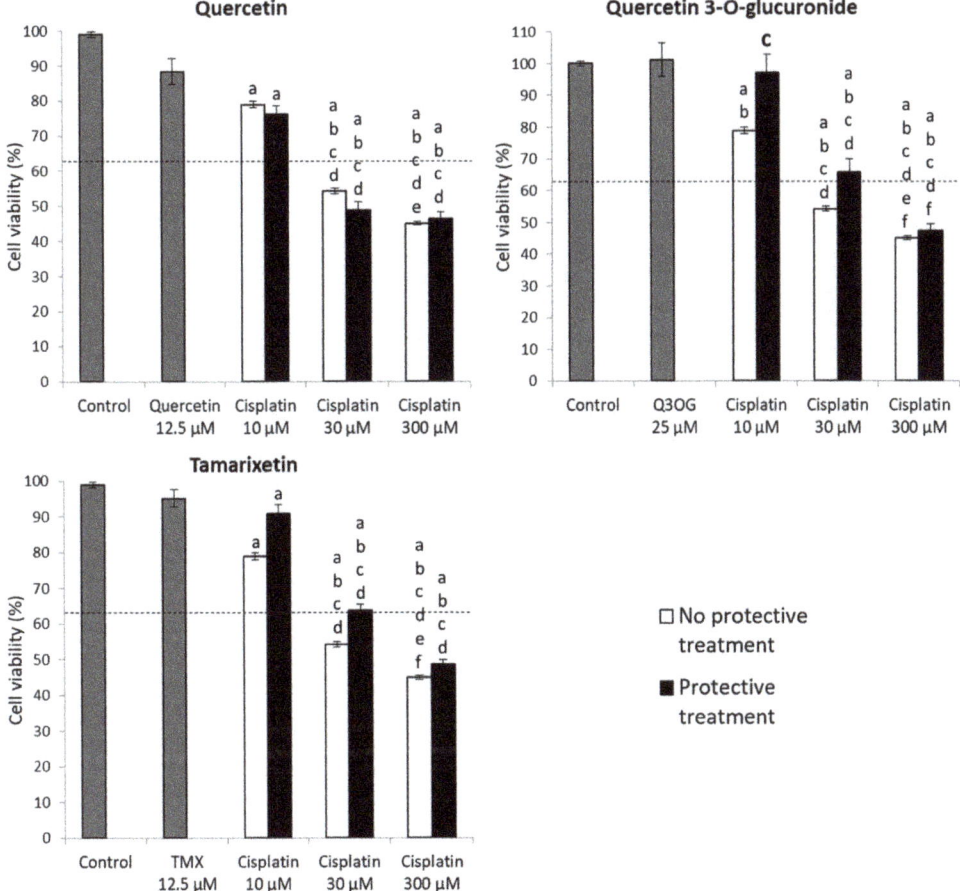

Figure 4. Quercetin, quercetin 3-O-glucuronide and tamarixetin efficacy assays in HK-2 cells for 24 h against cisplatin cytotoxicity. The cells were pretreated 6 h with the corresponding metabolite and subsequently cotreated for 18 h with the metabolite and cisplatin 10, 30 and 300 µM, respectively. The MTT technique was used to estimate cell viability. Values are expressed as the mean ± SEM. Significant differences ($p < 0.05$): a vs. Control; b vs. Metabolite alone (quercetin, quercetin 3-O-glucuronide or tamarixetin); c vs. 10 µM cisplatin; d vs. cisplatin 10 µM + protective treatment; e vs. 30 µM cisplatin; f vs. cisplatin 30 µM + protective treatment. The dotted line represents cell viability before treatment (time 0 h). Q3OG: quercetin-3-O-glucuronide; TMX: tamarixetin.

Regarding the methylated metabolites, the cotreatment of the cells with tamarixetin (12.5 µM) originated an increment on the cell viability profile in cells treated with cisplatin at the three doses tested. Moreover, photos of the cell culture were taken under light microscope, showing that cells cotreated with tamarixetin and cisplatin at doses 10 and 30 µM presented a better status than those treated with the same dose of cisplatin itself (Figure 5). However, these differences were not significant with respect to the single antineoplastic treatment at each of the concentrations (Figure 4). Instead, cotreatment with isorhamnetin (1.56 µM) did not protect from the decrease in cell viability produced by cisplatin; by contrast, an enhancing effect of isorhamnetin on the cytotoxicity of cisplatin was even observed, although the data were not statistically significant (Figure S2).

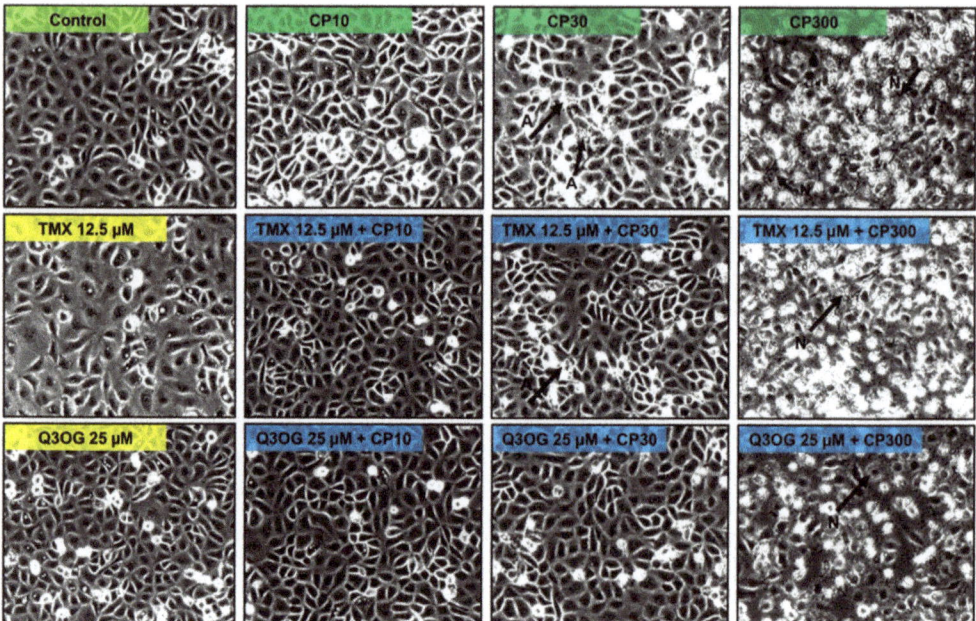

Figure 5. Representative images of HK-2 cells taken under the light microscope (10× magnification). First line: cells without treatment (Control) and treated with cisplatin 10 (CP10), 30 (CP30) and 300 (CP300) μM for 18 h. Second line: cells treated with tamarixetin (TMX) at a concentration of 12.5 μM for 24 h, and cells pretreated with TMX 12.5 μM for 6 h and subsequently cotreated 18 h with TMX 12.5 μM and CP10, CP30 and CP300, respectively. Third line: cells treated with quercetin 3-O-glucuronide (Q3OG) at a concentration of 25 μM for 24 h, and cells pre-treated with Q3OG 25 μM for 6 h and subsequently cotreated for 18 h with Q3OG 25 μM and CP10, CP30 and CP300, respectively. TMX: tamarixetin; G3OG: quercetin 3-O-glucuronide; CP: cisplatin; A: apoptosis; N: necrosis.

When the cells were pretreated with quercetin-3-O-glucuronide and, subsequently, cotreated with the metabolite and cisplatin at doses of 10 and 30 μM, an increase in cell viability of, respectively, 20 and 10% was observed, compared to treatments with cisplatin without metabolite (Figure 4). This result indicates that cotreatment with quercetin 3-O-glucuronide induces cellular protection, thus avoiding or reducing the damage caused by cisplatin. In fact, photos of the cell culture showed that cells cotreated with quercetin 3-O-glucuronide and cisplatin, at the three doses tested, presented a better status than those treated with the same dose of cisplatin itself (Figure 5). However, in the second case (cotreatment with cisplatin 30 μM) the results were not statistically significant, although a cytoprotection profile was observed.

2.4. Evaluation of the Protective Capacity of Quercetin and Its Metabolites against Cisplatin Cytotoxicity in NRK-52E Cells

Considering the results obtained in HK-2 cells for quercetin 3-O-glucuronide (see Section 2.3), it was decided to corroborate the cytoprotective effect of the metabolite in a different tubular cell line. Thus, the NRK-52E cell line was used for several reasons: first of all, they are cells from rats, which allowed us to evaluate the effect of quercetin 3-O-glucuronide in a different species (HK-2 cells come from human kidney). Second, they are renal epithelial cells [24], so they have a less specific origin than HK-2 cells, which are defined as proximal tubule cells [23].

The experiments with NRK-52E cells were carried out with only three compounds: quercetin (reference molecule in our experiments), quercetin 3-O-glucuronide and tamarixetin. The latter was introduced based on the good cytoprotective tendency observed in the experiments with HK-2 cells (see Section 2.3.3).

2.4.1. Titration of Cisplatin in NRK-52E Cells

In the same way, the cytotoxicity of cisplatin was determined in NRK-52E cells. The aim was to find the cisplatin concentrations that gave rise to the same phenotypes found in HK-2 cells (antiproliferative, apoptotic and necrotic effects) via MTT assay and light microscope photos, to subsequently perform cytoprotection experiments with these concentrations.

A decrease in the viability of NRK-52E cells was observed (Figure 6) as the cisplatin concentration increased, quite similar to what we had previously observed for HK-2 cells (see Section 2.3.1). At the 10 µM concentration of cisplatin, a decrease in cell viability of approximately 20% compared to cells without treatment was found, but the difference was not statistically significant compared to the control. From 30 µM, a statistically significant decrease in cell viability was observed, indicating the presence of cell damage at such concentrations of cisplatin for NRK-52E cells.

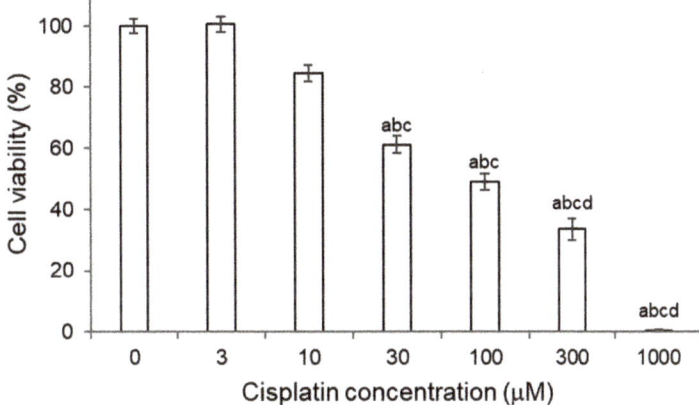

Figure 6. Titration of cisplatin cytotoxicity for 18 h of treatment in NRK-52E cells. Cell viability was determined using the MTT assay. Values are expressed as the mean ± SEM. Significant differences ($p < 0.05$): a vs. 0 µM, b vs. 3 µM; c vs. 10 µM; d vs. 30 µM.

The images of NRK-52E cells treated with different doses of cisplatin confirmed the presence of cellular damage, observing a similar phenotype to that of HK-2 cells at the same doses: cisplatin 10 µM concentration did not produce apparent changes in cell morphology, for which it seems to exert an antiproliferative effect according to MTT, while at a concentration of 30 µM of cisplatin, the formation of apoptotic bodies began to be observed. At 300 µM, death by necrosis was observed.

2.4.2. Titration of Quercetin and Its Metabolites in NRK-52E

In order to know the highest non-toxic concentrations for each metabolite in NRK-52E cells, cell safety experiments were assessed. Doses between 0.4–200 µM (Figure S3, Supplementary Material) were tested for each metabolite. The results are summarized in Table 2.

Table 2. Higher non-toxic concentrations of quercetin metabolites in NRK-52E cells. Data were obtained by cell viability assays (MTT assay; Figure S3, Supplementary Material).

	Maximum Non-Toxic Concentration in NRK-52E Cells (µM)
Quercetin	25
Quercetin 3-O-glucuronide	200
Tamarixetin	12.5

The data indicated that tamarixetin was the most toxic metabolite of those tested, while quercetin had an intermediate cellular safety range. On the other hand, quercetin 3-O-glucuronide was the compound that revealed the highest safety range, being 200 µM the highest non-toxic concentration of those evaluated. It is noteworthy that the safety range of the glucuronide was 4 times higher than in HK-2 cells.

2.4.3. Protection Assays of Quercetin and Its Metabolites against Cisplatin Cytotoxicity in NRK-52E Cells

Efficacy experiments were carried out in NRK-52E cells to check whether quercetin metabolites had a protective effect against cisplatin damage. Cytotoxic cisplatin concentrations were used according to the viability experiments (MTT) and the images obtained under the light microscope: antiproliferative (10 µM), apoptotic (30 µM) and necrotic (300 µM) effect. On the other hand, the highest non-toxic concentration of each metabolite was used for cotreatment with cisplatin, according to Table 2.

A 6-h pretreatment with quercetin/quercetin metabolite was performed, followed by an 18-h cotreatment with cisplatin and the corresponding compound. The results are presented in Figures 7 and 8.

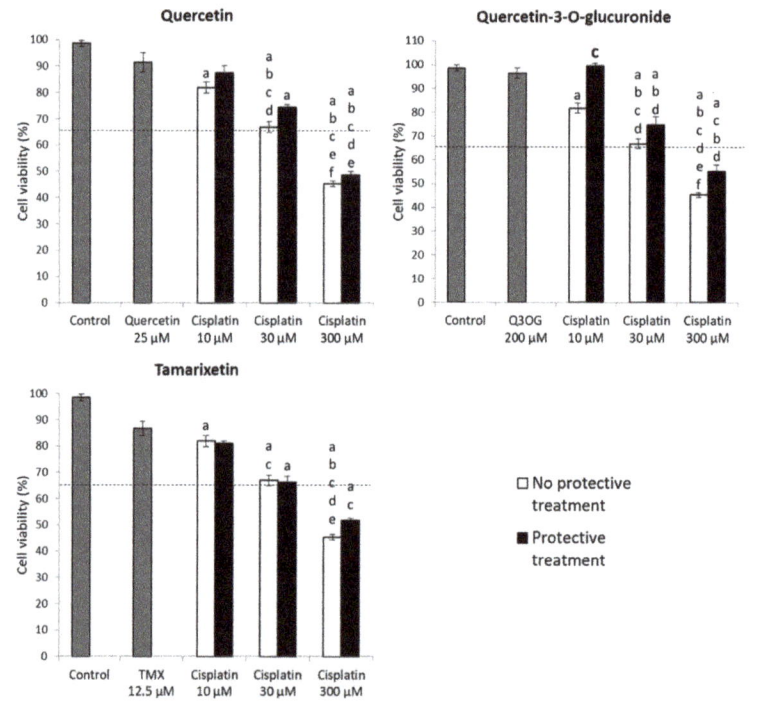

Figure 7. Quercetin, quercetin 3-O-glucuronide and tamarixetin efficacy assays in NRK-52E cells for

24 h against the cytotoxicity of cisplatin. The cells were pretreated for 6 h with the corresponding metabolite and subsequently co-treated for 18 h with the metabolite and 10, 30 and 300 µM cisplatin, respectively. The MTT technique was used to estimate cell viability. Values are expressed as the mean ± SEM. Significant differences ($p < 0.05$): a vs. Control; b vs. Protective treatment alone; c vs. 10 µM cisplatin; d vs. Cisplatin 10 µM + Protective treatment; e vs. 30 µM cisplatin; f vs. Cisplatin 30 µM + Protective treatment. The dotted line represents the state of the cells before treatment (time 0 h). Q3OG: quercetin-3-O-glucuronide; TMX: tamarixetin.

Cotreatment with quercetin 25 µM and cisplatin resulted in a mild increase in cell viability for the three concentrations of cisplatin tested, compared to the single treatment with the drug, although the differences were not significant. On the other hand, cotreatment with tamarixetin and cisplatin made no difference compared to treatment without the metabolite. Therefore, tamarixetin does not appear to protect against drug-induced damage in our experiments with NRK-52E cells (Figure 7).

As with HK-2 cells, in NRK-52E cells the most interesting results were found with quercetin 3-O-glucuronide. When cells are pretreated and subsequently cotreated with the metabolite in combination with cisplatin at concentrations of 10, 30 and 300 µM, an increase in cell viability of 20, 10 and 10%, respectively, was observed, in comparison with treatments without metabolite. This result indicates that glucuronide cotreatment at a concentration of 200 µM induces cellular protection, thus avoiding or reducing the damage caused by cisplatin. Nevertheless, the results were not statistically significant for the cotreatment with cisplatin 30 and 300 µM and glucuronide. Therefore, it can be stated that quercetin 3-O-glucuronide, according to our results, protects against cisplatin damage partially (Figures 7 and 8).

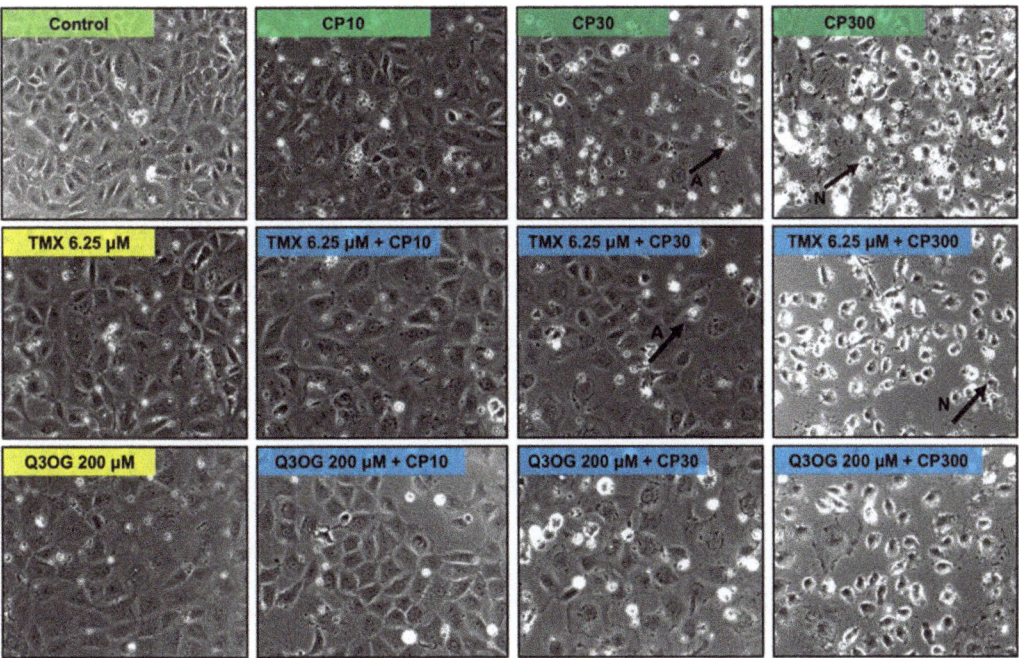

Figure 8. Representative images of NRK-52E cells taken under the light microscope (10× magnification). First line: cells without treatment (control) and treated with cisplatin 10 (CP10), 30 (CP30)

and 300 (CP300) μM for 18 h. Second line: cells treated with tamarixetin (TMX) at a concentration of 12.5 μM for 24 h, and cells pre-treated with TMX 12.5 μM for 6 h and subsequently cotreated 18 h with TMX 12.5 μM and CP10, CP30 and CP300, respectively. Third line: cells treated with Quercetin 3-*O*-glucuronide (Q3OG) at a concentration of 25 μM for 24 h, and cells pre-treated with Q3OG 25 μM for 6 h and subsequently cotreated for 18 h with Q3OG 25 μM and CP10, CP30 and CP300, respectively. TMX: tamarixetin; Q3OG: quercetin 3-*O*-glucuronide; CP: cisplatin; A: apoptosis; N: necrosis.

3. Discussion

In our experiments, quercetin 3-*O*-glucuronide induced cellular protection in the two tubular cell lines assessed, HK-2 and NRK-52E, since it was able to palliate the cytotoxic effect of cisplatin. In fact, the drug produced an antiproliferative effect at 10 μM concentration, which was reversed by the cotreatment with the quercetin metabolite. In contrast, at higher cisplatin concentrations (30 and 300 μM), no statistically significant results were obtained, although a cytoprotective trend was also observed at these concentrations. Therefore, the glucuronide appears to exert a moderate cytoprotective effect, preventing the cytotoxic effect caused by cisplatin at antiproliferative drug concentrations, and reducing the cell death caused by the drug at higher concentrations partially.

Based on those results, quercetin 3-*O*-glucuronide could be responsible, at least in part, for the nephroprotective effect of quercetin against cisplatin damage observed in vivo [3,15]. Furthermore, the protective effect of the metabolite could explain our in vivo results regarding to the absence of protection of oral P-quercetin, when the same formulation has previously been shown to be effective intraperitoneally [16]. The differences observed between the administration of P-quercetin by the intraperitoneal and oral vias cannot be explained by its absorption, since it seems to be very similar for both routes. However, plasma concentration of some quercetin metabolites was different, which might be explained by possible pharmacokinetic differences between both vias [25]. After i.p. administration, metabolism in enterocytes does not take place [12], while a first-pass effect occurs via p.o., with the compound being biotransformed before entering systemic circulation [26]. Therefore, the nephroprotection of quercetin against cisplatin could be related to the concentrations reached in the kidney by quercetin itself or its metabolites, which could be different depending on the administration route of the flavonoid. A relevant metabolite could be quercetin 3-*O*-glucuronide, according to our in vitro results.

Quercetin 3-*O*-glucuronide is one of the most common quercetin metabolites produced by phase II biotransformation in small intestine and liver cells, besides sulfated and methylated derivatives [27]. Once in the kidney, the entry of quercetin metabolites in tubular cells takes place mainly through influx transporters in the basolateral membrane and transporters in the apical membrane via tubular reabsorption [28]. It has been reported that glucuronide conjugates, such as quercetin 3'-*O*-glucuronide, have a high affinity for OAT3, whereas quercetin 3-*O*-glucuronide and quercetin 7-*O*-glucuronide are weak substrates of OAT1 and OAT3 [29]. Thus, it can be speculated that the use of quercetin 3'-*O*-glucuronide might still improve the cytoprotective effect observed for quercetin 3-*O*-glucuronide in the present work. It has also been suggested that these metabolites could be involved in the induction of antioxidant defense mechanisms through inducing the expression of antioxidant enzymes [12]. In addition, some published studies have demonstrated therapeutic effects of glucuronide metabolites [30,31]. In the kidney, an organ where metabolites accumulate to be eliminated, some glucuronides have been shown to have a biological effect. For example, they can reduce oxidative stress [32] or hypoxia signals in kidney cells [33]. On the other hand, other glucuronides behave as active compounds, increasing the prodrug toxicity [34,35]. Therefore, the biological activity of some glucuronide metabolites has been previously demonstrated. All in all, quercetin 3-*O*-glucuronide would reach kidney tissue through systemic distribution [36], and could exert a cytoprotective effect at the tubular level as indicated by our results. Nonetheless, the mechanisms by which the glucuronide exerts its protective effects from cisplatin toxicity must still be unveiled.

Cotreatment of quercetin aglycone, glycosylated derivatives (quercetin 3-O-glucoside and rutin), or methylated derivatives (tamarixetin and isorhamnetin) with cisplatin did not protect against drug cytotoxicity in our experiments. However, quercetin glycosides are present in human urine samples after ingesting foods rich in quercetin, whereas they are not present in plasma. This implies that they are formed in the kidney in situ based on local metabolism [12]. It should not be ruled out that glycosides are involved in nephroprotection, perhaps by action on other renal cells.

It either cannot be ruled out that aglycone could be responsible for the cytoprotective effect, since there are deconjugation processes in the tubular cell to give rise to aglycone [10]. Indeed, it has been reported that conjugation is a reversible process and glucuronides, but not sulfates, can be deconjugated at tissue level, yielding the parent aglycone, which could be the actual active form [37,38]. In fact, the tubular cells present all the enzymatic machinery to carry out a third biotransformation of quercetin in the kidney, which may contribute to the final concentrations reached in the kidney of quercetin and its metabolites. The high β-glucuronidase activity in tubular epithelial cells can, thus, be responsible for the deconjugation of glucuronides, to give rise to quercetin aglycone [12]. However, it has also been described that this deglucuronidation could be followed by instantaneous sulfation, and subsequent reglucuronidation [10]. This is consistent with the idea that glucuronide derivatives (and specifically quercetin 3-O-glucuronide) can be responsible, at least in part, for the protective effects of quercetin in the tubular cell.

The obtained results are very similar in both tubular cell lines, HK-2 and NRK-52E, despite the fact that, theoretically, there are some differences between them. HK-2 are defined as proximal tubule cells, while NRK-52E are renal epithelial cells, so they have a less specific origin. These differences could explain the results obtained with tamarixetin, a metabolite that seems to exert a certain cytoprotective action in the proximal tubule (HK-2), while this effect could be masked when carrying out the experiments with NRK-52E. However, this hypothesis needs to be analyzed more deeply.

Taking into account that the cytoprotective effect observed for quercetin 3-O-glucuronide in our experiments is partial and that quercetin metabolism is very complex [12], it is conceivable that the nephroprotective effect of quercetin observed in vivo could be the result of the joint action of several metabolites and on other targets in addition to the tubular cell. In fact, besides tubular protection, beneficial effects of quercetin have also been demonstrated at the hemodynamic level [39,40], which contribute to the blood supply of the kidney and the maintenance of the glomerular filtration rate when the flavonoid is coadministered with cisplatin [3]. In addition, according to in vitro studies, vasodilator effect can be exerted by both quercetin and isorhamnetin [41,42], which supports the idea that the metabolites could contribute to the in vivo effects of quercetin.

A strategy for future nephroprotection experiments could consist of administrating quercetin 3-O-glucuronide p.o. The administration of the glucuronide metabolite would facilitate access to the tubular cell, as well as its renal metabolism and accumulation. Although we are aware that quercetin glucuronide has shown a moderate cytoprotective effect, this strategy could be a further advance in the nephroprotective application of quercetin.

4. Materials and Methods

4.1. Animals and Bioethics

All procedures were approved by the Bioethics Committee of the University of Salamanca and the Regional Government of Castile and Leon, Ministry of Agriculture and Livestock (code: 0000075, 29 April 2016). Animals were handled according to the guidelines of the European Community Council Directive 2010/63/UE and to the current Spanish legislation for experimental animal use and care (RD 53/2013, 1 February 2013). Male Wistar rats (200–250 g) were maintained under controlled environmental conditions, with free access to water and standard chow.

4.2. Nephroprotection Study with Oral P-quercetin

A cisplatin nephrotoxicity model previously developed in our laboratory was used [16]. Rats were divided into three experimental groups (Figure 9): Control (n = 3), animals received vehicle (water) orally through a intragastric gavage (p.o.) for 9 days; CP (n = 5), animals received water p.o. for 9 days and a single nephrotoxic dose of cisplatin in NaCl 0.9% (6.5 mg/kg, i.p.) (Merck, Darmstadt, Germany) on day 3 of the experiment; and CP+PQor (n = 6) animals received a daily dose of P-quercetin (100 mg/kg, p.o., through a intragastric gavage (i.e., containing 50 mg/kg quercetin)) for 9 days and a dose of cisplatin in NaCl 0.9% (6.5 mg/kg, i.p.) on day 3. P-quercetin is a micellar formulation of quercetin, prepared as described previously [43].

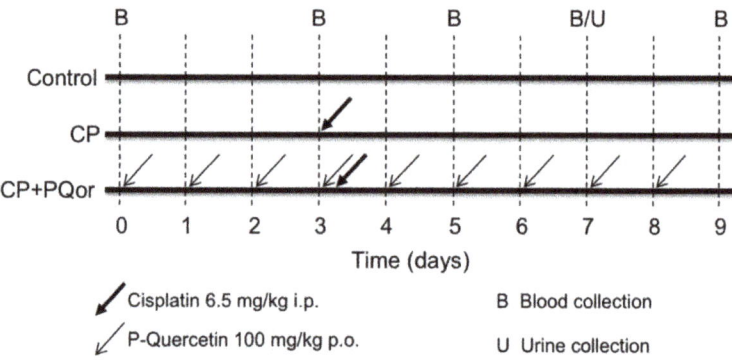

Figure 9. Scheme of the nephrotoxicity model. CP: cisplatin (6.5 mg/kg, i.p.) on day 3; CP + PQor: P-quercetin (100 mg/kg, p.o.) for 9 days and cisplatin (6.5 mg/kg, i.p.) on day 3.

Blood samples (150 µL) were collected on days 0, 3, 5, 7 and 9 in heparinized capillaries from a small incision in the tail tip. Plasma was separated by centrifugation (11,000 rpm for 3 min) and kept at −80 °C. On day 7 (time of maximum nephrotoxicity), 24 h urine was collected in metabolic cages, cleared by centrifugation (2000× g for 9 min) and stored at −80 °C. At the end of the experiment (day 9), rats were anesthetized and sacrificed by exsanguination.

4.3. Analysis of Quercetin and Its Metabolites in Blood and Urine Samples

A pilot assay was carried out in order to analyze the presence of quercetin and its metabolites in biological samples after the administration of P-quercetin orally and i.p. Two rats were treated with P-quercetin, p.o, at the dose of 100 mg/kg/day, for 3 days. Two other rats were simultaneously treated with the same dose, but administering the formulation i.p. On the fourth day, blood and urine (24 h) samples were collected and the rats were sacrificed by exsanguination. Blood and urine samples were processed as described in Section 4.2.

For HPLC analysis, 100 µL of plasma was taken and 300 µL of methanol was added. It was vortexed for one minute and centrifuged at 12,000 rpm for 10 min. Subsequently, the supernatant was collected, brought to dryness in the speedvac and the residue was redissolved in 120 µL of acetonitrile: 0.1% formic acid (20:80). In the case of urine samples, the protocol proposed by Mullen et al. [44] was followed, based on direct urine analysis. 200 µL of urine was taken and 10 µL of methanol was added in order to precipitate proteins. The samples were then shaken, centrifuged and injected into the HPLC system. Analysis were carried out by HPLC-DAD-ESI-MS as described elsewhere [45]. A Hewlett-Packard 1200 chromatograph (Agilent Technologies, Waldbronn, Germany) connected to an API 3200 138 Qtrap (Applied 139 Biosystems, Darmstadt, Germany) mass spectrometer was used. For detection, 280, 330 and 370 nm were selected as preferred wavelengths for the DAD and the MS was operated in the negative ion mode, spectra were recorded between *m/z* 100 and *m/z* 1500. The phenolic metabolites were identified by using data reported from literature and by comparison with our database library. All samples were analyzed in duplicate.

4.4. Renal Cells

Kidney tubular cells were used: HK-2 cells (human kidney 2), it is an immortalized proximal tubule cell line obtained from healthy human kidney (Ref: ATCC, CRL-2190), and NRK-52E (normal rat kidney), is an immortalized cell line of epithelial origin obtained from the rat (Ref: ATCC, CRL-1571). The culture medium for the HK-2 used was RPMI-1640 (Merck, Darmstadt, Germany), supplemented with fetal bovine serum (FBS) (10%) (ThermoFisher, Waltham, MA, USA), L-glutamine (1 mM) (Merck, Darmstadt, Germany) and penicillin-streptomycin (500 U/mL) (Merck, Darmstadt, Germany). For the NRK-52E, a DMEM medium was used: F12 with HEPES, glucose and L-glutamine (Lonza, Basel, Switzerland), supplemented with FBS (10%) and penicillin-streptomycin (500 U/mL).

All cell culture procedures were performed in a laminar flow hood using sterile material. The cells were incubated at 37 °C in a humid atmosphere, with 5% CO_2.

4.5. Quercetin Metabolites

Quercetin and its metabolites isorhamnetin, tamarixetin, quercetin 3-*O*-glucoside, rutin and quercetin 3-*O*-glucuronide (Figure 10), were used in the experiments carried out. All compounds were from Merck (Darmstadt, Germany) with the exception of quercetin (Acros Organics, Madrid, Spain) and tamarixetin (Cymit Química, Barcelona, Spain).

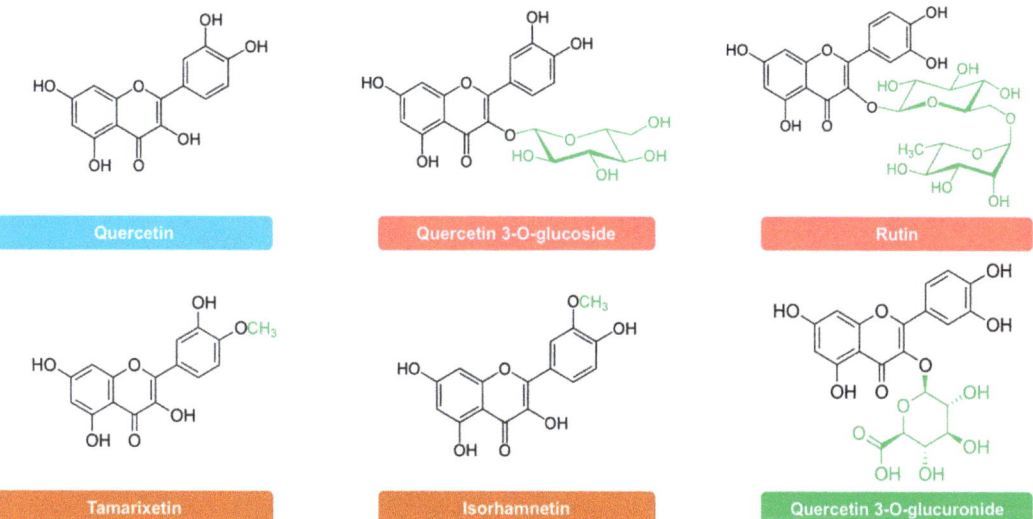

Figure 10. Chemical structure of quercetin and its metabolites used in this study.

All of them were dissolved in dimethylsulfoxide (DMSO) and stored at −80 °C, in order to have a 0.1 M stock solution.

4.6. Design of Experiments In Vitro

24-well plates (15.6 mm diameter) were used to carry out the experiments. After trypsinize and counting the cells, 35,000 cells were seeded per well. Once they had adhered and were at 70% confluence, the experiments were started. The final volume in each well was 500 μL. In each experiment, four duplicates (wells) were made for each experimental condition.

4.6.1. Titration of Cisplatin

Cisplatin (Merck, Darmstadt, Germany) titration in both cell types was performed as previously described [5]. Three experiments were performed for each tested cisplatin concentration (0–1000 µM). For the preparation of the solutions at different concentrations, a 1 M stock solution was started, and subsequently serial dilutions were made. After 18 h of treatment, the MTT test was carried out to determine cell viability (see Section 4.6).

4.6.2. Titration of Quercetin and Its Metabolites

In order to establish the highest non-toxic concentration of each metabolite, four titration experiments were carried out for each cell type and metabolite, adjusting the concentrations according to the results that were obtained. For the preparation of solutions at different concentrations of each metabolite, a 0.1 M stock solution (in DMSO) was started, and serial dilutions were made. The highest concentration tested for each metabolite was 200 µM. After 24 h of treatment, the 3-(4,5-dimethylthiazol-2-yl)-2,5-diphenyltetrazolium bromide (MTT) test was carried out to determine cell viability (see Section 4.6).

In the case of NRK-52E cells, the metabolites to be tested were selected based on the results obtained in HK-2 cells.

In each experiment, a plate of cells was seeded to which the MTT assay was performed at the time of starting the treatments (time 0).

4.6.3. Efficacy Experiments of Quercetin and Its Metabolites against Cisplatin Cytotoxicity

To check whether quercetin and/or its metabolites protected against the cytotoxicity of cisplatin, three concentrations of the drug were used: 10, 30 and 300 µM. These concentrations were chosen based on the results obtained in the titration experiments. Regarding quercetin and its metabolites, the highest non-toxic doses were selected for each cell type. Four experiments were performed for each cell type and metabolite.

A 6-h pretreatment was carried out with the quercetin metabolites and then, without removing the medium, cisplatin was added for 18 h. Once the incubations had elapsed, images were obtained by light microscope of each experimental condition and, subsequently, the MTT protocol was followed (see Section 4.6).

4.7. MTT Assay

As an index of cell viability, the MTT test was carried out. To each well was added, without removing the medium, 50 µL of MTT, (0.5 mg/mL) (Merck, Darmstadt, Germany) for 4 h at 37 °C (MTT concentration in the well 0.05 mg/mL). Subsequently, 500 µL of 10% SDS in 0.01 M HCl was added to dissolve the formed formazan crystals and the plates were kept overnight at 37 °C, 5% CO_2 and in the dark. The next day the absorbance of each well was measured at 595 nm. From this, 100% cell viability was considered that of the negative control (cells without treatment).

4.8. Statistical Analysis

Data are presented as mean ± standard error of the mean (SEM) of n animals/wells performed. Normal distribution of the data was evaluated using the Shapiro–Wilk ($n < 50$) or Kolmogorov–Smirnov ($n \geq 50$) test. Comparisons between groups were assessed by an ANOVA–Scheffé or a Kruskal–Wallis test. A value of $p < 0.05$ was considered significant. Statistical analysis was performed using IBM SPSS Statistics 20.0 software (International Business Machines, Armonk, NY, USA). Microsoft Office Excel and PowerPoint 2016 (Microsoft, Redmond, WA, USA) were used to create the artwork and illustrations.

5. Conclusions

P-quercetin protects against cisplatin damage via intraperitoneal, but not by the oral one, even though the formulation is absorbed by both routes. The nephroprotection of quercetin against cisplatin could be related to the concentration of some of its metabolites as they pass through the kidney. In our cytoprotection experiments, the compounds quercetin,

quercetin 3-*O*-glucoside, rutin and isorhamnetin did not appear to protect against cisplatin cytotoxicity in tubular cells. More experiments are needed for tamarixetin to check the cytoprotection profile observed. In contrast, quercetin 3-*O*-glucuronide exerts a moderate cytoprotective effect, preventing the cytotoxicity and antiproliferative effect of cisplatin in renal tubular cells, therefore being a possible candidate for future nephroprotection strategies in vivo.

Supplementary Materials: The following are available online, Figure S1: Quercetin metabolites safety assays in HK-2 cells for 24 h of treatment, Figure S2: Isorhamnetin, quercetin 3-*O*-glucoside and rutin efficacy assays in HK-2 cells for 24 h against the cytotoxicity of cisplatin, Figure S3: Quercetin, quercetin 3-*O*-glucuronide and tamarixetin safety assays in NRK-52E cells for 24 h of treatment. Table S1: Mass spectral data and tentative identification of the compounds detected in the analyzed urine samples.

Author Contributions: Conceptualization, A.I.M., F.J.L.-H. and M.P.; methodology, D.M.-R., A.G.C., A.M.G.-P. and Á.M.; investigation, D.M.-R., A.G.C., A.M.G.-P., Á.M. and M.P.; writing—original draft preparation, D.M.-R., A.G.C. and M.P.; writing—review and editing, D.M.-R., A.G.C., A.M.G.-P., Á.M.; C.S.-B., A.I.M., F.J.L.-H. and M.P.; supervision, C.S.-B., A.I.M., F.J.L.-H. and M.P.; funding acquisition, Á.M., C.S.-B. and A.I.M. All authors have read and agreed to the published version of the manuscript.

Funding: This research was funded by Fundación FUESCYL—Banco de Santander, (grant: Ed. 2014–2015 Desafío UNIV-EMP); Fundación General de la Universidad de Salamanca, Fondo Europeo de Desarrollo Regional (FEDER) y la Junta de Castilla y León (grant: Ed. 2015 Lanzadera TC); and Junta de Castilla y León, (grant: VA225U14); The GIP-USAL is financially supported by the Spanish Ministerio de Ciencia e Innovación (Project PID2019-106167RB-I00), and Consejería de Educación (Project SA093P20) and Strategic Research Programs for Units of Excellence from Junta de Castilla y León (ref. CLU-2018-04).

Institutional Review Board Statement: The study was conducted according to the guidelines of the Declaration of Helsinki, and approved by the Ethics Committee of the University of Salamanca and the Regional Government of Castile and Leon, Ministry of Agriculture and Livestock (code: 0000075, 29 April 2016).

Informed Consent Statement: Not applicable.

Data Availability Statement: The data presented in this study are available on request from the corresponding author.

Conflicts of Interest: The authors declare no conflict of interest.

Sample Availability: Not available.

References

1. Yao, X.; Panichpisal, K.; Kurtzman, N.; Nugent, K. Cisplatin Nephrotoxicity: A Review. *Am. J. Med. Sci.* **2007**, *334*, 115–124. [CrossRef] [PubMed]
2. Manohar, S.; Leung, N. Cisplatin Nephrotoxicity: A Review of the Literature. *J. Nephrol.* **2018**, *31*, 15–25. [CrossRef] [PubMed]
3. Sanchez-Gonzalez, P.D.; Lopez-Hernandez, F.J.; Perez-Barriocanal, F.; Morales, A.I.; Lopez-Novoa, J.M. Quercetin Reduces Cisplatin Nephrotoxicity in Rats without Compromising Its Anti-Tumour Activity. *Nephrol. Dial. Transplant.* **2011**, *26*, 3484–3495. [CrossRef] [PubMed]
4. Kuhlmann, M.; Burkhardt, G.; Kohler, H. Insights into Potential Cellular Mechanisms of Cisplatin Nephrotoxicity and Their Clinical Application. *Nephrol. Dial. Transplant.* **1997**, *12*, 2478–2480. [CrossRef]
5. Sancho-Martínez, S.M.; Piedrafita, F.J.; Cannata-Andía, J.B.; López-Novoa, J.M.; López-Hernández, F.J. Necrotic Concentrations of Cisplatin Activate the Apoptotic Machinery but Inhibit Effector Caspases and Interfere with the Execution of Apoptosis. *Toxicol. Sci.* **2011**, *122*, 73–85. [CrossRef]
6. Pabla, N.; Dong, Z. Cisplatin Nephrotoxicity: Mechanisms and Renoprotective Strategies. *Kidney Int.* **2008**, *73*, 994–1007. [CrossRef]
7. Dasari, S.; Bernard Tchounwou, P. Cisplatin in Cancer Therapy: Molecular Mechanisms of Action. *Eur. J. Pharmacol.* **2014**, *740*, 364–378. [CrossRef]
8. DeHaan, R.D.; Yazlovitskaya, E.M.; Persons, D.L. Regulation of P53 Target Gene Expression by Cisplatin-Induced Extracellular Signal-Regulated Kinase. *Cancer Chemother. Pharm.* **2001**, *48*, 383–388. [CrossRef]

9. Gluba, A.; Banach, M.; Hannam, S.; Mikhailidis, D.P.; Sakowicz, A.; Rysz, J. The Role of Toll-like Receptors in Renal Diseases. *Nat. Rev. Nephrol.* **2010**, *6*, 224–235. [CrossRef]
10. Wang, W.; Sun, C.; Mao, L.; Ma, P.; Liu, F.; Yang, J.; Gao, Y. The Biological Activities, Chemical Stability, Metabolism and Delivery Systems of Quercetin: A Review. *Trends Food Sci. Technol.* **2016**, *56*, 21–38. [CrossRef]
11. Russo, G.L.; Russo, M.; Spagnuolo, C. The Pleiotropic Flavonoid Quercetin: From Its Metabolism to the Inhibition of Protein Kinases in Chronic Lymphocytic Leukemia. *Food Funct.* **2014**, *5*, 2393–2401. [CrossRef] [PubMed]
12. Muñoz-Reyes, D.; Morales, A.I.; Prieto, M. Transit and Metabolic Pathways of Quercetin in Tubular Cells: Involvement of Its Antioxidant Properties in the Kidney. *Antioxidants* **2021**, *10*, 909. [CrossRef] [PubMed]
13. González-Paramás, A.M.; Ayuda-Durán, B.; Martínez, S.; González-Manzano, S.; Santos-Buelga, C. The Mechanisms Behind the Biological Activity of Flavonoids. *Curr. Med. Chem.* **2019**, *26*, 6976–6990. [CrossRef]
14. Yang, H.; Song, Y.; Liang, Y.; Li, R. Quercetin Treatment Improves Renal Function and Protects the Kidney in a Rat Model of Adenine-Induced Chronic Kidney Disease. *Med. Sci. Monit.* **2018**, *24*, 4760–4766. [CrossRef] [PubMed]
15. Sánchez-González, P.D.; López-Hernández, F.J.; Dueñas, M.; Prieto, M.; Sánchez-López, E.; Thomale, J.; Ruiz-Ortega, M.; López-Novoa, J.M.; Morales, A.I. Differential Effect of Quercetin on Cisplatin-Induced Toxicity in Kidney and Tumor Tissues. *Food Chem. Toxicol.* **2017**, *107*, 226–236. [CrossRef]
16. Casanova, A.G.; Prieto, M.; Colino, C.I.; Gutiérrez-Millán, C.; Ruszkowska-Ciastek, B.; de Paz, E.; Martín, Á.; Morales, A.I.; López-Hernández, F.J. A Micellar Formulation of Quercetin Prevents Cisplatin Nephrotoxicity. *Int. J. Mol. Sci.* **2021**, *22*, 729. [CrossRef] [PubMed]
17. Bellomo, R.; Ronco, C.; Kellum, J.A.; Mehta, R.L.; Palevsky, P. Acute Renal Failure—Definition, Outcome Measures, Animal Models, Fluid Therapy and Information Technology Needs: The Second International Consensus Conference of the Acute Dialysis Quality Initiative (ADQI) Group. *Crit Care* **2004**, *8*, R204. [CrossRef]
18. Mehta, R.L.; Kellum, J.A.; Shah, S.V.; Molitoris, B.A.; Ronco, C.; Warnock, D.G.; Levin, A. Acute Kidney Injury Network: Report of an Initiative to Improve Outcomes in Acute Kidney Injury. *Crit Care* **2007**, *11*, R31. [CrossRef] [PubMed]
19. Khwaja, A. KDIGO Clinical Practice Guidelines for Acute Kidney Injury. *Nephron* **2012**, *120*, c179–c184. [CrossRef]
20. De Corte, W.; Vanholder, R.; Dhondt, A.W.; De Waele, J.J.; Decruyenaere, J.; Danneels, C.; Claus, S.; Hoste, E.A.J. Serum Urea Concentration Is Probably Not Related to Outcome in ICU Patients with AKI and Renal Replacement Therapy. *Nephrol. Dial. Transplant.* **2011**, *26*, 3211–3218. [CrossRef]
21. Diskin, C.J. Creatinine and Glomerular Filtration Rate: Evolution of an Accommodation. *Ann. Clin. Biochem.* **2007**, *44*, 16–19. [CrossRef] [PubMed]
22. Sánchez-González, P.D.; López-Hernández, F.J.; López-Novoa, J.M.; Morales, A.I. An Integrative View of the Pathophysiological Events Leading to Cisplatin Nephrotoxicity. *Crit. Rev. Toxicol.* **2011**, *41*, 803–821. [CrossRef] [PubMed]
23. Murphy, R.; Stafford, R.; Petrasovits, B.; Boone, M.; Valentovic, M. Establishment of HK-2 Cells as a Relevant Model to Study Tenofovir-Induced Cytotoxicity. *Int. J. Mol. Sci.* **2017**, *18*, 531. [CrossRef] [PubMed]
24. Lash, L.H.; Putt, D.A.; Matherly, L.H. Protection of NRK-52E Cells, a Rat Renal Proximal Tubular Cell Line, from Chemical-Induced Apoptosis by Overexpression of a Mitochondrial Glutathione Transporter. *J. Pharm. Exp.* **2002**, *303*, 476–486. [CrossRef] [PubMed]
25. Galindo, P.; González-Manzano, S.; Zarzuelo, M.J.; Gómez-Guzmán, M.; Quintela, A.M.; González-Paramás, A.; Santos-Buelga, C.; Pérez-Vizcaíno, F.; Duarte, J.; Jiménez, R. Different Cardiovascular Protective Effects of Quercetin Administered Orally or Intraperitoneally in Spontaneously Hypertensive Rats. *Food Funct.* **2012**, *3*, 643. [CrossRef]
26. Almeida, A.F.; Borge, G.I.A.; Piskula, M.; Tudose, A.; Tudoreanu, L.; Valentová, K.; Williamson, G.; Santos, C.N. Bioavailability of Quercetin in Humans with a Focus on Interindividual Variation: Variability in Quercetin Bioavailability. *Compr. Rev. Food Sci. Food Saf.* **2018**, *17*, 714–731. [CrossRef]
27. Nemeth, K.; Piskula, M.K. Food Content, Processing, Absorption and Metabolism of Onion Flavonoids. *Crit. Rev. Food Sci. Nutr.* **2007**, *47*, 397–409. [CrossRef]
28. Semenova, S.; Rozov, S.; Panula, P. Distribution, Properties, and Inhibitor Sensitivity of Zebrafish Catechol-O-Methyl Transferases (COMT). *Biochem. Pharmacol.* **2017**, *145*, 147–157. [CrossRef]
29. Wong, C.C.; Botting, N.P.; Orfila, C.; Al-Maharik, N.; Williamson, G. Flavonoid Conjugates Interact with Organic Anion Transporters (OATs) and Attenuate Cytotoxicity of Adefovir Mediated by Organic Anion Transporter 1 (OAT1/SLC22A6). *Biochem. Pharmacol.* **2011**, *81*, 942–949. [CrossRef]
30. Baral, S.; Pariyar, R.; Kim, J.; Lee, H.-S.; Seo, J. Quercetin-3-O-Glucuronide Promotes the Proliferation and Migration of Neural Stem Cells. *Neurobiol. Aging* **2017**, *52*, 39–52. [CrossRef]
31. Chen, X.; Wu, B.; Wang, P. Glucuronides in Anti-Cancer Therapy. *Curr. Med. Chem.-Anti-Cancer Agents* **2003**, *3*, 139–150. [CrossRef] [PubMed]
32. Deiana, M.; Incani, A.; Rosa, A.; Atzeri, A.; Loru, D.; Cabboi, B.; Paola Melis, M.; Lucas, R.; Morales, J.C.; Assunta Dessì, M. Hydroxytyrosol Glucuronides Protect Renal Tubular Epithelial Cells against H_2O_2 Induced Oxidative Damage. *Chem.-Biol. Interact.* **2011**, *193*, 232–239. [CrossRef] [PubMed]
33. Asai, H.; Hirata, J.; Watanabe-Akanuma, M. Indoxyl Glucuronide, a Protein-Bound Uremic Toxin, Inhibits Hypoxia-Inducible Factor–dependent Erythropoietin Expression through Activation of Aryl Hydrocarbon Receptor. *Biochem. Biophys. Res. Commun.* **2018**, *504*, 538–544. [CrossRef] [PubMed]

34. Iwamura, A.; Watanabe, K.; Akai, S.; Nishinosono, T.; Tsuneyama, K.; Oda, S.; Kume, T.; Yokoi, T. Zomepirac Acyl Glucuronide Is Responsible for Zomepirac-Induced Acute Kidney Injury in Mice. *Drug Metab. Dispos.* **2016**, *44*, 888–896. [CrossRef] [PubMed]
35. Steffen, P.; Keller, F. Analgesic Drug Therapy in Kidney Patients. *Dtsch. Med. Wochenschr.* **2021**, *146*, 1009–1015. [CrossRef]
36. Yang, L.-L.; Xiao, N.; Li, X.-W.; Fan, Y.; Alolga, R.N.; Sun, X.-Y.; Wang, S.-L.; Li, P.; Qi, L.-W. Pharmacokinetic Comparison between Quercetin and Quercetin 3-O-β-Glucuronide in Rats by UHPLC-MS/MS. *Sci. Rep.* **2016**, *6*, 35460. [CrossRef]
37. Perez-Vizcaino, F.; Duarte, J.; Santos-Buelga, C. The Flavonoid Paradox: Conjugation and Deconjugation as Key Steps for the Biological Activity of Flavonoids: The Flavonoid Paradox. *J. Sci. Food Agric.* **2012**, *92*, 1822–1825. [CrossRef]
38. Menendez, C.; Dueñas, M.; Galindo, P.; González-Manzano, S.; Jimenez, R.; Moreno, L.; Zarzuelo, M.J.; Rodríguez-Gómez, I.; Duarte, J.; Santos-Buelga, C.; et al. Vascular Deconjugation of Quercetin Glucuronide: The Flavonoid Paradox Revealed? *Mol. Nutr. Food Res.* **2011**, *55*, 1780–1790. [CrossRef]
39. Perez, A.; Gonzalez-Manzano, S.; Jimenez, R.; Perez-Abud, R.; Haro, J.M.; Osuna, A.; Santos-Buelga, C.; Duarte, J.; Perez-Vizcaino, F. The Flavonoid Quercetin Induces Acute Vasodilator Effects in Healthy Volunteers: Correlation with Beta-Glucuronidase Activity. *Pharmacol. Res.* **2014**, *89*, 11–18. [CrossRef]
40. Galindo, P.; Rodriguez-Gómez, I.; González-Manzano, S.; Dueñas, M.; Jiménez, R.; Menéndez, C.; Vargas, F.; Tamargo, J.; Santos-Buelga, C.; Pérez-Vizcaíno, F.; et al. Glucuronidated Quercetin Lowers Blood Pressure in Spontaneously Hypertensive Rats via Deconjugation. *PLoS ONE* **2012**, *7*, e32673. [CrossRef]
41. Duarte, J.; Pérez-Palencia, R.; Vargas, F.; Angeles Ocete, M.; Pérez-Vizcaino, F.; Zarzuelo, A.; Tamargo, J. Antihypertensive Effects of the Flavonoid Quercetin in Spontaneously Hypertensive Rats: Quercetin and Hypertension. *Br. J. Pharmacol.* **2001**, *133*, 117–124. [CrossRef] [PubMed]
42. Perez-Vizcaino, F.; Duarte, J.; Andriantsitohaina, R. Endothelial Function and Cardiovascular Disease: Effects of Quercetin and Wine Polyphenols. *Free Radic. Res.* **2006**, *40*, 1054–1065. [CrossRef] [PubMed]
43. Fraile, M.; Buratto, R.; Gómez, B.; Martín, Á.; Cocero, M.J. Enhanced Delivery of Quercetin by Encapsulation in Poloxamers by Supercritical Antisolvent Process. *Ind. Eng. Chem. Res.* **2014**, *53*, 4318–4327. [CrossRef]
44. Mullen, W.; Edwards, C.A.; Crozier, A. Absorption, Excretion and Metabolite Profiling of Methyl-, Glucuronyl-, Glucosyl- and Sulpho-Conjugates of Quercetin in Human Plasma and Urine after Ingestion of Onions. *Br. J. Nutr.* **2006**, *96*, 107. [CrossRef]
45. Gil-Sánchez, I.; Ayuda-Durán, B.; González-Manzano, S.; Santos-Buelga, C.; Cueva, C.; Martín-Cabrejas, M.A.; Sanz-Buenhombre, M.; Guadarrama, A.; Moreno-Arribas, M.V.; Bartolomé, B. Chemical Characterization and in Vitro Colonic Fermentation of Grape Pomace Extracts. *J. Sci. Food Agric.* **2017**, *97*, 3433–3444. [CrossRef]

Article

Carrageenan of Red Algae *Eucheuma gelatinae*: Extraction, Antioxidant Activity, Rheology Characteristics, and Physicochemistry Characterization

Hoang Thai Ha [1], Dang Xuan Cuong [1,2,3,*], Le Huong Thuy [4,*], Pham Thanh Thuan [5], Dang Thi Thanh Tuyen [6], Vu Thi Mo [2,3] and Dinh Huu Dong [1]

1 Department of Food Technology, Ho Chi Minh City University of Food Industry, Ho Chi Minh 700000, Vietnam; haht@hufi.edu.vn (H.T.H.); dongdh@hufi.edu.vn (D.H.D.)
2 Department of Biology, Graduate University of Science and Technology, VAST, Ha Noi 100000, Vietnam; thaonguyenxanh1607@gmail.com
3 Department of Organic Material from Marine Resource, Nha Trang Institute of Technology Research and Application, VAST, Nha Trang 650000, Vietnam
4 Institute of Biotechnology and Food Technology, Industrial University of Ho Chi Minh City, Ho Chi Minh 700000, Vietnam
5 General Surgery Department, Ninh Thuan Provincial General Hospital, Phan Rang 59000, Vietnam; phamthanhthuan2015@gmail.com
6 Department of Food Science, Nha Trang University, Nha Trang 650000, Vietnam; thanhtuyen151809@gmail.com
* Correspondence: xuancuong@nitra.vast.vn (D.X.C.); lehuongthuy@iuh.edu.vn (L.H.T.); Tel.: +84-905-239-482 (D.X.C.); +84-932-082-199 (L.H.T.)

Abstract: Carrageenan is an anionic sulfated polysaccharide that accounts for a high content of red seaweed *Eucheuma gelatinae*. This paper focused on the extraction, optimization, and evaluation of antioxidant activity, rheology characteristics, and physic-chemistry characterization of β-carrageenan from *Eucheuma gelatinae*. The extraction and the optimization of β-carrageenan were by the maceration-stirred method and the experimental model of Box-Behken. Antioxidant activity was evaluated to be the total antioxidant activity and reducing power activity. The rheology characteristics of carrageenan were measured to be gel strength and viscosity. Physic-chemistry characterization was determined, including the molecular weight, sugar composition, function groups, and crystal structure, through GCP, GC-FID, FTIR, and XRD. The results showed that carrageenan possessed antioxidant activity, had intrinsic viscosity and gel strength, corresponding to 263.02 cps and 487.5 g/cm², respectively. Antioxidant carrageenan is composed of rhamnose, mannose, glucose, fucose, and xylose, with two molecular weight fractions of 2.635×10^6 and 2.58×10^6 g/mol, respectively. Antioxidant carrageenan did not exist in the crystal. The optimization condition of antioxidant carrageenan extraction was done at 82.35 °C for 115.35 min with a solvent-to-algae ratio of 36.42 (v/w). At the optimization condition, the extraction efficiency of carrageenan was predicted to be 87.56 ± 5.61 (%), the total antioxidant activity and reducing power activity were predicted to 71.95 ± 5.32 (mg ascorbic acid equivalent/g DW) and 89.84 ± 5.84 (mg FeSO₄ equivalent/g DW), respectively. Purity carrageenan content got the highest value at 42.68 ± 2.37 (%, DW). Antioxidant carrageenan from *Eucheuma gelatinae* is of potential use in food and pharmaceuticals.

Keywords: β-carrageenan; antioxidant activity; Box-Behken; extraction; *Eucheuma gelatinae*; physic-chemistry; rheology

1. Introduction

Eucheuma gelatinae belongs to the *Solieriaceae* family, specifically the Rhodophyta division, and is a commonly popular marine plant acting as the material for processing β-carrageenan that is widely used in food, functional foods, and pharmaceuticals. The

Eucheuma gelatinae species is a small individual size and lives in dead coral areas. Currently, the demand for carrageenan increases more in the commercial rhodophytes and plays a vital role in the world [1–3]. The *Eucheuma* species was the most farmed red algae, with 10.2 million tonnes in 2015, cultured in Korea, the Philippines, Malaysia, and China as the main algae species [4]. There were about 30,000 tons of cottonii (*Eucheuma alvarezii* Doty), 6000 tons of spinosum (*Eucheuma denticulatum* [Burman] Collins & Hervey), and 100 tons of gelatinae (*Eucheuma gelatinae* [Esper] J. Agardh) farmed for producing κ-carrageenan, ι-carrageenan, and a mixture of γ-, β- and κ-carrageenans [5], respectively.

Carrageenan is a galactan polysaccharidesand exists in the intercellular matrix of red algae. Carrageenan possesses numerous various bioactivities, for instance, anticoagulant [6], antiviral [7], antithrombotic [8], antibacterial [9], cholesterol-lowering [10], antitumor [11], immunomodulatory [12], and antihyperlipidemic [13]. It is also used in the treatment of stomach ulcers [14], and as an antioxidant. The bioactivity of carrageenan was clearly both in vitro and in vivo, and led to the potential promising in developing therapeutic agents. Among those biological activities, the antioxidant activity of carrageenan is most remarkable because antioxidants will eliminate free radicals and contribute to improving resistance and minimizing diseases in the human body. Therefore, carrageenan has potential in functional foods and pharmaceuticals, and is commonly applied in the food and pharmaceutical industries, and is used primarily for drug delivery (tablets, suppositories, fast-dissolving insert, beads, pellets, films, oral suspensions, micro/nanoparticles, floating model, intranasal system, wafers, hydrogel, and tissue engineering (bone or cartilage, and 3-D bioprinting applications) [15] in the latter. Most publications are mainly on *kappa* carrageenan from *Eucheuma denticulatum* and *Kappaphycus alverazii*.

Nowadays, there are numerous different methods for carrageenan extraction, for example, enzyme-assisted extraction [16], maceration [17], stirring soak [18], pressurized-assisted maceration [19], ultrasound-assisted [20], microwave-assisted [21] and extraction optimization [22]. The results on carrageenan from *Eucheuma gelatinae* were less and did not present the content, antioxidant activity, rheological and physicochemical properties of carrageenan from *Eucheuma gelatinae*, especially species grown in Vietnam. Carrageenan application development in functional foods and pharmaceuticals, the control of extraction conditions, the extraction optimization of multi-objective functions including refined carrageenan content, the antioxidant activity, rheology, and the physicochemical properties of carrageenan all demonstrate its essential role.

Therefore, the study focuses on the extraction and optimization of antioxidant carrageenan extracting from *Eucheuma gelatinae* grown in Vietnam, and the evaluation of its antioxidant activity, rheology characteristics, and physiochemistry properties.

2. Results

2.1. Extraction of Antioxidant Carrageenan

2.1.1. Purity Carrageenan Content

The purity of carrageenan content varied from 23.41 ± 1.27 to 41.94 ± 3.05 (%, DW) in the range of the extraction condition as described in Section 4.3 (Table 1). Solvent (pH 7) caused the highest purity carrageenan content, compared to other pH solvents at the same condition. The significant difference in the purifying carrageenan content ($p < 0.05$) occurred as a pH solvent over 8. Pure carrageenan content was the highest at 100 °C, compared to others. However, using an extracting temperature from 80 to 100 °C affected non-significantly the pure carrageenan content ($p > 0.05$), except for the temperature, which lowered to 80 °C. Purity carrageenan content was in the range of 33.77 ± 1.55 to 39.62 ± 1.82 (%, DW) as surveying the extracting temperature. Pure carrageenan content got 32.45 ± 1.62 (%, DW) and 39.52 ± 2.11 (%, DW) at the extracting time of 30 min and 120 min, respectively. The difference in purity carrageenan content did not occur while the extraction time was from 60 to 120 min ($p > 0.05$) (Table 1). The highest purity carrageenan content was 41.02 ± 3.52 (%, DW) for the extracting time of 90 min. The solvent-to-algae ratio was significantly affected purity carrageenan content ($p < 0.05$) as lower than 30/1

(v/w). The purity carrageenan content of 42.68 ± 2.37 was found at the solvent-to-algae ratio of 40/1 (v/w), and this was the highest value compared to other conditions.

Table 1. Effect of the extraction condition on purity carrageenan content and its antioxidant activity.

Std	pH	Extracting Temperature (°C)	Extracting Time (min)	Solvent-to-algae Ratio (v/w)	Purity Carrageenan Content (%, DW)	Total Antioxidant Activity (mg Ascorbic Acid Equivalent/g DW)	Reducing Power Activity (mg FeSO$_4$ Equivalent/g DW)
I	7	80	60	30/1	36.09 ± 1.21 [a]	24.58 ± 1.15 [a]	26.75 ± 1.98 [a]
	8	80	60	30/1	34.57 ± 1.42 [a]	19.73 ± 2.01 [b]	23.29 ± 2.44 [ac]
	9	80	60	30/1	27.28 ± 1.03 [b]	15.62 ± 2.19 [b]	27.84 ± 2.71 [a]
	10	80	60	30/1	23.41 ± 1.27 [c]	10.04 ± 1.68 [c]	20.61 ± 2.52 [bc]
II	7	70	60	30/1	33.77 ± 1.55 [a]	19.32 ± 1.92 [a]	20.57 ± 2.03 [a]
	7	80	60	30/1	36.09 ± 1.47 [ab]	24.58 ± 1.15 [b]	26.75 ± 1.98 [b]
	7	90	60	30/1	38.58 ± 1.69 [b]	27.29 ± 2.75 [bc]	29.52 ± 3.01 [b]
	7	100	60	30/1	39.62 ± 1.82 [bc]	29.31 ± 2.48 [c]	30.59 ± 3.26 [b]
III	7	90	30	30/1	32.45 ± 1.62 [a]	19.05 ± 2.23 [a]	23.17 ± 1.78 [a]
	7	90	60	30/1	38.58 ± 1.69 [b]	27.29 ± 2.75 [b]	29.52 ± 3.01 [b]
	7	90	90	30/1	41.02 ± 3.52 [b]	28.58 ± 3.10 [b]	30.09 ± 2.67 [b]
	7	90	120	30/1	39.52 ± 2.11 [b]	30.76 ± 3.21 [b]	32.68 ± 3.24 [b]
IV	7	90	90	20/1	30.14 ± 2.38 [a]	20.37 ± 1.92 [a]	25.73 ± 2.75 [a]
	7	90	90	30/1	41.02 ± 3.52 [b]	28.58 ± 3.10 [b]	30.09 ± 2.67 [ab]
	7	90	90	40/1	42.68 ± 2.37 [b]	29.72 ± 3.22 [b]	34.67 ± 2.78 [bc]
	7	90	90	50/1	41.94 ± 3.05 [b]	30.04 ± 2.57 [b]	36.25 ± 3.01 [cd]

Note: Std I, II, III, and IV included four lines. Letters a, b, c, and d in each column exhibited a significant difference in the column of each Std with $p < 0.05$, $n = 3$.

2.1.2. Antioxidant Activity

Total Antioxidant Activity

The total antioxidant activity was in the range of 10.04 ± 1.68 to 30.76 ± 3.21 (mg ascorbic acid equivalent/g DW) when the survey of the extraction conditions and the change of total antioxidant activity were significant ($p < 0.05$). The difference in solvent pH led to the difference in total antioxidant activity ($p < 0.05$), except for solvents pH 8 and 9. The total antioxidant activity got the highest value of 24.58 ± 1.15 (mg ascorbic acid equivalent/g DW) at solvent pH 7 when compared to other solvents. The extracting temperature affected total antioxidant activity ($p < 0.05$) when the temperature increased from 70, 80 °C to 100 °C. The non-significant difference in total antioxidant activity occurred when the extracting temperature increased from 80 to 90 °C and 90 to 100 °C. However, the total antioxidant activity was still evaluated highest with the value of 29.31 ± 2.48 (Table 1).

The extracting time of carrageenan from 60 to 120 min did not significantly affect total antioxidant activity ($p < 0.05$); the extracting time of 30 min impacted to significant ($p < 0.05$), compared to other the extracting time. The highest total antioxidant activity was for 120 min compared to other extraction conditions.

Under the impact of the solvent-to-algae ratio, total antioxidant activity varied from 20.37 ± 1.92 to 30.04 ± 2.57 (mg ascorbic acid equivalent/g DW), corresponding to the solvent-to-algae ratio of 20/1 and 50/1 (v/w). However, the difference in total antioxidant activity only occurred ($p < 0.05$) when the solvent-to-algae ratio was lower than 30/1 (v/w) in comparison to other solvent-to-algae ratios.

Reducing Power Activity

Reducing power activity changed from 20.57 ± 2.03 to 36.25 ± 3.01 (mg FeSO$_4$ equivalent/g DW) under the impact of other extraction conditions. Reducing power activity got the highest value of 27.84 ± 2.71 (mg FeSO$_4$ equivalent/g DW) at solvent pH 9, but was not a significant difference from solvent pH 7 and 8. Reducing power activity was the lowest, corresponding to 20.61 ± 2.52 (mg $FeSO_4$ equivalent/g DW) at solvent pH 10. The extracting temperature only led to a significant difference ($p < 0.05$) in reducing power activity when the temperature was below 80 °C, compared to other temperatures. The reducing power activity of 30.59 ± 3.26 (mg $FeSO_4$ equivalent/g DW) was the highest value, compared to 70 to 90 °C. The reducing power activity (32.68 ± 3.24, mg $FeSO_4$ equivalent/g DW) was highest for the extracting time of 120 min compared to other times (Table 1). The reducing power activity was not significantly different, compared at 60 and 90 min. The lowest reducing power activity (23.17 ± 1.78, mg $FeSO_4$ equivalent/g DW) exhibited a significant difference ($p < 0.05$) in comparison to other extracting times.

Under the impact of the solvent-to-algae ratios, the reducing power activity varied from 25.73 ± 2.75 to 36.25 ± 3.01 (mg $FeSO_4$ equivalent/g DW). A significant difference did not occur ($p > 0.05$) between the solvent-to-algae ratio of 20/1 and 30/1 (v/w), 30/1 and 40/1 (v/w), and 40/1 and 50/1 (v/w). The reducing power activity reached the highest and lowest value when the solvent-to-algae ratio was 50/1 and 20/1 (v/w), respectively.

2.1.3. Correlation between Carrageenan Content and Antioxidant Activity

Purity carrageenan content strongly correlated to total antioxidant activity and weakly to reducing power activity, corresponding to 0.97 and 0.41 when the impact survey of solvent pH on purity carrageenan content and antioxidant activity was added. There is a strong correlation between purity carrageenan content and antioxidant activity ($R^2 > 0.9$), especially total antioxidant activity (0.99), and reducing power activity (0.97) as with the extracting temperatures survey. The total antioxidant activity and reducing power activity strongly correlated to the purity carrageenan content, corresponding to 0.94 and 0.91, respectively, as the extracting time survey. A strong correlation was found between purity carrageenan content and antioxidant activity, as the survey of the solvent-to-algae ratio corresponded to 0.99 (total antioxidant activity) and 0.88 (reducing power activity), respectively.

2.2. Optimization of Antioxidant Carrageenan

2.2.1. Analysis of Optimization Model

According to the study type of response surface with the design type of Box-Behnken on the quadratic model and randomized subtype, the results showed the distribution of the target functions were focusing on the centre of the survey interval, compared with the boundary region. The fit optimization model of Y_1 function was the quadratic model ($p = 0.0001 < 0.05$) with non-significant lack-of-fit ($p = 0.08 > 0.05$) and adjusted R^2 (0.95). Response surface Y_1 had a standard deviation (SD) of 5.61 and coefficient of variation (C.V%) of 13.51. Response surface Y_2 was the model quadratic ($p = 0.0002 < 0.05$), compared to the model of 2FI, linear, and cubic. SD and C.V% of Y_2 were 2.53 and 15.8, respectively. The lack-of-fit of model Y_2 ($p = 0.43 > 0.05$) was non-significant, and its adjusted R^2 got 0.94. Sequential p-value and adjusted R^2 of the response surface Y_3 corresponded to the one of response surface Y_1. The lack of fit of model Y_3 had a p-value of 0.1. The quadratic model Y_3 got a C.V% of 13.74 and SD of 2.53.

The response surface Y_1 was in the range of 15.82 to 88.24 (%) and got the average value of 41.52 ± 5.61 (%). The response surface Y_2 changed in the range of values (5.11 to 36.20 mg ascorbic acid equivalent/g DW), and its average value got 16.03 ± 2.53 (mg ascorbic acid equivalent/g DW). The value range of response surface Y_3 varied from 5.94 to 39.53 (mg $FeSO_4$ equivalent/g DW), with the average value of 18.43 ± 2.53 (mg $FeSO_4$ equivalent/g DW) (Table 2).

Table 2. The experiment results of the optimization design according to the Box-Behnken model.

Std	X_1	X_2	X_3	Y_1	Y_2	Y_3
1	70	30	35	23.56	9.02	10.48
2	100	30	35	27.74	10.60	12.30
3	70	120	35	30.40	11.61	13.48
4	100	120	35	38.24	14.59	16.94
5	70	75	20	22.18	8.47	9.83
6	100	75	20	24.49	9.35	10.87
7	70	75	50	13.40	5.11	5.94
8	100	75	50	49.79	19.01	22.11
9	85	30	20	15.82	6.04	7.00
10	85	120	20	41.01	15.66	18.19
11	85	30	50	31.42	12.00	13.92
12	85	120	50	45.28	17.30	20.07
13	85	75	35	88.24	33.71	37.43
14	85	75	35	86.97	31.72	39.53
15	85	75	35	84.30	36.20	38.32

Note: Values expressed as mean value, $n = 3$.

After ANOVA analysis for three response models, the results showed the solvent-to-material ratio had a non-significant effect on all response surfaces because their p-value was 0.9527, 0.9606, and 0.9521, corresponding to surface Y_1, Y_2, and Y_3, respectively. Interaction factors of x_1x_2 and x_2x_3 possessed a p-value higher than 0.05, showing that these interaction factors did not affect the response surfaces (Table 3). The results also showed the coding variable equation of response surface Y_1, Y_2, and Y_3, as follows:

$$Y_1 = 74.17 + 37.02x_1 - 44.98x_2 - 0.1796x_3 + 2.03x_1x_2 + 11.36x_1x_3 - 4.72x_2x_3 - 55.50x_1^2 - 70.28x_2^2 - 27.82x_3^2 \quad (1)$$

$$Y_2 = 28.93 + 14.51x_1 - 18.11x_2 - 0.0675x_3 + 0.7778x_1x_2 + 4.34x_1x_3 - 1.80x_2x_3 - 21.94x_1^2 - 28.00x_2^2 - 11.05x_3^2 \quad (2)$$

$$Y_3 = 32.94 + 16.44x_1 - 20.04x_2 - 0.0821x_3 + 0.9111x_1x_2 + 5.04x_1x_3 - 2.10x_2x_3 - 24.65x_1^2 - 31.28x_2^2 - 12.37x_3^2 \quad (3)$$

Table 3. The basic parameters of the response surface equation.

Source	Response Surface Y_1			Response Surface Y_2			Response Surface Y_3		
	p-Value	CE	SE	p-Value	CE	SE	p-Value	CE	SE
Model	0.0007	74.17	3.02	0.0014	28.93	1.36	0.0008	32.94	1.36
$x_1 - X_1$	0.0004	37.02	4.47	0.0008	14.51	2.02	0.0004	16.44	2.02
$x_2 - X_2$	0.0018	−44.98	7.45	0.0030	−18.11	3.36	0.0019	−20.04	3.36
$x_3 - X_3$	0.9527	−0.1796	2.88	0.9606	−0.0675	1.30	0.9521	−0.0821	1.30
x_1x_2	0.7575	2.03	6.23	0.7932	0.7778	2.81	0.7591	0.9111	2.81
x_1x_3	0.0288	11.36	3.74	0.0499	4.34	1.69	0.0305	5.04	1.69
x_2x_3	0.3590	−4.72	4.68	0.4325	−1.80	2.11	0.3652	−2.10	2.11
x_1^2	0.0001	−55.50	5.19	0.0002	−21.94	2.34	0.0001	−24.65	2.34
x_2^2	0.0003	−70.28	8.11	0.0006	−28.00	3.66	0.0004	−31.28	3.66
x_3^2	0.0002	−27.82	2.92	0.0004	−11.05	1.32	0.0002	−12.37	1.32

SE: Standard error; CE: Coefficient estimate.

The importance of response surfaces was equal, and the extraction optimization of antioxidant carrageenan via the software Design-Expert version 13 showed the optimization point of 82.35 °C, 115.35 min, and 36.42 (v/w) with the overlay figure of response surfaces (Figure 1d). Antioxidant carrageenan was the white color and yarn type in the optimization point (Figure 1e,f). At the optimization condition, response surfaces Y_1, Y_2, and Y_3 were predicted to get the average value of 87.56 ± 5.61 (%), 71.95 ± 5.32 (mg ascorbic acid equivalent/g DW), and 89.84 ± 5.84 (mg FeSO$_4$ equivalent/g DW), respectively. Response surfaces were the spherical surface (Figure 1a–c).

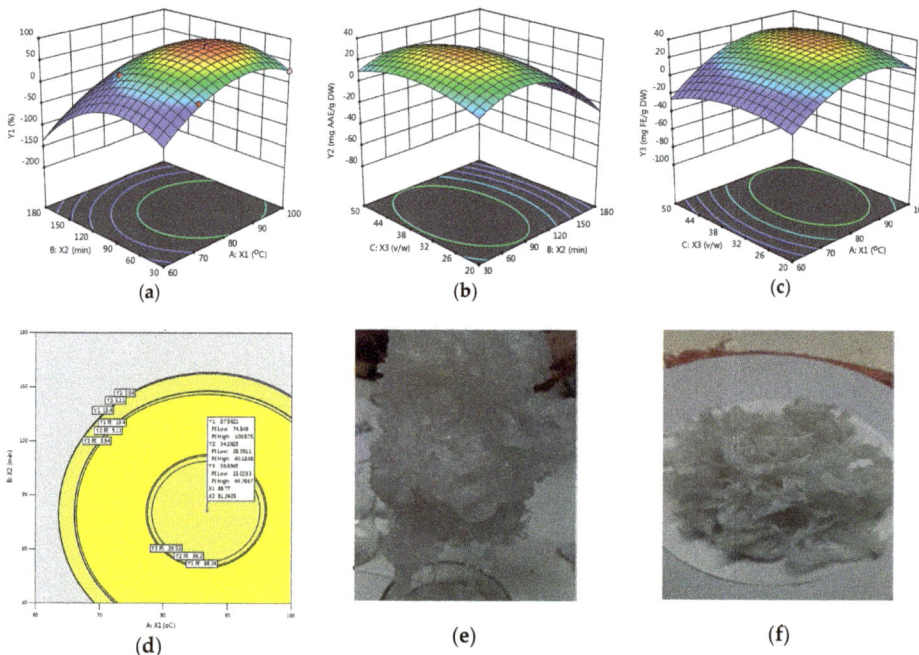

Figure 1. Response surface, overlay surface and antioxidant carrageenan: (**a**) Response surface of Y_1; (**b**) Response surface of Y_2; (**c**) Response surface of Y_3; (**d**) Overlay surface of Y_1, Y_2, and Y_3; (**e**) Carrageenan after precipitation using 96% ethanol; (**f**) Carrageenan after drying.

2.2.2. Test of Optimization Model by the Experiment

Following the experiment on the optimization condition and the correlation analysis between the actual target functions and the predicted target function, the results showed the strong correlation between the experiment and the prediction (Figure 2). The experiment value of the target functions corresponded to 86.52 (Y_1) (Figure 2a), 87.69 (Y_2) (Figure 2b), and 85.73% (Y_3) (Figure 2c) when compared to the predicted target functions by the software Design Expert version 13.

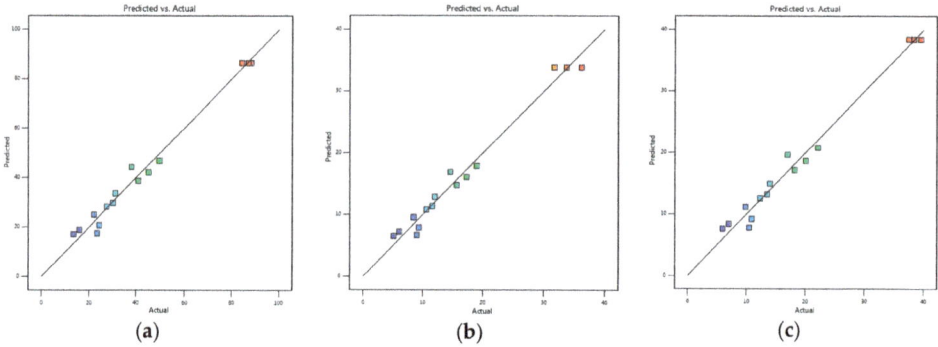

Figure 2. The correlation between predicted target function and actual target function: (**a**) Y_1, (**b**) Y_2, (**c**) Y_3, respectively.

2.3. Characteristics of Rheology and Physical-Chemistry of Antioxidant Carrageenan

2.3.1. Rheological Characteristic of Antioxidant Carrageenan

Intrinsic viscosity and gel strength of antioxidant carrageenan had a value of 263.02 cps and 487.5 g/cm^2, respectively.

2.3.2. Physical-Chemistry Characteristics of Antioxidant Carrageenan

Different sugars were found in antioxidant carrageenan, such as rhamnose, mannose, glucose, fucose, and xylose. These sugars got the value of 59.16, 52.63, 78.20, 20.24, and 96.98, respectively. Galactose was not detected in antioxidant carrageenan (Figure 3).

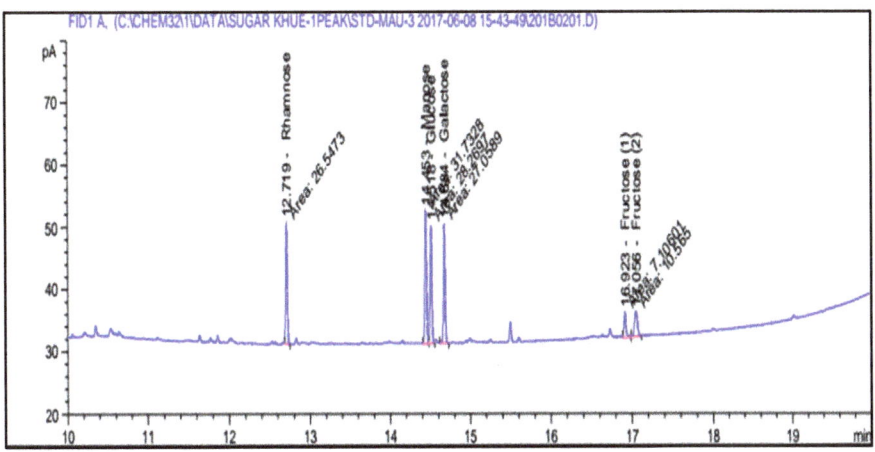

Figure 3. Sugar composition of antioxidant carrageenan.

The FTIR method utilizes the material's light absorption by manipulating how different molecular compounds respond to infrared light to determine the analyzed material's structure. This method is also known as absorption spectroscopy. It is applied in various ways, including light beams of a limited frequency group or using monochromatic light. This technique exploits that the fact that each frequency responds differently to the material and works by using more than one different frequency in the beam. In this way, the composition of unknown material is precisely determined. FTIR spectroscopy offers the advantage of measuring a small sample (a few milligrams or milliliters) in the least amount of time. FTIR analysis in the spectrum range of 580–3420 cm^{-1} showed different peaks, such as 3416.64, 2926.42, 1722.49, 1643.63, 1417.96, 1376.90, 1264.34, 1161.71, 1071.64, 845.86, and 582.00 cm^{-1} occurring in the FTIR of antioxidant carrageenan extracting from *Eucheuma gelatinae* grown in Vietnam. The things showed the functional groups, for example, -OH, C-H, C=O, NH$_2$ deformation, C-O-H stretch or C-O/C-H bending, C-O or CH$_3$ deformation, alkyl ketone or C-O-C stretch, alkylamine, sulphation of C4 of the /3–1,3-linked residue, and S-O-S bending (Figure S1). The peak of 842.89 cm^{-1} exhibited the C-O-SO$_4$ group on C$_4$ of galactose (Figure 4). The peak of 927.76 was the characteristic for *k*-carrageenan without *μ*-carrageenan and presented 3,6-anhydro-D-galactose group (Figure 4). The stretching vibration of the entire anhydro-glucose ring of antioxidant carrageenan was presented at the peak of 574.79 cm^{-1}, 769.60 cm^{-1}, 891.11 cm^{-1}, and 927.76 cm^{-1} (Figure 4), respectively. The peak of 891.11 cm^{-1} and 842.89 cm^{-1} were the properties of *β*-carrageenan and *j*-carrageenan, respectively (Figure 4).

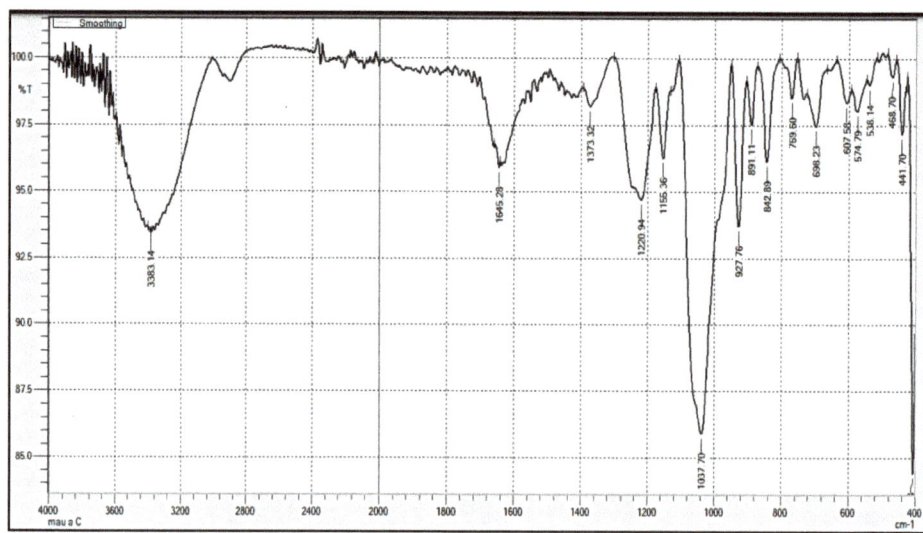

Figure 4. FTIR spectroscopy of antioxidant carrageenan.

GPC is a new solution for the effortless analysis of polymeric compounds in nature and facilitates molecular mass analysis; carrageenan is dissolved in an alkaline medium. Figure 5 shows that antioxidant carrageenan, which was extracted from *Eucheuma gelatinae*, possessed two fractions, with the average molecular weight of 2.635×10^6 and 2.58×10^6 g/mol, respectively.

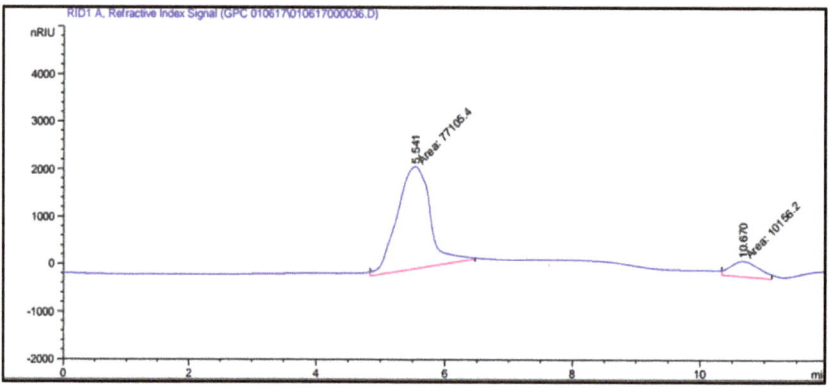

Figure 5. GPC spectroscopy of antioxidant carrageenan.

X-ray diffraction (XRD) is the sole laboratory technique that equips structural information such as chemical composition, crystal structure, crystal size, strain, and layer thickness. As a result, materials researchers use XRD to examine a wide range of materials, from powder X-ray diffraction (XRPD) to solids, thin films, and nanomaterials. In the current study, X-ray diffraction was used to analyze the crystal structure of carrageenan. Antioxidant carrageenan had a high purity degree and did not form a crystal structure, exhibited in Figure 6.

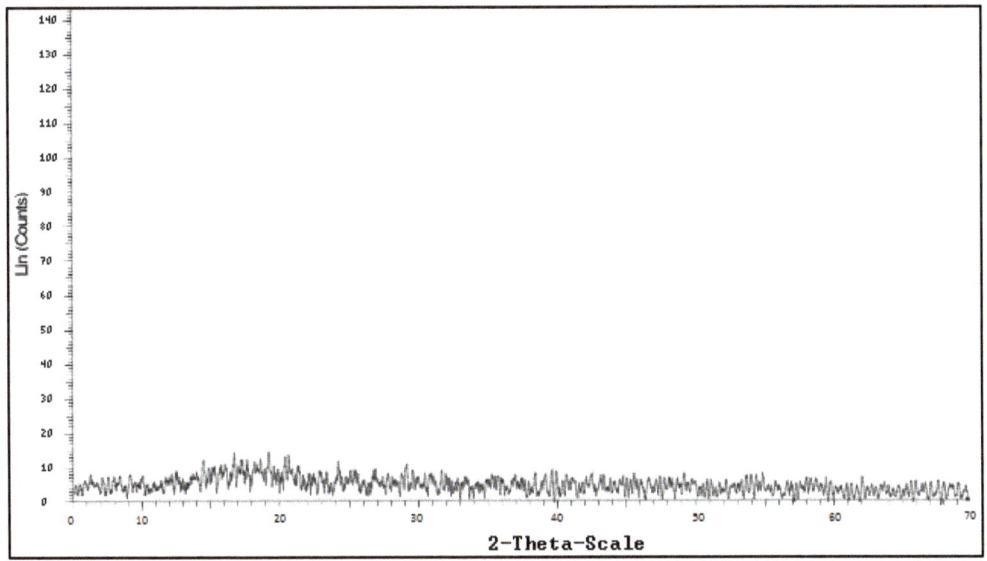

Figure 6. GPC spectroscopy of antioxidant carrageenan.

3. Discussion

Purity carrageenan content was collected using solvent pH 7 at 1.54 and 1.32 times, compared to solvent pH 10 and 9, respectively. The highest drop in purity carrageenan content occurred when solvent pH increased from 8 to 9, corresponding to 21.09%. In the range of the extracting temperature from 80–100 °C, the temperature increased by 10 °C, and purity carrageenan content increased by 6.8%. At the extracting temperature of 70 °C, the purity carrageenan content was 0.94, 0.88, and 0.85 times, compared to 80, 90, and 100 °C, respectively. The impact of solvent pH for purity carrageenan content was the descend linear model when solvent pH increased, however, one of the temperatures was inversed. The change of purity carrageenan content was the quadratic model trend with the maximum peak at 90 min. When the extracting time increased from 30 to 60 min, purity carrageenan content also increased by 18.89%. Increasing of the time from 60 to 90 min, the following increase of purity carrageenan content was only 6.32%. After the extraction at 90 min, purity carrageenan content decreased. The increase ratio of solvent-to-ratio from 20 to 40 (v/w) led to the increasing purity carrageenan content of 1.42 times. Purity carrageenan content tended parallel to the horizontal axis when the solvent-to-algae ratio increased from 30/1 to 50/1 (v/w). The difference in algae species and extraction methods caused different carrageenan-extracted content, for example, using ohmic heating for carrageenan extraction from *Eucheuma spinosum*, needing the temperature (95 °C), the time (240 min), and the solvent-to-algae ratio of 45/1 (v/w) [17]. The study of Andi et al. (2021) [17] exhibited the extraction condition of carrageenan from *Eucheuma spinosum* higher than in the current study.

The total antioxidant activity and reducing power activity changed according to the increasing solvent pH linear model trend, similar to purity carrageenan content under the impact of solvent pH. The increase of solvent pH was from 9 to 10, and total antioxidant activity and reducing power activity decreased by 55.58 and 35.08%, respectively. The antioxidant change of total antioxidant activity and reducing power activity was similar to purity carrageenan content change under the impact of the extracting temperature. It showed that antioxidant activity was proportional to the purity carrageenan content. The rate of increase in antioxidant activity decreased when the temperature increased. When the extracting temperature increased from 70 °C to 80, 90, and 100 °C, the ratio of total antioxidant activity increased by 27.23, 11.03, and 7.4%, the increasing ratio of reducing power activity corresponded to 30.04, 10.36, and 3.63%. The change of antioxidant activity composed of total antioxidant activity and reducing power activity was proportional to the purity carrageenan content change according to the linear model as the increase of extraction time. When the extracting time was 60, 90, and 120 min, total antioxidant activity increased by 43.26, 50.03, and 61.47%, and the increase of reducing power activity was 27.41, 29.86, and 41.05%, compared to the extracting time for 30 min. The impact of the solvent-to-algae ratio on purity carrageenan content and the antioxidant was similar to that of the extracting time. However, the effect of the raw material solvent ratio on the refined carrageenan content was only evident when changing the raw material solvent ratio from 20 to 30/1 (v/w). The solvent-to-algae ratio impacted antioxidant activity more clearly than the purity carrageenan content. It meant that carrageenan from *Eucheuma gelatinae* possessed antioxidant activity, and short-chain carrageenan had higher antioxidant activity than long-chain carrageenan. Previous studies had only shown the antioxidant capacity of κ-carrageenan. For carrageenan from *Eucheuma gelatinae*, its antioxidant activity was only shown based on the DPPH and ABTS method, and a correlation between carrageenan and antioxidant activity [23] was not found. In the current study, the antioxidant activity of beta carrageenan is evaluated based on total antioxidant activity and reducing power activity, also described as the correlation between them. The impact of different extraction conditions on antioxidant carrageenan from *Eucheuma gelatinae* was not presented in previous studies, but found in the current study.

Previous studies on optimization of carrageenan extraction mainly focused on the objective functions of extraction yield, gel strength, and viscosity of carrageenan on *Kappaphycus alverazii* and *Eucheuma spinosum* with alkaline solvent. The input factors were mainly studied to be temperature and time [24,25]. There was only one Chinese notice of *Eucheuma gelatinae* pretreatment optimization for carrageenan extraction. However, this publication did not address the antioxidant activity of carrageenan. The optimization of the target functions, such as extraction efficiency and antioxidant activity of carrageenan from *Eucheuma gelatinae*, the survey of solvent pH, temperature, time and solvent-to-material ratio in the determination of optimization domain, and the optimization of three input factors (solvent-to-material ratio, temperature, and time of extraction) in the optimization were presented in the current study. A Box-Behnken design model with a sphere response model had also not been found in studies of carrageenan extraction.

Antioxidant carrageenan from *Eucheuma gelatinae* in Vietnam possessed the characteristics of gel strength and viscosity higher than carrageenan from *Eucheuma sp.* in previous studies [24,26]. Sugar compositions of antioxidant carrageenan were noticed in the current study and were one of the few publications on sugar compositions of carrageenan and the only one on the sugar content of the antioxidant carrageenan from *Eucheuma gelatinae*. The average molecular weight of antioxidant carrageenan in the current study was higher than the results obtained by Nishinaria and Watase (1992) [27] and Ahmed et al. (2018) [28].

The results showed the existence of β-carrageenan in peak 891 cm^{-1} in the spectrum range of 580–3420 cm^{-1}, [29], and the appearance of peak 820 cm^{-1} related to γ-carrageenan. The results were interesting compared to the previous studies. Algae species differently led in structure and functional groups composition in carrageenan [30], for example, ι-carrageenan in *Eucheuma serra* [31], and κ-carrageenan in *Eucheuma cottonii* [24]. The peak of 842 cm^{-1} and 925 cm^{-1} are C-O-SO_4 group presence on C4 of galactose and 3,6-anhydro-D-galactose [18]. The entire anhydro-glucose ring achieved the stretching vibration at 573 cm^{-1}, 760 cm^{-1}, 858 cm^{-1}, and 928 cm^{-1} [32], respectively. The spectra of 848 cm^{-1} and 891 cm^{-1} appeared as C-4 sulfate of the ι-carrageenan and β-carrageenan, respectively [33].

4. Materials and Methods

4.1. Source of the Plant Material

Red algae *Eucheuma gelatinae* was collected from Ninh Hai district, Ninh Thuan province, Vietnam in April 2017.

4.2. Sample Preparation

Eucheuma gelatinae was selected, washed and dried until the humidity was lower than 8%. Next, the seaweed was crushed to a size of 2–3 cm for further study.

4.3. Extraction of Carrageenan

Carrageenan was extracted by agitating maceration with classical experimental design, fixing an independent variable and running the remaining variables.

For surveying solvent pH from 7 to 10 with a jump (δ) 1, temperature, time, and the ratio of solvent-to-algae of extraction were fixed, corresponding to 80 °C, 60 min, and 30/1 (v/w), respectively.

For surveying the extracting temperature from 70 to 100 °C with δ 10 °C, time and the ratio of solvent-to-algae of extraction were similar to the study on solvent pH and collection of solvent pH from the above results.

For surveying the extracting time from 30 to 120 min with δ 30 min, the ratio of solvent-to-algae of extraction was 30/1 (v/w). pH solvent and the extracting temperatures were from the above results.

For surveying the solvent-to-algae ratio from 20 to 50 (v/w) with δ 10 (v/w), three independent variables (solvent pH, temperature, and time of extraction) were from the above results.

All extracts were filtered through Whatman No 1. and precipitated using 96% ethanol. The residues were dried at 40 °C for carrageenan collection. Carrageenan was analysed for the purity content and antioxidant activity.

4.4. Optimization of Carrageenan Extraction

The optimization of carrageenan extraction was based on the experiment design model of Box-Behnken with four input factors, including temperature (X_1, °C), time (X_2, minutes), the solvent-to-material ratio (X_3, v/w), and three target functions such as, for example, carrageenan purity degree (Y_1, %), total antioxidant (Y_2, mg ascorbic acid equivalent/g DW), and reducing power (Y_3, mg FeSO$_4$/g DW). The optimal experimental domain of four input factors and the independent-to-coding variable conversion are presented in Table 4. The experimental design included 12-factor experiments and three replicate experiments at the center of the plan (Table 5). The target functions were selected and based on the combined results of independent variables in Table 5. The experiments were randomly carried out to minimize the effects of unusual changes in the observations. The variables are coded according to the following equation:

$$x = \frac{(X_i - X_0)}{\Delta X} \quad (4)$$

wherein x is the code variable, X_i is the real variable, X_0 is the central experiment variable, and ΔX is the difference between the maximum value of the real variable and the value of X_0. The mathematical equation corresponding to the Box-Behnken experimental model is as follows:

$$Y = \beta_0 + \sum_{i=1}^{3} \beta_i X_i + \sum_{i=1}^{3} \beta_{ii} X_i^2 + \sum_{i=1}^{2} \sum_{j=1+1}^{3} \beta_{ij} X_i X_j + \varepsilon \quad (5)$$

Requirement of objective functions:

Y_1: extraction efficiency of carrageenan (%): max
Y_2: total antioxidant activity (mg ascorbic acid equivalent/g DW): max
Y_3: reducing power activity (mg FeSO$_4$ equivalent/g DW): max

Table 4. The conversion between code variables and reality variables.

Input Factor (Independent Variable)	Code Variable		
	−1	0	1
Extracting temperature (X_1, °C)	70	85	100
Extracting time (X_2, minutes)	30	75	120
Solvent-to-material ratio (X_3, v/w)	20	35	50

Carrageenan was collected in the condition of optimization extraction and analyzed for rheological (gel strength and intrinsic viscosity) and physical-chemistry characteristics (molecular weight, functional groups, crystal structure, and sugar composition).

Table 5. Experiment design and the results.

Std	Actual Variable			Coded Variable			Target Function		
	X_1	X_2	X_3	x_1	x_2	x_3	Y_1	Y_2	Y_3
1	70	30	35	−1	−1	0			
2	100	30	35	+1	−1	0			
3	70	120	35	−1	+1	0			
4	100	120	35	+1	+1	0			
5	70	75	20	−1	0	−1			
6	100	75	20	+1	0	−1			
7	70	75	50	−1	0	+1			
8	100	75	50	+1	0	+1		Y_{ij}	
9	85	30	20	0	−1	−1			
10	85	120	20	0	+1	−1			
11	85	30	50	0	−1	+1			
12	85	120	50	0	+1	+1			
13	85	75	35	0	0	0			
14	85	75	35	0	0	0			
15	85	75	35	0	0	0			

4.5. Determination of the Content and the Extraction Efficiency of Carrageenan

4.5.1. Purity Carrageenan Content

Purity carrageenan content is calculated based on the following equation:

$$\text{Purity carrageenan content } (C_P, \%) = \frac{C_R - H_C - A_C}{W_A} (\%) \qquad (6)$$

where in C_R (%) is the dry weight of carrageenan after extraction as described in Section 4.3. The humidity content of C_R is H_c, which is determined according to the drying method at 105 °C until constant weight. W_A is the dried algae powder. The ash content of C_R is A_c, which is calculated and based on the white ash weight of the material after being calcined at 650 °C.

4.5.2. Extraction Efficiency of Carrageenan

The extraction efficiency of carrageenan was determined according to the following equation:

$$EEC\ (\%) = \frac{C_p}{\text{Carrageenan in initial algae}} (\%) \qquad (7)$$

$$\text{Carrageenan in initial algae} = W_A - H_A - A_A - \text{Protein}_A - \text{Lipid}_A - \text{Cellulose}_A \qquad (8)$$

where in:

W_A: the dried algae powder (g);

H_A: humidity content of algae powder (g); dried at 105 °C.

A_A: ash content of algae powder (g); calcined at 650 °C.

Protein$_A$: protein content of algae powder (g); determined based on the method of Lowry.

Lipid$_A$: lipid content of algae powder (g); determined based on the soxhlet method.

Cellulose$_A$: cellulose content of algae powder (g);

4.6. Detemination of Antioxidant Activity

4.6.1. Total Antioxidant Activity

100 µL extract, in turn, was added to 900 µL of distilled water and solution A (0.6 M H_2SO_4, 28 mM sodium phosphate and 04 mM ammonium molybdate). The mixture was vortexed and kept for 90 min at 95 °C and then measured at the wavelength of 695 nm with an ascorbic acid standard [34].

4.6.2. Reducing Power Activity

The reducing power activity was determined according to the method of Zhu et al. (2002) [35]. Firstly, 0.5 mL phosphate buffer at pH 7.2 was added to 500 µL extract. Secondly, 0.2 mL of 1% $K_3[Fe(CN)_6]$ was added to the compound. The compound was kept at 50 °C for 20 min. Thirdly, 500 µL of 10% CCl_3COOH with 300 µL distilled water and 80 µL of 0.1% $FeCl_3$ were added. Finally, the compound was measured at 655 nm with the standard substance $FeSO_4$.

4.7. Determination of Rheological Characteristics

4.7.1. Gell Strength

Carrageenan gel strength was measured with a Brookfield rheometer and the maintenance of samples at 20 °C. Samples were prepared by dissolving 1.7 g of carrageenan powder in 98.3 mL of distilled water at 80 °C to form a 1.5% carrageenan solution. KCl was then added until reaching 0.1% KCl solution and soaking to 20 °C for 2.5 h. Carrageenan gel was continuously cut into slices with a thickness of 1.5 cm and put into the rheometer for measurement.

4.7.2. Intrinsic Viscosity

The viscosity of 1.5% carrageenan solution at 80 °C was measured with a Brookfield rheometer.

4.8. Determination of Physic-Chemistry Characterization

4.8.1. Sugar Compositions

The sugar composition determination of antioxidant carrageenan was according to the GC-FID method on Agilent's 6890 N gas chromatograph (USA) that was composed of an automatic sample injector, an injection chamber, a column furnace, a flame ionization detector (FID), and an HP5 MS column (30 m × 0.25 m × 0.25 m). The chamber temperature was set at 280 °C with the line split ratio of 0/1. The program temperature column was set at 100 °C with a 20 °C/min rate for getting 325 °C and kept for 10 min, and then the probe temperature at 300 °C and carried gas speed at 01 mL/min. Derivation process: The sample was hydrolyzed in 1.5 M HCl, then poured to 10 mL of the cylinder by acetic anhydride, and the compound was finally injected into the GC system.

4.8.2. Molecular Weight

The average molecular weight of carrageenan was measured with gel permeation chromatography, using a model YL9100 GPC.

4.8.3. Functional Groups

The samples and anhydrous KBr were mixed according to the ratio of KBr-to-sample of 98:2 (w/w). The mixture was then measured on a Bruker FT-IR spectrometer ALPHA with a wavelength range from 4000 to 500 cm^{-1}. Finally, the results were analysed on OPUS 7.0 software (Bruker, Ettlingen, Germany).

4.8.4. Crystal Structure

The crystal structure of carrageenan was measured on a Brucker-Germany instrument, model D8 Advance, which met the ISO 9001:2000 international standards, in addition to standards for radiation safety and European CE standards for electrical safety. 0.5 g of carrageenan was finely ground in an agate mortar and pestle and placed in a special tray of a polycrystalline powder diffraction apparatus. Samples in the tray were flat and spread evenly over the tray surface. The sample holder mounted on an incident beam (narrow, monochromatic, parallel X-ray beam projected onto the sample). X-ray was rotated at an angle theta (θ) for the incident ray, and obtained the dispersion of X-ray by the detector–right. The sodium iodide (NaI) flicker detector would rotate 2θ from 5–70°. One sample rotation was 0.03° and the single point diffraction time was one second. The measuring temperature of samples was at 25 °C.

4.9. Statistical Analysis

All experiments were undertaken in triplicate and exhibited under mean ± SD with a significant level ($p < 0.05$). Analysis of statistics, ANOVA, and regression was calculated using the software MS. Excel 2013 and Design Expert 13. Duncan method was used for the movement of the non-normal value.

5. Conclusions

In summary, surveying and condition optimization of carrageenan extraction from *Eucheuma gelatinae* with the target functions such as purity carrageenan content, total antioxidant activity, and reducing power activity were performed in this study. Antioxidant carrageenan extracted at the optimal condition was evaluated, including the rheology and physicochemical properties. Antioxidant carrageenan contained rhamnose, mannose, glucose, fucose, and xylose. The molecular weight of carrageenan reached an average value of $2.635e^6$ and $2.58e^6$ g/mol. The solvent-to-algae ratio had the least effect on the objective functions. Antioxidant carrageenan from *Eucheuma gelatinae* had intrinsic viscosity (263.02 cps) and gel strength (487.5 g/cm^2), and did not exist in the crystal. At the optimization condition (82.35 °C for 115.35 min with the solvent-to-algae ratio of 36.42 (v/w)), target functions were predicted such as carrageenan yield of extraction (87.56 ± 5.61, %), total antioxidant activity (71.95 ± 5.32, mg ascorbic acid equivalent/g DW), and reducing power activity (89.84 ± 5.84, mg FeSO$_4$ equivalent/g DW)). The highest value of purity carrageenan content is 42.68 ± 2.37 (%, DW), and it has potential in the food and pharmaceutical industries.

Supplementary Materials: The following supporting information can be downloaded online. Figure S1. FTIR spectroscopy of antioxidant carrageenan.

Author Contributions: Conceptualization, H.T.H., D.X.C. and L.H.T.; methodology, D.X.C. and P.T.T.; software, D.H.D.; validation, V.T.M. and D.T.T.T.; formal analysis, D.T.T.T. and L.H.T.; investigation, H.T.H. and D.H.D.; resources, D.X.C. and D.T.T.T.; data curation, H.T.H. and L.H.T.; writing—original draft preparation, P.T.T. and V.T.M.; writing—review and editing, D.X.C. and V.T.M.; visualization, P.T.T. and H.T.H.; supervision, D.X.C. and D.T.T.T.; project administration, L.H.T. and D.H.D.; funding acquisition, H.T.H., P.T.T. and V.T.M. All authors have read and agreed to the published version of the manuscript.

Funding: This research received no external funding.

Institutional Review Board Statement: Not applicable.

Informed Consent Statement: Not applicable.

Data Availability Statement: Datas are available from the authors.

Acknowledgments: Thankful for the support of Tran Thi Thanh Van in the current study.

Conflicts of Interest: The authors declare that they have no conflict of interest.

References

1. Bixler, H.; Porse, H. A decade of change in the seaweed hydrocolloids industry. *J. Appl. Phycol.* **2011**, *23*, 321–335. [CrossRef]
2. FAO. The state of the world fisheries and aquaculture 2020. *Sustain. Action Rome* **2020**, 244. [CrossRef]
3. Brakel, J.; Sibonga, R.C.; Dumilag, R.V.; Montalescot, V.; Campbell, I.; Cottier-Cook, E.J.; Ward, G.; Le, M.V.; Liu, T.; Msuya, F.E.; et al. Exploring, harnessing and conserving marine genetic resources towards a sustainable seaweed aquaculture. *Plants People Planet* **2021**, *3*, 337–349. [CrossRef]
4. FAO. The global status of seaweed production, trade and utilization. *Globefish Res. Programme* **2018**, *24*, 124.
5. Available online: https://www.fao.org/3/x5819e/x5819e06.htm (accessed on 30 November 2016).
6. Gómez-Ordóñez, E.; Jiménez-Escrig, A.; Rupérez, P. Bioactivity of sulfated polysaccharides from the edible red seaweed *Mastocarpus stellatus*. *Bioact. Carbohydr. Diet. Fibre* **2014**, *3*, 29–40. [CrossRef]
7. Besednova, N.; Zaporozhets, T.; Kuznetsova, T.; Makarenkova, I.; Fedyanina, L.; Kryzhanovsky, S.; Malyarenko, O.; Ermakova, S. Metabolites of seaweeds as potential agents for the prevention and therapy of influenza infection. *Mar. Drugs* **2019**, *17*, 373. [CrossRef]
8. Necas, J.; Bartosikova, L. Carrageenan: A review. *Vet. Med.* **2013**, *58*, 187–205. [CrossRef]
9. Mingjin, Z.; Liming, G.; Yongbo, L.; Yaxin, Zi.; Xinying, Li.; Defu, L.; Changdao, M. Preparation, characterization and antibacterial activity of oxidized κ-carrageenan. *Carbohydr. Polym.* **2017**, *174*, 1051–1058. [CrossRef]
10. Ratih, P.; Se-Kwon, K. Biological Activities of Carrageenan. *Adv. Food Nutr.* **2014**, *72*, 113–124. [CrossRef]
11. Maxim, K.; Vladlena, T.; Aleksandra, K.; Maria, B.; Rodion, K.; Ekaterina, L.; Igor, B.; Yuri, K. Antitumor potential of carrageenans from marine red algae. *Carbohydr. Polym.* **2020**, *246*, 116568. [CrossRef]
12. Eduardas, C.; Aleksandra, A.; Kalitnik, Y.A.K.; Manoj, S.G.M.R.; Anant, A.; Anna, O.K. Immunomodulating properties of carrageenan from *Tichocarpus crinitus*. *Inflammation* **2020**, *43*, 1387–1396.
13. Xia, Q.; Wenwen, Z. Antihyperglycemic and antihyperlipidemic effects of low-molecular-weight carrageenan in rats. *Open Life Sci.* **2018**, *13*, 379–384. [CrossRef]
14. Tobacman, J.K. Review of harmful gastrointestinal effects of carrageenan in animal experiments. *Environ. Health Perspect.* **2001**, *109*, 983–994. [CrossRef]
15. Edisson-Mauricio, P.-Q.; Roberto, R.-C.; María-Dolores, V. Carrageenan: Drug delivery systems and other biomedical applications. *Mar. Drugs* **2020**, *18*, 583–622. [CrossRef]
16. Tarman, U.S.; Joko, S.; Linawati, H. Carrageenan and its enzymatic extraction. In *Kustiariyah Encyclopedia of Marine Biotechnology: Five Volume Set*, 1st ed.; John Wiley & Sons Ltd.: Hoboken, NJ, USA, 2020.
17. Andi, H.; Meta, M.; Amran, L.; Metusalach, M. Extraction of carrageenan from *Eucheuma spinosum* using ohmic heating: Optimization of extraction conditions using response surface methodology. *Food Sci. Technol.* **2021**, *41*, 928–937.
18. Zainab, M.A.-N.; Ahmed, A.-A.; Insaaf, A.-M. The effect of extraction conditions on chemical and thermal characteristics of kappa-carrageenan extracted from *Hypnea bryoides*. *J. Mar. Sci.* **2019**, *2019*, 5183261. [CrossRef]
19. Siti, M.; Widiyastuti, W.; Hideki, K.; Sugeng, W.; Motonobu, G. Pressurized hot water extraction of carrageenan and phenolic compounds from *Eucheuma cottonii* and *Gracilaria* sp.: Effect of extraction conditions. *ARPN J. Eng. Appl.* **2019**, *14*, 3113–3123.
20. Mariel, G.T.; Lucille, V.A.; Virgilio, D.E.; Drexel, H.C. Ultrasound-assisted depolymerization of kappa-carrageenan and characterization of degradation product. *Ultrason. Sonochem.* **2021**, *73*, 105540. [CrossRef]
21. Vázquez-Delfín, E.; Robledo, D.; Freile-Pelegrín, Y. Microwave-assisted extraction of the Carrageenan from *Hypnea musciformis* (*Cystocloniaceae*, Rhodophyta). *J. Appl. Phycol.* **2014**, *26*, 901–907. [CrossRef]
22. Deng, C.-M.; Wu, Z.-J.; He, L.-Z.; Zhang, G.-G.; Wu, Y.-L.; Wen, Y.-M.; Kang, X.-H. Technological optimization of alkali pretreatment in the carrageenan extraction from *Eucheuma gelatinae*. *Sci. Technol. Food Ind.* **2017**, *22*, 178–183. [CrossRef]
23. Tran, T.T.V.; Vo, M.N.H.; Cao, T.T.H.; Phan, T.H.T.; Tran, M.D.; Quach, T.M.T. Structural characteristics and biological activity of sulfated polysaccharide from red algae *Betaphycus gelatinus*. *Vietnam J. Sci. Technol.* **2020**, *58*, 252–260.
24. Vanessa, W.; Sabrina, M.D.C.; Paulo, J.O.; Leila, H.; Pedro, L.M.B. Optimization of the extraction of carrageenan from *Kappaphycus alvarezii* using response surface methodology. *Food Sci. Technol.* **2012**, *32*, 812–818.
25. Chen, F.; Peng, J.; Lei, D.; Liu, J.; Zhao, G. Optimization of genistein solubilization by κ-carrageenan hydrogel using response surface methodology. *Food Sci. Hum. Wellness* **2013**, *2*, 124–131. [CrossRef]
26. Joseph, W.; Bolton, J.J.; Derek, K.; Lincoln, R. Seasonal changes in carrageenan yield and gel properties in three commercial eucheumoids grown in southern Kenya. *Bot. Mar.* **2006**, *49*, 208–215.
27. Nishinari, K.; Watase, M. Effects of sugars and polyols on the gel-sol transition of kappa-carrageenan gels. *Thermochim. Acta* **1992**, *206*, 149–162. [CrossRef]
28. Ahmed, A.-A.; Pothiraj, C.; Abdullah, A.-M.; Insaaf, A.-M.; Mohammad, S.R. Characterization of red seaweed extracts treated by water, acid and alkaline solutions. *Int. J. Food Eng.* **2018**, *14*, 1–9.
29. Pereira, L. Identification of phycocolloids by vibrational spectroscopy. In *World Seaweed Resources—An Authoritative Reference System*; Critchley, A.T., Ohno, M., Largo, D.B., Eds.; ETI Information Services Ltd., UNESCO: Paris, France, 2006.
30. Greer, C.W.; Yaphe, W. Characterization of hybrid (Beta-Kappa-Gamma) carrageenan from *Eucheuma gelatinae* J. Agardh (Rhodophyta, Solieriaceae) using carrageenases, infrared and 13C-nuclear magnetic resonance spectroscopy. *Bot. Mar.* **1984**, *27*, 473–478. [CrossRef]

31. Li-Hwa, L.; Masakuni, T.; Fujiya, H. Isolation and characterization of ι-carrageenan from *Eucheuma serra* (Togekirinsai). *J. Appl. Glycosci.* **2000**, *47*, 303–310.
32. Fang, J.; Fowler, P.; Sayers, C.; Williams, P. The chemical modification of a range of starches under aqueous reaction conditions. *Carbohydr. Polym.* **2004**, *55*, 283–289. [CrossRef]
33. Aimei, W.; Nahidul, I.M.; Xiaojuan, Q.; Hongxin, W.; Yaoyao, P.; Chaoyang, M. Purification, identification, and characterization of D-galactose-6-sulfurylase from marine algae (*Betaphycus gelatinus*). *Carbohydr. Res.* **2014**, *388*, 94–99.
34. Prieto, P.; Pineda, M.; Aguilar, M. Spectrophotometric quantitation of antioxidant capacity through the formation of a phosphomolybdenum complex: Specific application to the determination of vitamin E. *Anal. Biochem.* **1999**, *269*, 337–341. [CrossRef]
35. Zhu, Q.Y.; Hackman, R.M.; Ensunsa, J.L.; Holt, R.R.; Keen, C.L. Antioxidative activities of oolong tea. *J. Agric. Food Chem.* **2002**, *50*, 6929–6934. [CrossRef]

Article

Molecular Docking and Dynamics Investigations for Identifying Potential Inhibitors of the 3-Chymotrypsin-like Protease of SARS-CoV-2: Repurposing of Approved Pyrimidonic Pharmaceuticals for COVID-19 Treatment

Amin Osman Elzupir

College of Science, Deanship of Scientific Research, Imam Mohammad Ibn Saud Islamic University (IMSIU), Riyadh 11623, Saudi Arabia; aoalamalhuda@imamu.edu.sa

Abstract: This study demonstrates the inhibitory effect of 42 pyrimidonic pharmaceuticals (PPs) on the 3-chymotrypsin-like protease of SARS-CoV-2 ($3CL^{pro}$) through molecular docking, molecular dynamics simulations, and free binding energies by means of molecular mechanics–Poisson Boltzmann surface area (MM-PBSA) and molecular mechanics–generalized Born surface area (MM-GBSA). Of these tested PPs, 11 drugs approved by the US Food and Drug Administration showed an excellent binding affinity to the catalytic residues of $3CL^{pro}$ of His41 and Cys145: uracil mustard, cytarabine, floxuridine, trifluridine, stavudine, lamivudine, zalcitabine, telbivudine, tipiracil, citicoline, and uridine triacetate. Their percentage of residues involved in binding at the active sites ranged from 56 to 100, and their binding affinities were in the range from -4.6 ± 0.14 to -7.0 ± 0.19 kcal/mol. The molecular dynamics as determined by a 200 ns simulation run of solvated docked complexes confirmed the stability of PP conformations that bound to the catalytic dyad and the active sites of $3CL^{pro}$. The free energy of binding also demonstrates the stability of the PP–$3CL^{pro}$ complexes. Citicoline and uridine triacetate showed free binding energies of -25.53 and -7.07 kcal/mol, respectively. Therefore, I recommend that they be repurposed for the fight against COVID-19, following proper experimental and clinical validation.

Keywords: coronavirus SARS-CoV-2; COVID-19; 3-chymotrypsin-like protease; pyrimidonic pharmaceuticals; molecular dynamics simulations; binding free energy

Citation: Elzupir, A.O. Molecular Docking and Dynamics Investigations for Identifying Potential Inhibitors of the 3-Chymotrypsin-like Protease of SARS-CoV-2: Repurposing of Approved Pyrimidonic Pharmaceuticals for COVID-19 Treatment. *Molecules* **2021**, *26*, 7458. https://doi.org/10.3390/molecules26247458

Academic Editors: Giovanni Ribaudo and Laura Orian

Received: 30 September 2021
Accepted: 29 November 2021
Published: 9 December 2021

Publisher's Note: MDPI stays neutral with regard to jurisdictional claims in published maps and institutional affiliations.

Copyright: © 2021 by the author. Licensee MDPI, Basel, Switzerland. This article is an open access article distributed under the terms and conditions of the Creative Commons Attribution (CC BY) license (https://creativecommons.org/licenses/by/4.0/).

1. Introduction

Over a year has passed since the COVID-19 pandemic began. Some vaccines, such as those by Pfizer and Moderna, and some drugs, such as remdesivir, have been approved for use in therapy. The efforts by governments, health organizations, and other sectors to stem the alarmingly increasing numbers of deaths and cases were unprecedented [1–6]. However, SARS-CoV-2 continues to threaten the world, with over four million deaths and 227 million cases as of 16 September 2021 (https://www.worldometers.info/coronavirus/ accessed on 27 November 2021). COVID-19 was declared a pandemic by the World Health Organization on 11 March 2020. Today, the new SARS-CoV-2 virus, the causative agent of COVID-19, has been detected in almost every country on the planet [5,7–9].

Coronaviruses are positive-stranded RNA viruses with the largest viral genomes ever known, ranging from 16 to 32 kb. The 3-chymotrypsin-like protease ($3CL^{pro}$) produced by SARS-CoV-2 is a cysteine protease encoded as nonstructural protein 3 in the polyprotein. $3CL^{pro}$ is responsible for the cleavage of 11 specific sites of polyproteins (pp1a, pp1ab) produced by the 229E gene. These polyproteins are involved in the production of a functional polypeptide essential for viral replication and transcription. Further, the specificity of $3CL^{pro}$ is dissimilar to that of human host-cell protease. Thus, $3CL^{pro}$ has become the focus of drug repurposing and development programs to combat the COVID-19 pandemic [10–13].

Recent and ongoing research has reported that some pharmaceutical, synthetic, and natural products can act as 3CL^{pro} inhibitors or against SARS-CoV-2 in general. These include selenium-containing heterocyclic compounds, chloroquine phosphate, indinavir, darunavir, lopinavir, eravacycline, naproxen, salix cortex, antioxidants, chiral phytochemicals from *Opuntia ficus-indica*, elbasvir, valrubicin, favipiravir isoflavone, and myricitrin [6,14–22]. Although some of these have entered human clinical trials or were even approved, more studies are still needed. The importance of the pyridone ring was highlighted in synthetic materials and drugs containing pyridone [11,23]. The pyrimidone ring has the exact shape of pyridine but is more functionalized and electron-deficient. Herein, we screened the inhibitory activity of 42 approved pyrimidonic pharmaceuticals (PPs) against 3CL^{pro} using a combination of molecular docking analyses, molecular dynamics simulations, and calculations of the MM-PBSA and MM-GBSA binding free energies. The sites of action of active inhibitors were investigated, discussed, and explored.

2. Materials and Methods
2.1. The Pyrimidonic Pharmaceuticals (PPs)

The PPs were selected using the search engine of the drug bank database. The search uncovered 46 PPs; the pharmaceuticals containing caffeine were entirely excluded as all except enprofylline have previously been studied. In addition, the macropolymeric drug mipomersen, drugs composed of a mixture of drugs, and withdrawn drugs were not included in this study. The chosen drugs were classified into four categories according to their structures. 1PPs have only one heterocycle, 2aPPs have two, 2bPPs have two heterocycles with a pyrimidone ring having an extra carbonyl group, and 3PPs have three or more heterocycles (Table 1).

Table 1. Structures of pyrimidonic pharmaceuticals and their classification according to the number of rings.

1PPs	2aPPs	2bPPs	3PPs
Cidofovir	Gemcitabine	Idoxuridine	Riboflavin
Fluorouracil	Lamivudine	Floxuridine	Flavin adenine dinucleotide
Uracil mustard	Emtricitabine	Trifluridine	Alogliptin

Table 1. *Cont.*

1PPs	2aPPs	2bPPs	3PPs
Flucytosine	Zalcitabine	Telbivudine	Flavin mononucleotide
-	Cytarabine	Zidovudine	Trametinib
-	Capecitabine	Stavudine	Dasabuvir
-	Citicoline	Brivudine	Relugolix
-	Sulfacytine	Tegafur	Elagolix

Table 1. Cont.

1PPs	2aPPs	2bPPs	3PPs
-	-	Uridine triacetate	Sofosbuvir
-	-	Tipiracil	-
-	-	Enprofylline	-

2.2. Generation and Energy Minimization of the PPs and 3CL^{pro}

The 3D structures of the selected PPs were downloaded from the PubChem website as SDF files; their energy was minimized for 10,000 steepest descent steps at 5000 conjugate gradient steps using antechamber plugin UCSF Chimera [24,25]. For alogliptin, the 3D structure was obtained by utilizing OpenBabel converter tools and ChemSkech [26]. The crystal structure of SARS-CoV-2 3CL^{pro} was obtained from the Protein Data Bank database website (PDB ID: 6Y2E). For analysis, water was removed from the 3CL^{pro} structure, and the energy was then minimized for 1000 steepest descent steps at 20 conjugate gradient steps.

2.3. Molecular Docking

Blind molecular docking experiments were performed using the AutoDock Vina tool implemented with the interactive visualization and analysis program UCSF Chimera. The default parameter values were adopted with a grid box ($-15 \times -25 \times 15$) Å, centered at (35, 65, 65) Å. The predicted affinity values of the score were observed using the View Dock tool. The binding between ligands and 3CL^{pro} active sites and the images were processed and visualized using UCSF Chimera [24–28].

2.4. Molecular Dynamics Simulations

MD simulations were performed as previously described [29]. The PP ligands were separated from the docked complexes using UCSF Chimera. The missed hydrogens were added and saved as PDB files using AMBER's large-structure serial numbering. Topology files and parameters of the receptor and the ligands were made using leap and antechamber of Amber Tools 21 [30,30], utilizing Amber force fields of GAFF2 [31] and ff14SB [32] to assign inhibitors and 3CL^{pro} structure, respectively. The systems were solvated with TIP3P water molecules [33] and were neutralized via sodium ions. Subsequently, molecular dynamics (MD) simulations were performed by means of the Nanoscale Molecular Dynamics

(NAMD) Simulation 2.6 program [34]. Each system was minimized for 1 ps at 273.15 K using the NVE ensemble. The temperature was gradually increased to 310 K using the NVT ensemble in a protocol consisting of 1600 minimization steps. Then, each system was minimized for 10 ps at 310 K followed by 200 ns of MD simulation control using the NVT ensemble at 310 K and a time step of 2 fs. In order to calculate electrostatic interactions, the particle mesh Ewald process and periodic boundary conditions were applied [35,36]. The root-mean-square fluctuation (RMSF) and the root-mean-square deviation (RMSD) for each system were obtained by analyzing the trajectory using the VMD 1.8 program [37].

2.5. The Binding Free Energies

The binding free energies of the PP-3CL^{pro} complexes were calculated by means of molecular mechanics–Poisson Boltzmann surface area (MM-PBSA) and molecular mechanics–generalized Born surface area (MM-GBSA) using the MMPBSA.py module of Amber Tools 21 [38]. The MD simulation over 200 ns provided several conformations sampled after equilibrium, using the last frames to lessen the computational cost. CPPTRAJ was used to obtain the snapshots [39]. The conformational changes were evaluated through quasi-harmonic entropy approximation [40]. The free energy of the binding interaction between inhibitors and 3CL^{pro} complexes can be obtained via the following equations:

$$\Delta G = \Delta H - T\Delta S \quad (1)$$

$$\Delta H = \Delta G_{gas} + \Delta G_{sol} \quad (2)$$

$$\Delta G_{gas} = E_{vdw} + E_{elec} \quad (3)$$

$$\Delta G_{sol} = E_{pb/gb} + E_{np} \quad (4)$$

where ΔH represents enthalpy change, $T\Delta S$ represents the entropic contribution, E_{vdw} represents the van der Waals interaction energy, E_{ele} represents the electrostatic interaction energy, ΔG_{sol} represents the polar solvation energy, and E_{np} represents the nonpolar solvation energy.

3. Results and Discussion

The results of the molecular docking are tabulated in Tables S1–S4. Figure 1 shows the catalytic dyad and the active sites of 3CL^{pro}. The crucial residues HIS 41, GLY 143, SER 144, and CYS 145 forming the S1' site are shown in black. Then, PHE 140, LUE 141, ASN 142, HIS 163, GLU 166 (magenta), and the *N*-terminal amino acid residues (blue) are involved in the formation of the S1 subsite of the substrate-binding pocket. The MET 49, TYR 54, HIS 164, ASP 187, and ARG R188 residues form the S2 site (green). MET 165, LEU 167, GLN 189, THR 190, and GLN 192 comprise the S4 site (cyan). The SER 284, ALA 285, and LEU 286 residues (yellow) are a result of genetic mutation leading to an increase in the SASR-CoV-2 3CL^{pro} activity of 3.6 fold over that of the 3CL^{pro} predecessor of SARS-CoV [12,41].

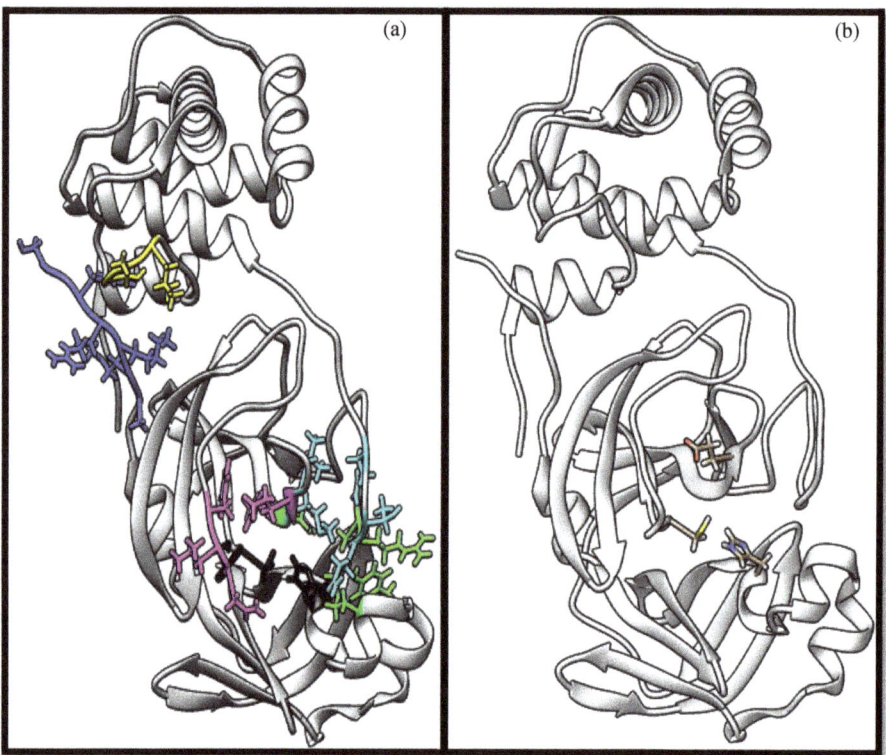

Figure 1. The crystal structure of chymotrypsin-like protease of SARS-CoV-2 (PDB ID: 6Y2E) and its active residues. (**a**) Color indicates the residues involved in the formation of the S1 site (shown in magenta), S1 site from the other promotor (blue), S2 site (green), S4 site (cyan), and S1' site (black), in addition to SER 284, ALA 285, and LEU 286 (yellow). (**b**) Only the catalytic dyad and GLU 166 residues.

3.1. Molecular Docking

The docked complexes of the top 11 candidates are depicted in Figure 2. Their binding affinities to the active sites of 3CL^{pro} are shown in Table 2. The 3PPs showed significant interactions with the residues LEU 286, SER 284, and ALA 285, and a relatively lower interaction ratio to the catalytic dyad, in contrast to the other groups. Of the 3PPs, alogliptin and flavin mononucleotide were found to have the highest binding percentage with the catalytic dyad and to form hydrogen bonds with the S1 and S'1 sites. These were followed by riboflavin and sofosbuvir with an advantage in binding to the LEU 286 residue (Table S1). Flavin adenine dinucleotide showed excellent binding affinity to LEU 286 but not with the catalytic dyad. Zidovudine and gemcitabine demonstrated similar activity to alogliptin and flavin mononucleotide (Tables S2 and S3).

Among the 2bPPs, anti-hepatitis B infection telbivudine, anti-orotic aciduria uridine triacetate, and anticancer tipiracil were found to have the highest binding to 3CL^{pro} active sites, followed by antimetabolite floxuridine, anti-herpesvirus trifluridine, and anti-HIV stavudine. Here, it is worth noting the importance of the molecular structure, as this set differed from the previous one by its increased ability to bind to the 3CL^{pro} catalytic dyad. The 2aPPs showed similar activity to that of the 2bPPs. Anticancer cytarabine, antiglaucoma citicoline, and anti-HIV drugs lamivudine and zalcitabine showed promising inhibitory activity (Table S3).

Finally, but very importantly, of the 1PPs, the chemotherapy drug uracil mustard showed binding to the catalytic dyad with all of its simulated conformations, followed by anti-cytomegalovirus cidofovir (Table S4).

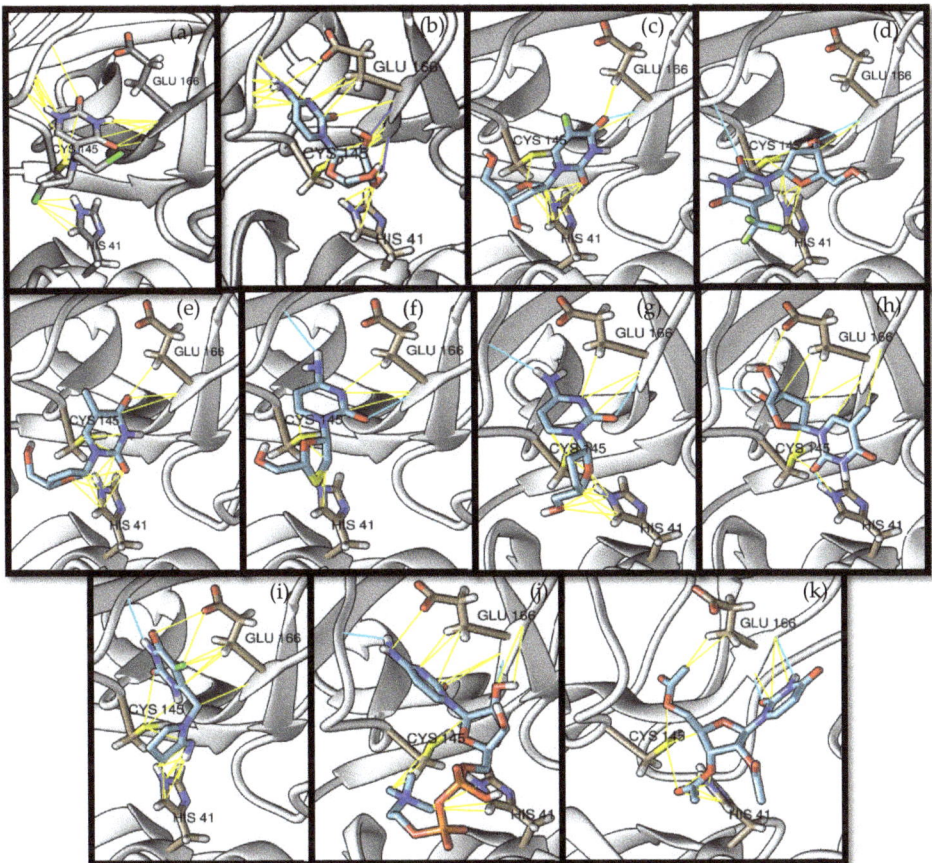

Figure 2. The PPs docked with 3CL^{pro}, focusing on contacts with HIS 41, CYS 145, and GLU 166. (**a**) uracil mustard, (**b**) cytarabine, (**c**) floxuridine, (**d**) trifluridine, (**e**) stavudine, (**f**) lamivudine, (**g**) zalcitabine, (**h**) telbivudine, (**i**) tipiracil, (**j**) citicoline, (**k**) uridine triacetate. The hydrocarbon skeleton is shown in cyan, nitrogen atoms are blue, and oxygens are red. Hydrogen bonds are represented by blue lines; van der Waals forces are represented in yellow.

Table 2. The binding affinities of the potential pyrimidonic pharmaceuticals with 3-chymotrypsin-like protease (3CL^{pro}).

Pharmaceutical Name	Binding Percentage [a]	Score ± SD (kcal/mol) [b]	RMSD	Hydrogen Bond (Number of Bonds/Number of Conformations)	Van Der Waals (Distance) (Number of Bonds/Number of Conformations)
Uracil mustard	a. 100 * c. 89 d. 100 e. 100 All. 100	a. −4.6 ± 0.14 c. −4.6 ± 0.14 d. −4.6 ± 0.14 e. −4.6 ± 0.14	a. 0.00–7.13 c. 0.00–7.13 d. 0.00–7.13 e. 0.00–7.13	a. HIS 163 (5/5), GLU 166 (3/3), LEU 141 (3/3), ASN 142 (1/1)	a. HIS 163 (31/5), GLU 166 (40/8), LEU 141 (19/5), ASN 142 (50/7), PHE 140 (8/5) c. MET 49 (52/7), HIS 164 (12/4) d. GLN 189 (32/9), MET 165 (28/8) e. HIS 41 (53/9), SER 144 (10/5), GLY 143 (4/1), CYS 145 (28/9)
Cytarabine	a. 67 c. 67 d. 67 e. 67 All. 67	a. −5.4 ± 0.26 c. −5.4 ± 0.26 d. −5.4 ± 0.26 e. −5.4 ± 0.26	a. 0.00–6.07 c. 0.00–6.07 d. 0.00–6.07 e. 0.00–6.07	a. GLU 166 (3/3), LEU 141 (2/2), HIS 163 (1/1), PHE 140 (2/2), ASN 142 (1/1) c. HIS 164 (2/2) d. GLN 189 (1/1) e. GLY 143 (1/1)	a. GLU 166 (44/5), LEU 141 (18/3), HIS 163 (18/4), PHE 140 (19/3), ASN 142 (27/6) c. HIS 164 (10/4), MET 49 (20/6) d. MET 165 (28/4), GLN 189 (4/2) e. GLY 143 (11/2), SER 144 (8/2), CYS 145 (15/6), HIS 41 (21/5)
Floxuridine	a. 44 c. 44 d. 44 e. 44 All. 56	a. 5.5 ± 0.13 c. 5.6 ± 0.25 d. 5.6 ± 0.25 e. 5.6 ± 0.19	a. 3.24–8.28 c. 0.00–8.28 d. 0.00–8.28 e. 0.00–8.28	a. ASN 142 (1/1), HIS 163 (2/2), GLU 166 (2/2), PHE 140 (1/1), LEU 141 (1/1) e. HIS 41 (1/1)	a. GLU 166 (23/4), LEU 141 (6/2), PHE 140 (8/2), HIS 163 (5/2), ASN 142 (19/4) c. MET 49 (9/2), HIS 164 (3/3) d. MET 165 (5/3), GLN 189 (4/1) e. GLY 143 (6/1), SER 144 (4/1), CYS 145 (5/3), HIS 41 (16/3)
Trifluridine	a. 44 c. 44 d. 44 e. 44 f. 11 All. 56	a. −6.03 ± 0.17 c. −6.03 ± 0.17 d. −6.03 ± 0.17 e. −6.03 ± 0.17 f. −5.7	a. 0.00–5.58 c. 0.00–5.58 d. 0.00–5.58 e. 0.00–5.58 f. 28.34–30.21	a. GLU 166 (2/2), ASN 142 (1/1). c. HIS 164 (2/2) e. GLY 143 (1/1)	a. GLU 166 (18/4), ASN 142 (15/3), HIS 163 (2/1), LEU 141 (2/1) b. c. HIS 164 (10/3), MET 49 (17/4) d. MET 165 (19/4), GLN 189 (1/1). e. CYS 145 (13/4), GLY 143 (10/2), HIS 41 (10/4) f. SER 284 (9/1)
Stavudine	a. 44 c. 56 d. 44 e. 56 All. 56	a. −5.6 ± 0.28 c. −5.6 ± 0.28 d. −5.6 ± 0.28 e. −5.6 ± 0.28	a. 27.12–32.34 c. 27.12–35.03 d. 27.12–32.34 e. 27.12–35.03	a. GLU 166 (1/1)	a. ASN 142 (13/3), GLU 166 (14/4), HIS 163 (2/1), LEU 141 (1/1) c. HIS 164 (6/3), MET 49 (20/5) d. MET 165 (12/4), GLN 189 (1/1) e. HIS 41 (39/5), GLY 143 (9/2), CYS 145 (7/3)
Lamivudine	a. 56 c. 44 d. 44 e. 56 f. 11 All. 67	a. −5.4 ± 0.24 c. −5.4 ± 0.28 d. −5.4 ± 0.25 e. −5.4 ± 0.24 f. −5.2	a. 0.00–4.77 c. 0.00–3.38 d. 0.00–4.77 e. 0.00–4.77 f. 26.28–28.62	a. HIS 163 (3/3), ASN 142 (1/1), PHE 140 (3/3), LEU 141 (2/2), GLU 166 (2/2) d. GLN 189 (1/1) e. SER 144 (2/2)	a. HIS 163 (26/5), ASN 142 (19/2), PHE 140 (31/5), LEU 141 (15/5), GLU 166 (46/5) c. MET 49 (12/4), HIS 164 (1/1) d. GLN 189 (12/2), MET 165 (10/4), LEU 167 (1/1) e. SER 144 (18/4), HIS 41 (5/2), CYS 145 (7/3) f. LEU 286 (1/1)

Table 2. Cont.

Pharmaceutical Name	Binding Percentage [a]	Score ± SD (kcal/mol) [b]	RMSD	Hydrogen Bond (Number of Bonds/Number of Conformations)	Van Der Waals (Distance) (Number of Bonds/Number of Conformations)
Zalcitabine	a. 44 b. 11 c. 33 d. 44 e. 44 f. 22 All. 67	a. −5.5 ± 0.29 b. −5.1 c. −5.4 ± 0.35 d. −5.5 ± 0.29 e. −5.5 ± 0.29 f. −5.1 ± 0.00	a. 0.00–6.39 b. 28.02–29.48 c. 0.00–6.39 d. 0.00–6.39 e. 0.00–6.39 f. 28.02–31.88	a. PHE 140 (3/3), LEU 141 (1/1), GLU 166 (3/3), ASN 142 (1/1) c. GLN 189 (1/1)	a. HIS 163 (15/3), PHE 140 (16/3), LEU 141 (9/3), GLU 166 (25/4), ASN 142 (11/3) b. LYS 5 (6/1), ARG 4 (12/1), PHE 3 (4/1) c. MET 49 (16/2), HIS 164 (5/3) d. GLN 189 (6/2), MET 165 (15/4) e. SER 144 (7/2), HIS 41 (13/2), CYS 145 (6/3) f. SER 284 (4/1), LEU 286 (4/1)
Telbivudine	a. 56 c. 44 d. 44 e. 56 f. 11 All. 67	a. −5.6 ± 0.38 c. −5.7 ± 0.42 d. −5.7 ± 0.42 e. −5.6 ± 0.38 f. −5.3	a. 0.00–7.50 c. 0.00–7.50 d. 0.00–7.50 e. 0.00–7.50 f. 22.85–24.17	a. ASN 142 (2/1), HIS 163 (3/3), GLU 166 (1/1), PHE 140 (2/2), LEU 141 (1/1) c. HIS 164 (1/1) d. GLN 189 (1/1) e. HIS 41 (1/1)	a. GLU 166 (36/4), HIS 163 (11/3), PHE 140 (11/3), ASN 142 (27/5), LEU 141 (9/3) c. HIS 164 (6/3), MET 49 (27/3) d. GLN 189 (10/3), MET 165 (14/4), e. CYS 145 (11/4), SER 144 (8/2), GLY 143 (3/1), HIS 41 (12/5) f. LEU 286 (5/1)
Tipiracil	a. 56 b. 11 c. 44 d. 44 e. 44 f. 11 All. 67	a. −5.8 ± 0.16 b. −5.7 c. −5.9 ± 0.17 d. −5.8 ± 0.08 e. −5.9 ± 0.17 f. −5.7	a. 26.28–29.69 b. 19.12–20.08 c. 26.28–29.69 d. 26.67–29.69 e. 26.28–29.69 f. 19.12–20.08	a. HIS 163 (2/2), GLU 166 (2/1), PHE 140 (2/2), ASN 142 (1/1) b. LYS 5 (1/1) c. HIS 164 (1/1) e. GLY 143 (1/1)	a. HIS 163 (14/3), GLU 166 (40/4), PHE 140 (8/3), ASN 142 (22/4), LEU 141 (10/3) b. PHE 3 (6/1), LYS 5 (9/1), ARG 4 (5/1) c. HIS 164 (8/4), MET 49 (22/3) d. MET 165 (9/3), GLN 189 (13/3) e. GLY 143 (7/2), HIS 41 (26/4), SER 144(2/1), CYS 145 (13/4) f. SER 284 (7/1), LEU 286 (2/1)
Citicoline	a. 56 * c. 56 d. 56 e. 56 All. 56	a. −7.0 ± 0.19 c. −7.0 ± 0.19 d. −7.0 ± 0.19 e. −7.0 ± 0.19	a. 0.00–8.01 c. 0.00–8.01 d. 0.00–8.01 e. 0.00–8.01	a. PHE 140 (2/2), GLU 166 (4/4), HIS 163 (3/3), ASN 142 (2/1), LEU 141 (1/1) e. SER 144 (1/1)	a. PHE 140 (24/5), GLU 166 (57/5), HIS 163 (21/5), ASN 142 (34/5), LEU 141 (21/5) c. MET 49 (23/5), HIS 164 (1/1) d. MET 165 (29/5), GLN 189 (6/3) e. SER 144 (12/3), GLY 143 (4/1), CYS 145 (9/5), HIS 41 (21/5)
Uridine triacetate	a. 56 c. 56 d. 56 e. 56 f. 11 All. 67	a. −6.2 ± 0.26 c. −6.2 ± 0.26 d. −6.2 ± 0.26 e. −6.2 ± 0.26 f. −6.4	a. 0.00–6.79 c. 0.00–6.79 d. 0.00–6.79 e. 0.00–6.79 f. 24.97–28.01	a. HIS 163 (2/2), GLU 166 (3/2) e. HIS 41 (1/1)	a. HIS 163 (5/2), GLU 166 (31/5), ASN 142 (30/5), PHE 140 (4/2), LEU 141 (6/2) c. MET 49 (18/5), HIS 164 (5/2) d. GLN 189 (15/4), MET 165 (18/4) e. CYS 145 (16/5), HIS 41 (35/5), SER 144 (4/2), GLY 143 (2/1) f. LEU 286 (1/1)

[a] Binding percentage was calculated based on the number of conformations attached to the active sites of the CL^{pro} (nine conformations in total). [b] SD based on the other score energies of conformations. * Alphabetical order indicates the type of active site involved in bonding: a. S1 site, b. S1 site from the other promotor, c. S2 site, d. S4 site, e. S'1 site, and f. SER 284, ALA 285, and LEU 286 residues. When letters are missing, this means no interactions were observed at that site.

3.2. Molecular Dynamics Simulations

MD was performed on the hole complexes of the top 11 PPs candidates. Based on the conformer score energy from docking, the complex with the conformer with the lowest value and interacting with the 3CL^{pro} active site was selected. The RMSDs were computed along the trajectories using the initial structure as a reference. Figure 3 shows that the binding of PPs significantly affected the equilibration states of 3CL^{pro}, as the majority of the tested systems reached their equilibrium at around 100 ns. The PP-3CLpro complexes revealed relatively lower average values for the RMSDs, between 0.41 and 0.52 Å, throughout the simulation, clarifying their good behavior in forming stable complexes. Moreover, the fluctuations in the 3CL^{pro} backbone residues were analyzed by means of the RMSF (Figure 4). The 3CL^{pro}/PPs exhibited lower fluctuations, particularly at the active site. The fluctuations at the catalytic dyad and GLU 166 were minor, demonstrating the loss of flexibility at these regions upon binding to the PPs. Table 3 shows the superior stability of the PP-3CL^{pro} complexes formed throughout the production runs; these results support the use of these PPs as 3CL^{pro} inhibitors.

Table 3. The binding interactions of the potential pyrimidonic pharmaceuticals/3-chymotrypsin-like protease 3CL^{pro} complexes at different times throughout the production runs.

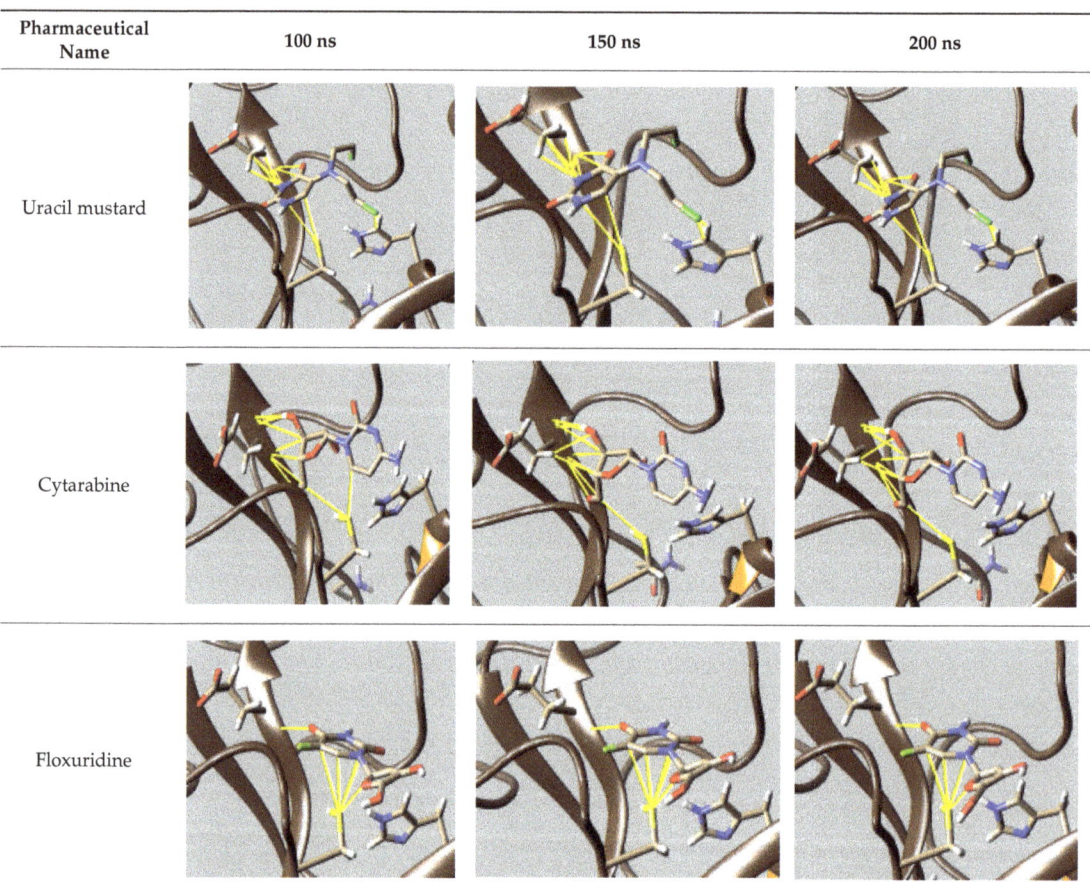

Pharmaceutical Name	100 ns	150 ns	200 ns
Uracil mustard			
Cytarabine			
Floxuridine			

Table 3. Cont.

Pharmaceutical Name	100 ns	150 ns	200 ns
Trifluridine			
Stavudine			
Lamivudine			
Zalcitabine			
Telbivudine			

Table 3. *Cont.*

Pharmaceutical Name	100 ns	150 ns	200 ns
Tipiracil			
Citicoline			
Uridine triacetate			

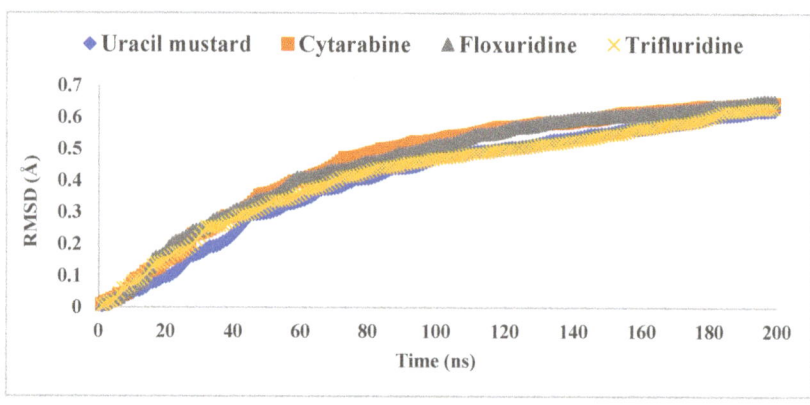

Figure 3. *Cont.*

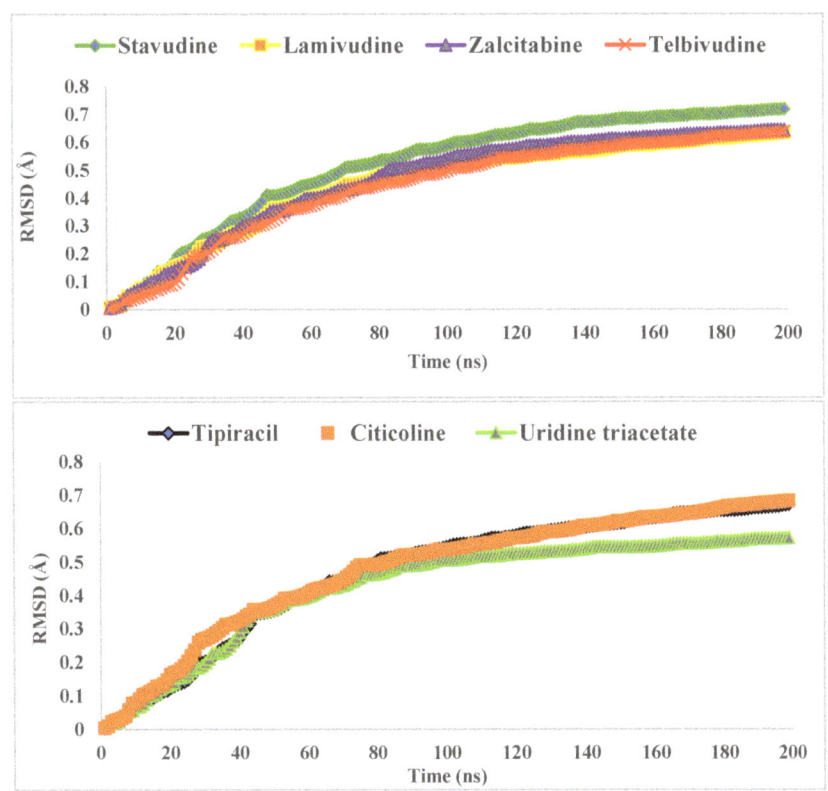

Figure 3. The RMSD values of the simulated PP-3CL^{pro} complexes throughout the 200 ns production runs.

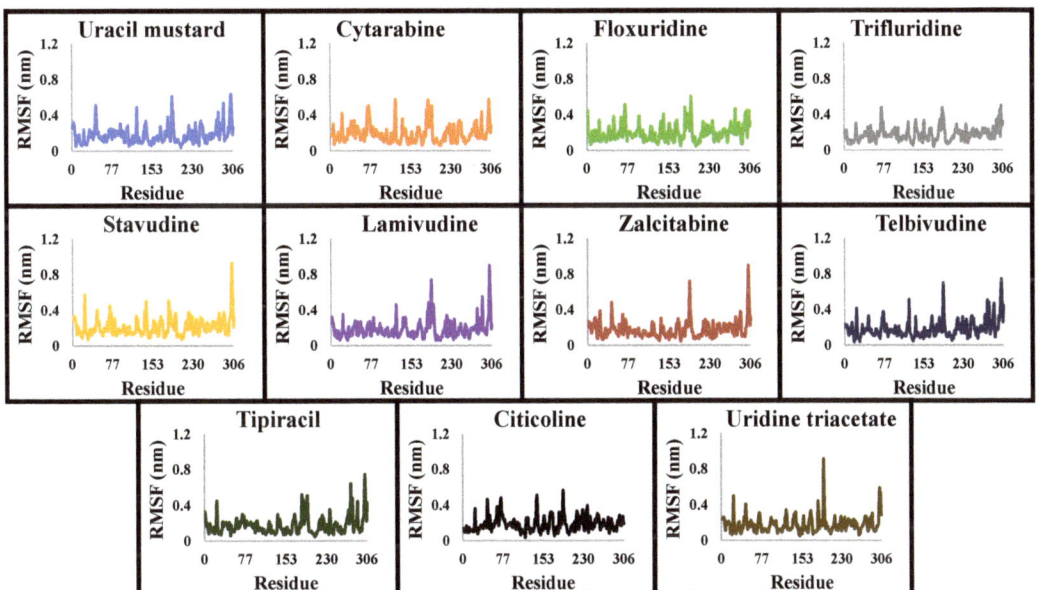

Figure 4. The RMSF values of the simulated PP-3CL^{pro} complexes throughout the 200 ns production runs.

3.3. The Binding Free Energies

The data on the binding free energies of the PP-3CL^{pro} complexes are tabulated in Table 4. In both MM-GBSA and MM-PBSA, van der Waals and electrostatic interactions acted as driving forces for the PP ligands to bind to 3CL^{pro}, contrasting the solvation energies. The MM-GBSA and MM-PBSA results suggest that the PPs have an excellent ability to inhibit 3CL^{pro}. Of these PPs, citicoline revealed the most promising inhibitory activity, followed by uridine triacetate (Table 4).

Table 4. The MMPBSA and MMGBSA data for the binding of pyrimidone containing-pharmaceuticals to 3CL^{pro} of SARS-CoV-2.

3CL^{pro} Complex Type	$-T\Delta S$	E_{vdw}	MMGBSA			MMPBSA		
			E_{alac}	E_{sol}	Δg (kcal/mol)	E_{alac}	E_{sol}	Δg (kcal/mol)
Uracil mustard	22.58	−23.76	−16.08	20.46	3.20	−0.80	0.94	−1.05
Cytarabine	22.39	−19.25	−33.50	28.39	−1.96	−1.67	1.09	2.55
Floxuridine	22.42	−24.54	0.00	3.70	1.58	0.00	0.39	−1.73
Trifluridine	22.96	−29.48	0.00	4.20	−2.32	0.00	0.41	−6.11
Stavudine	22.25	−24.41	−17.67	17.90	−1.94	−0.88	0.93	−2.12
Lamivudine	22.26	−20.58	−35.11	28.94	−4.50	−1.76	1.06	0.98
Zalcitabine	22.12	−24.34	−30.10	26.50	−5.81	−1.50	1.19	2.53
Telbivudine	22.48	−25.29	−46.88	42.53	−7.16	−2.34	1.61	−3.54
Tipiracil	22.54	−28.79	0.00	5.07	−1.18	0.00	0.44	−5.81
Citicoline	24.17	−54.50	0.00	4.80	**−25.53**	0.00	0.64	**−29.69**
Uridine triacetate	23.60	−32.98	−31.09	33.40	**−7.07**	−1.55	1.43	**−9.51**

Among the challenges of discovering 3CL^{pro} inhibitors for COVID-19 treatment, these inhibitors must be highly bioavailable inside the cytosol [13]. The 2aPP and 2bPP structures contain a primidone heterocycle and ribose ring. These heterocycles increase their hydrophilicity and solubility in plasma. Thus, they can satisfy the requirement of bioavailability. Citicoline, the most promising inhibitor among the PPs investigated, has high hydrophilicity and good ADME properties [42,43]. The 2aPPs and 2bPPs also have an intermediate structure size among the PP groups. This sheds light on the importance of the size and general structural features of PPs acting as 3CL^{pro} inhibitors.

Recent reports suggest a general hypothesis that 3CL^{pro} inhibitors comprise electrophilic sites such as Michael acceptors [12]. That the pyrimidone ring is highly electron-deficient clarifies and confirms this hypothesis. The pyrimidone ring plays an essential role in PPs' inhibitory activity and has a high tendency to form hydrogen bonds, particularly with the GLU 166 residue. This may preclude the formation of the S1 pocket. Contacts between the pyrimidone ring and HIS 41 were observed in floxuridine, stavudine, and telbivudine. In all these cases, HIS 41 interacts with the oxygen of the pyrimidone group. Further, the electrophilic carbon, nitrogen, and oxygen in the pyrimidone ring were attracted to bind with the sulfur of the CYS 145 residue.

Interestingly, most PPs investigated were previously studied against SARS-CoV-2. For example, flavin mononucleotide and flavin adenine dinucleotide have been suggested as good 3CL^{pro} and RNA-dependent RNA polymerase inhibitors, respectively [22,44]. Riboflavin and sofosbuvir were shown to be suitable inhibitors of the spike protein S1 domain/ACE2 and RNA-dependent RNA polymerase [45–47]. Alogliptin was also suggested as a 3CL^{pro} inhibitor; however, the enzymatic assay demonstrated its inactivity against 3CL^{pro} [48]. Compelling clues have been found regarding the use of zidovudine and gemcitabine against spike protein/human ACE2 and in the inhibition SARS-CoV-2 in cell culture [49–53]. Gemcitabine and cidofovir were reported to inhibit SARS-CoV and SARS-CoV-2 proteins with IC$_{50}$ values of 4.95 μM and 36 μM [54,55]. Further, telbivudine, tipiracil, cytarabine, and citicoline were recommended as 3CL^{pro} inhibitors [56–61]. This transitory literature scanning confirms these pharmaceuticals' activity against 3CL^{pro} of SARS-CoV-2, as demonstrated in the present study.

To conclude, the inhibitory effect on 3CL^{Pro} by PPs was investigated based on their ability to form hydrogen bonds and van der Waals interactions with the 3CL^{Pro} active side through molecular docking, MD simulations, and the calculation of binding free energy. The overall analysis revealed 11 candidates from the initial set of 42 investigated PPs are promising 3CL^{Pro} inhibitors. These include citicoline and uridine triacetate as the best choices, followed by telbivudine, trifluridine, lamivudine, cytarabine, stavudine, zalcitabine, tipiracil, floxuridine, and flavin mononucleotide. The interactions of PPs with the catalytic dyad and the active sides of 3CL^{Pro} of SARS-CoV-2 were comprehensively and thoroughly investigated. The pyrimidone ring was found to play an essential role in the PPs' inhibitory activity.

Supplementary Materials: The following are available online, Table S1: The binding affinities of the pyrimidonic pharmaceuticals (group 3PPs) with 3-chymotrypsin-like protease (3CLpro), Table S2: The binding affinities of the pyrimidonic pharmaceuticals (group 2bPPs) with 3-chymotrypsin-like protease (3CLpro), Table S3: The binding affinities of the pyrimidonic pharmaceuticals (group 2aPPs) with 3-chymotrypsin-like protease (3CLpro), Table S4: The binding affinity of the pyrimidone containing-pharmaceuticals (group 1PCPs) with 3-chymotrypsinlike protease (3CLpro).

Funding: This research and the APC were funded by the Deanship of Scientific Research, Imam Mohammad Ibn Saud Islamic University (IMSIU), Saudi Arabia, grant number [21-13-18-040].

Institutional Review Board Statement: Not applicable.

Informed Consent Statement: Not applicable.

Data Availability Statement: The data will be available upon request.

Conflicts of Interest: There is no conflict of interest to be reported.

Sample Availability: Not applicable.

References

1. Ledford, H.; Cyranoski, D.; Van Noorden, R. The UK has approved a COVID vaccine-here's what scientists now want to know. *Nature* **2020**, *588*, 205–206. [CrossRef]
2. Mahase, E. Covid-19: UK approves Moderna vaccine to be given as two doses 28 days apart. *BMJ* **2021**, *372*. [CrossRef] [PubMed]
3. Mahase, E. Covid-19: Moderna applies for US and EU approval as vaccine trial reports 94.1% efficacy. *BMJ* **2020**, *371*. [CrossRef] [PubMed]
4. Lamb, Y.N. Remdesivir: First approval. *Drugs* **2020**, *80*, 1355–1363. [CrossRef]
5. Rubin, D.; Chan-Tack, K.; Farley, J.; Sherwat, A. FDA approval of remdesivir—A step in the right direction. *N. Engl. J. Med.* **2020**, *383*, 2598–2600. [CrossRef]
6. Rivero-Segura, N.A.; Gomez-Verjan, J.C. In Silico Screening of Natural Products Isolated from Mexican Herbal Medicines against COVID-19. *Biomolecules* **2021**, *11*, 216. [CrossRef] [PubMed]
7. World Health Organization. Coronavirus Disease 2019 (COVID-19): Situation Report, 94. 2020. Available online: https://apps.who.int/iris/handle/10665/331865 (accessed on 27 November 2021).
8. Zhou, P.; Yang, X.-L.; Wang, X.-G.; Hu, B.; Zhang, L.; Zhang, W.; Si, H.-R.; Zhu, Y.; Li, B.; Huang, C.-L. A pneumonia outbreak associated with a new coronavirus of probable bat origin. *Nature* **2020**, *579*, 270–273. [CrossRef]
9. World Health Organization. COVID-19 Weekly Epidemiological Update. 2020. Available online: https://www.who.int/emergencies/diseases/novel-coronavirus-2019/situation-reports (accessed on 2 December 2021).
10. Ullrich, S.; Nitsche, C. The SARS-CoV-2 main protease as drug target. *Bioorganic Med. Chem. Lett.* **2020**, *30*, 127377. [CrossRef]
11. Zhang, L.; Lin, D.; Sun, X.; Curth, U.; Drosten, C.; Sauerhering, L.; Becker, S.; Rox, K.; Hilgenfeld, R. Crystal structure of SARS-CoV-2 main protease provides a basis for design of improved α-ketoamide inhibitors. *Science* **2020**, *368*, 409–412. [CrossRef]
12. Amin, S.A.; Banerjee, S.; Gayen, S.; Jha, T. Protease targeted COVID-19 drug discovery: What we have learned from the past SARS-CoV inhibitors? *Eur. J. Med. Chem.* **2021**, 113294. [CrossRef]
13. Steuten, K.; Kim, H.; Widen, J.C.; Babin, B.M.; Onguka, O.; Lovell, S.; Bolgi, O.; Cerikan, B.; Neufeldt, C.J.; Cortese, M. Challenges for targeting SARS-CoV-2 proteases as a therapeutic strategy for COVID-19. *ACS Infect. Dis.* **2021**. [CrossRef] [PubMed]
14. Beck, B.R.; Shin, B.; Choi, Y.; Park, S.; Kang, K. Predicting commercially available antiviral drugs that may act on the novel coronavirus (SARS-CoV-2) through a drug-target interaction deep learning model. *Comput. Struct. Biotechnol. J.* **2020**. [CrossRef] [PubMed]
15. Qamar, M.T.U.; Alqahtani, S.M.; Alamri, M.A.; Chen, L.-L. Structural basis of SARS-CoV-2 3CLpro and anti-COVID-19 drug discovery from medicinal plants. *J. Pharm. Anal.* **2020**, *10*, 313–319. [CrossRef]

16. Gomez, C.R.; Espinoza, I.; Faruke, F.S.; Hasan, M.; Rahman, K.M.; Walker, L.A.; Muhammad, I. Therapeutic intervention of COVID-19 by natural products: A population-specific survey directed approach. *Molecules* **2021**, *26*, 1191. [CrossRef] [PubMed]
17. Wang, J. Fast Identification of Possible Drug Treatment of Coronavirus Disease-19 (COVID-19) through Computational Drug Repurposing Study. *J. Chem. Inf. Model.* **2020**, *60*, 3277–3286. [CrossRef]
18. Chang, Y.-C.; Tung, Y.-A.; Lee, K.-H.; Chen, T.-F.; Hsiao, Y.-C.; Chang, H.-C.; Hsieh, T.-T.; Su, C.-H.; Wang, S.-S.; Yu, J.-Y. Potential therapeutic agents for COVID-19 based on the analysis of protease and RNA polymerase docking. *Preprints* **2020**. [CrossRef]
19. Terrier, O.; Dilly, S.; Pizzorno, A.; Chalupska, D.; Humpolickova, J.; Bouřa, E.; Berenbaum, F.; Quideau, S.; Lina, B.; Fève, B. Antiviral Properties of the NSAID Drug Naproxen Targeting the Nucleoprotein of SARS-CoV-2 Coronavirus. *Molecules* **2021**, *26*, 2593. [CrossRef]
20. Le, N.P.K.; Herz, C.; Gomes, J.V.D.; Förster, N.; Antoniadou, K.; Mittermeier-Kleßinger, V.K.; Mewis, I.; Dawid, C.; Ulrichs, C.; Lamy, E. Comparative Anti-Inflammatory Effects of Salix Cortex Extracts and Acetylsalicylic Acid in SARS-CoV-2 Peptide and LPS-Activated Human In Vitro Systems. *Int. J. Mol. Sci.* **2021**, *22*, 6766. [CrossRef]
21. Akhter, S.; Batool, A.I.; Selamoglu, Z.; Sevindik, M.; Eman, R.; Mustaqeem, M.; Akram, M.S.; Kanwal, F.; Lu, C.; Aslam, M. Effectiveness of Natural Antioxidants against SARS-CoV-2? Insights from the In-Silico World. *Antibiotics* **2021**, *10*, 1011.
22. Wu, C.; Liu, Y.; Yang, Y.; Zhang, P.; Zhong, W.; Wang, Y.; Wang, Q.; Xu, Y.; Li, M.; Li, X. Analysis of therapeutic targets for SARS-CoV-2 and discovery of potential drugs by computational methods. *Acta Pharm. Sin. B* **2020**. [CrossRef]
23. Elzupir, A.O. Inhibition of SARS-CoV-2 main protease 3CLpro by means of α-ketoamide and pyridone-containing pharmaceuticals using in silico molecular docking. *J. Mol. Struct.* **2020**, *1222*, 128878. [CrossRef]
24. Pettersen, E.F.; Goddard, T.D.; Huang, C.C.; Couch, G.S.; Greenblatt, D.M.; Meng, E.C.; Ferrin, T.E. UCSF Chimera—A visualization system for exploratory research and analysis. *J. Comput. Chem.* **2004**, *25*, 1605–1612. [CrossRef] [PubMed]
25. Wang, J.; Wang, W.; Kollman, P.A.; Case, D.A. Automatic atom type and bond type perception in molecular mechanical calculations. *J. Mol. Graph. Model.* **2006**, *25*, 247–260. [CrossRef] [PubMed]
26. O'Boyle, N.M.; Banck, M.; James, C.A.; Morley, C.; Vandermeersch, T.; Hutchison, G.R. Open Babel: An open chemical toolbox. *J. Cheminform.* **2011**, *3*, 1–14. [CrossRef] [PubMed]
27. Shapovalov, M.V.; Dunbrack Jr, R.L. A smoothed backbone-dependent rotamer library for proteins derived from adaptive kernel density estimates and regressions. *Structure* **2011**, *19*, 844–858. [CrossRef]
28. Trott, O.; Olson, A.J. AutoDock Vina: Improving the speed and accuracy of docking with a new scoring function, efficient optimization, and multithreading. *J. Comput. Chem.* **2010**, *31*, 455–461. [CrossRef]
29. Elzupir, A.O. Caffeine and caffeine-containing pharmaceuticals as promising inhibitors for 3-chymotrypsin-like protease of SARS-CoV-2. *J. Biomol. Struct. Dyn.* **2020**, 1–8. [CrossRef]
30. Da Silva, A.W.S.; Vranken, W.F. ACPYPE-Antechamber python parser interface. *BMC Res. Notes* **2012**, *5*, 1–8. [CrossRef]
31. Wang, J.; Wolf, R.M.; Caldwell, J.W.; Kollman, P.A.; Case, D.A. Development and testing of a general amber force field. *J. Comput. Chem.* **2004**, *25*, 1157–1174. [CrossRef]
32. Maier, J.A.; Martinez, C.; Kasavajhala, K.; Wickstrom, L.; Hauser, K.E.; Simmerling, C. ff14SB: Improving the accuracy of protein side chain and backbone parameters from ff99SB. *J. Chem. Theory Comput.* **2015**, *11*, 3696–3713. [CrossRef]
33. Jorgensen, W.L.; Chandrasekhar, J.; Madura, J.D.; Impey, R.W.; Klein, M.L. Comparison of simple potential functions for simulating liquid water. *J. Chem. Phys.* **1983**, *79*, 926–935. [CrossRef]
34. Nelson, M.T.; Humphrey, W.; Gursoy, A.; Dalke, A.; Kalé, L.V.; Skeel, R.D.; Schulten, K. NAMD: A parallel, object-oriented molecular dynamics program. *Int. J. Supercomput. Appl. High Perform. Comput.* **1996**, *10*, 251–268. [CrossRef]
35. De Leeuw, S.W.; Perram, J.W.; Smith, E.R. Simulation of electrostatic systems in periodic boundary conditions. I. Lattice sums and dielectric constants. *Proc. R. Soc. Lond. A Math. Phys. Sci.* **1980**, *373*, 27–56.
36. Essmann, U.; Perera, L.; Berkowitz, M.L.; Darden, T.; Lee, H.; Pedersen, L.G. A smooth particle mesh Ewald method. *J. Chem. Phys.* **1995**, *103*, 8577–8593. [CrossRef]
37. Humphrey, W.; Dalke, A.; Schulten, K. VMD: Visual molecular dynamics. *J. Mol. Graph.* **1996**, *14*, 33–38. [CrossRef]
38. Miller, B.R., III; McGee, T.D., Jr.; Swails, J.M.; Homeyer, N.; Gohlke, H.; Roitberg, A.E. MMPBSA.py: An efficient program for end-state free energy calculations. *J. Chem. Theory Comput.* **2012**, *8*, 3314–3321. [CrossRef]
39. Roe, D.R.; Cheatham, T.E., III. PTRAJ and CPPTRAJ: Software for processing and analysis of molecular dynamics trajectory data. *J. Chem. Theory Comput.* **2013**, *9*, 3084–3095. [CrossRef] [PubMed]
40. Numata, J.; Wan, M.; Knapp, E.-W. Conformational entropy of biomolecules: Beyond the quasi-harmonic approximation. *Genome Inform.* **2007**, *18*, 192–205. [PubMed]
41. Lim, L.; Shi, J.; Mu, Y.; Song, J. Dynamically-driven enhancement of the catalytic machinery of the SARS 3C-like protease by the S284-T285-I286/A mutations on the extra domain. *PLoS ONE* **2014**, *9*, e101941. [CrossRef] [PubMed]
42. Saver, J.L. Citicoline: Update on a promising and widely available agent for neuroprotection and neurorepair. *Rev. Neurol. Dis.* **2008**, *5*, 167–177.
43. Dinsdale, J.; Griffiths, G.; Rowlands, C.; Castelló, J.; Ortiz, J.; Maddock, J.; Aylward, M. Pharmacokinetics of 14C CDP-choline. *Arzneim. Forsch.* **1983**, *33*, 1066–1070.
44. Anwar, M.U.; Adnan, F.; Abro, A.; Khan, M.R.; Rehman, A.U.; Osama, M.; Javed, S.; Baig, A.; Shabbir, M.R.; Assir, M.Z. Combined Deep Learning and Molecular Docking Simulations Approach Identifies Potentially Effective FDA Approved Drugs for Repurposing Against SARS-CoV-2. *Comput. Biol. Med.* **2021**. [CrossRef]

45. Prajapat, M.; Shekhar, N.; Sarma, P.; Avti, P.; Singh, S.; Kaur, H.; Bhattacharyya, A.; Kumar, S.; Sharma, S.; Prakash, A. Virtual screening and molecular dynamics study of approved drugs as inhibitors of spike protein S1 domain and ACE2 interaction in SARS-CoV-2. *J. Mol. Graph. Model.* **2020**, *101*, 107716. [CrossRef]
46. Elfiky, A.A. Ribavirin, Remdesivir, Sofosbuvir, Galidesivir, and Tenofovir against SARS-CoV-2 RNA dependent RNA polymerase (RdRp): A molecular docking study. *Life Sci.* **2020**, *253*, 117592. [CrossRef]
47. Chien, M.; Anderson, T.K.; Jockusch, S.; Tao, C.; Li, X.; Kumar, S.; Russo, J.J.; Kirchdoerfer, R.N.; Ju, J. Nucleotide analogues as inhibitors of SARS-CoV-2 polymerase, a key drug target for COVID-19. *J. Proteome Res.* **2020**, *19*, 4690–4697. [CrossRef]
48. Nar, H.; Schnapp, G.; Hucke, O.; Hardman, T.C.; Klein, T. Action of dipeptidyl peptidase-4 inhibitors on SARS-CoV-2 main protease. *ChemMedChem* **2021**, *16*, 1425–1426. [CrossRef]
49. Zhou, L.; Wang, J.; Liu, G.; Lu, Q.; Dong, R.; Tian, G.; Yang, J.; Peng, L. Probing antiviral drugs against SARS-CoV-2 through virus-drug association prediction based on the KATZ method. *Genomics* **2020**, *112*, 4427–4434. [CrossRef]
50. Zhang, Y.-N.; Zhang, Q.-Y.; Li, X.-D.; Xiong, J.; Xiao, S.-Q.; Wang, Z.; Zhang, Z.-R.; Deng, C.-L.; Yang, X.-L.; Wei, H.-P. Gemcitabine, lycorine and oxysophoridine inhibit novel coronavirus (SARS-CoV-2) in cell culture. *Emerg. Microbes Infect.* **2020**, *9*, 1170–1173. [CrossRef] [PubMed]
51. Zheng, Z.; Groaz, E.; Snoeck, R.; De Jonghe, S.; Herdewijn, P.; Andrei, G. Influence of 4′-Substitution on the Activity of Gemcitabine and Its ProTide against VZV and SARS-CoV-2. *ACS Med. Chem. Lett.* **2020**, *12*, 88–92. [CrossRef] [PubMed]
52. Verma, A.K.; Aggarwal, R. Repurposing potential of FDA-approved and investigational drugs for COVID-19 targeting SARS-CoV-2 spike and main protease and validation by machine learning algorithm. *Chem. Biol. Drug Des.* **2021**, *97*, 836–853. [CrossRef] [PubMed]
53. Borbone, N.; Piccialli, G.; Roviello, G.N.; Oliviero, G. Nucleoside analogs and nucleoside precursors as drugs in the fight against SARS-CoV-2 and other coronaviruses. *Molecules* **2021**, *26*, 986. [CrossRef]
54. Jo, S.; Kim, S.; Yoo, J.; Kim, M.-S.; Shin, D.H. A Study of 3CLpros as Promising Targets against SARS-CoV and SARS-CoV-2. *Microorganisms* **2021**, *9*, 756. [CrossRef] [PubMed]
55. Pillaiyar, T.; Meenakshisundaram, S.; Manickam, M. Recent discovery and development of inhibitors targeting coronaviruses. *Drug Discov. Today* **2020**, *25*, 668–688. [CrossRef]
56. Maurya, S.K.; Maurya, A.K.; Mishra, N.; Siddique, H.R. Virtual screening, ADME/T, and binding free energy analysis of anti-viral, anti-protease, and anti-infectious compounds against NSP10/NSP16 methyltransferase and main protease of SARS CoV-2. *J. Recept. Signal Transduct.* **2020**, *40*, 605–612. [CrossRef]
57. Kandeel, M.; Al-Nazawi, M. Virtual screening and repurposing of FDA approved drugs against COVID-19 main protease. *Life Sci.* **2020**, *251*, 117627. [CrossRef]
58. Choi, R.; Zhou, M.; Shek, R.; Wilson, J.W.; Tillery, L.; Craig, J.K.; Salukhe, I.A.; Hickson, S.E.; Kumar, N.; James, R.M. High-throughput screening of the ReFRAME, Pandemic Box, and COVID Box drug repurposing libraries against SARS-CoV-2 nsp15 endoribonuclease to identify small-molecule inhibitors of viral activity. *PLoS ONE* **2021**, *16*, e0250019. [CrossRef]
59. Sharanya, C.; Abhithaj, J.; Sadasivan, C. Drug repurposing to identify therapeutics against COVID 19 with SARS-Cov-2 spike glycoprotein and main protease as targets: An in silico study. *ChemRxiv* **2020**. [CrossRef]
60. Fadlalla, M. COVID19 Approved Drug Repurposing: Pocket Similarity Approach. *ChemRxiv* **2020**. [CrossRef]
61. Rahman, N.; Basharat, Z.; Yousuf, M.; Castaldo, G.; Rastrelli, L.; Khan, H. Virtual screening of natural products against type II transmembrane serine protease (TMPRSS2), the priming agent of coronavirus 2 (SARS-CoV-2). *Molecules* **2020**, *25*, 2271. [CrossRef] [PubMed]

Article

Selective Oxidation of Clopidogrel by Peroxymonosulfate (PMS) and Sodium Halide (NaX) System: An NMR Study

Everaldo F. Krake and Wolfgang Baumann *

Leibniz-Institut für Katalyse e.V., Albert-Einstein-Straße 29a, 18059 Rostock, Germany; everaldokiko@gmail.com
* Correspondence: wolfgang.baumann@catalysis.de

Abstract: A selective transformation of clopidogrel hydrogen sulfate (CLP) by reactive halogen species (HOX) generated from peroxymonosulfate (PMS) and sodium halide (NaX) is described. Other sustainable oxidants as well as different solvents have also been investigated. As result of this study, for each sodium salt the reaction conditions were optimized, and four different degradation products were formed. Three products were halogenated at C-2 on the thiophene ring and have concomitant functional transformation, such as N-oxide in the piperidine group. A halogenated endo-iminium product was also observed. With this condition, a fast preparation of known endo-iminium clopidogrel impurity (new counterion) was reported as well. The progress of the reaction was monitored using nuclear magnetic resonance spectroscopy as an analytical tool and all the products were characterized by 1D-, 2D-NMR and HRMS.

Keywords: clopidogrel; NMR study; oxone; peroxymonosulfate; sodium halide; thienopyridine

Citation: Krake, E.F.; Baumann, W. Selective Oxidation of Clopidogrel by Peroxymonosulfate (PMS) and Sodium Halide (NaX) System: An NMR Study. *Molecules* **2021**, *26*, 5921. https://doi.org/10.3390/molecules26195921

Academic Editors: Giovanni Ribaudo and Laura Orian

Received: 20 August 2021
Accepted: 25 September 2021
Published: 29 September 2021

Publisher's Note: MDPI stays neutral with regard to jurisdictional claims in published maps and institutional affiliations.

Copyright: © 2021 by the authors. Licensee MDPI, Basel, Switzerland. This article is an open access article distributed under the terms and conditions of the Creative Commons Attribution (CC BY) license (https://creativecommons.org/licenses/by/4.0/).

1. Introduction

The direct insertion of halogens in (hetero)aromatic drugs, in a selective way, has been the object of much interest by the synthetic community [1]. The inclusion of a new C–X bond in these bioactive heterocyclic compounds can improve their physical and biological properties, increase potency, and be used as a handle in the further design and construction of pharmaceuticals [2]. Thiophene rings are five-membered heterocycles bearing sulfur atoms in their structure. Connected to a halogen, halothiophenes represent a class of bioactive molecules with extraordinary pharmacological properties [3,4], including the FDA-approved drugs Avatrombopag, Tioconazole, Lornoxicam, Rivaroxaban, and Brotizolam (Figure 1A).

In recent years, several methods of direct activation of halogens in organic compounds have been reported using safe halogen sources such as HX, NH$_4$X and NaX (X = Cl, Br and I). To sustainably transform these halides into more reactive species, the use of oxidizing agents that conform to the principles of Green Chemistry is essential [1].

Like other green oxidants such as O$_2$ and H$_2$O$_2$ [5], peroxymonosulfate (PMS, Oxone ®) has been widely used: (1) in the academia to develop new synthetic protocols [6]; (2) in pharmaceutical companies to promote oxidative stress testing of active pharmaceutical ingredients (API) to predict their degradation [7]; and (3) in hypersaline industrial wastewaters to remove organic contaminants [8–10]. This safe, sustainable, and inexpensive oxidant has shown extraordinary reactivity with alkaline metal halide salts [11–27] and with hydrogen halides (HCl, HBr and HI) [28–30].

In our previous work, we documented the reactive sequence of the oxidation–chlorination of Ticlopidine hydrochloride using PMS. We observed the formation of the reactive intermediate species (**DP-1**) containing a chlorine group at the C-2 carbon of the thiophene and the oxidation of cyclic amine to N-oxide in the piperidinic structure (Figure 1B) [31].

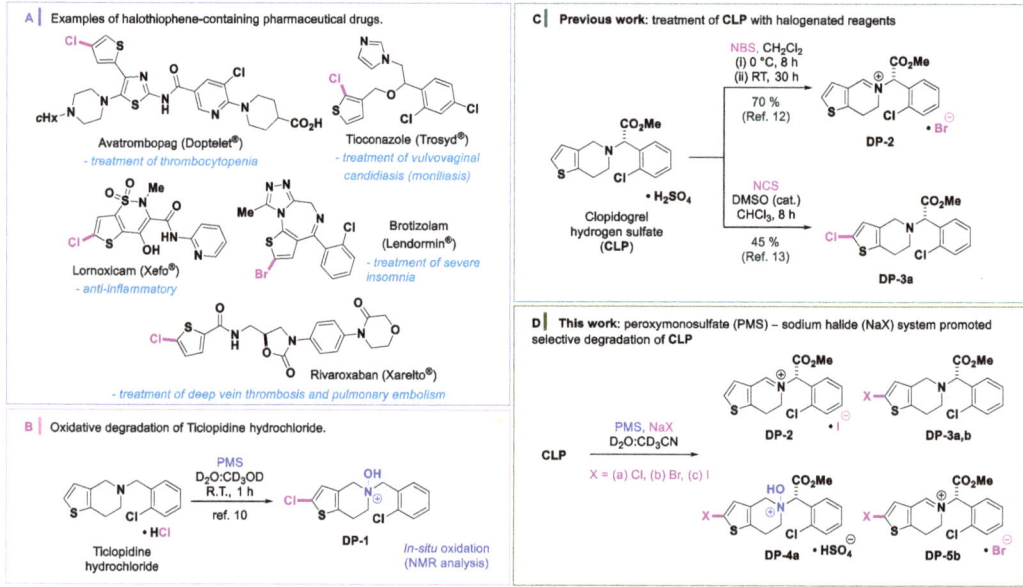

Figure 1. Halothiophene-containing pharmaceutical drugs and representative oxidation of **CLP**. (**A**) Examples of halothiophene-containing pharmaceutical drugs. (**B**) Our previous work: oxidative degradation of ticlopidine hydrochloride with PMS. (**C**) Previous work: treatment of **CLP** with halogenated succinimides. (**D**) This work: PMS-halide oxidation of **CLP**, anions may be halide or hydrogen sulfate.

Clopidogrel hydrogen sulfate (**CLP**, Plavix ^{TM}) is another thienopyridine drug that has powerful antiplatelet properties, and it plays an important role in the treatment of coronary, peripheral vascular and cerebrovascular diseases [32–34]. Several researchers have reported functional transformations of **CLP** using halogenated succinimide reagents (Figure 1C). Padi and coworkers have used N-bromosuccinimide (NBS) for the preparation of endo-iminium impurity **DP-2** on a large scale [35]. Jiao and coworkers have developed an efficient method using N-chlorosuccinimide (NCS) and dimethyl sulfoxide (DMSO) as a catalyst for the preparation of 2-Cl-clopidogrel **DP-3a** [36].

Continuing our efforts in the development and optimization of sustainable prediction methods for degradation of active pharmaceutical ingredients, we wanted to understand the oxidative halogenation reaction of heterocycles containing non-hydrohalic acids with safe halogen sources. Here, we show the oxidation of **CLP** using PMS/sodium halides (NaX, X = Cl, Br and I; Figure 1D). We observed the formation of four interesting classes of products that can predict the degradation of thienopyridines under high salinity media. To the best of our knowledge, detection, characterization, and selective preparation of the products in this oxidative method are first reported herein.

2. Results and Discussion

2.1. Impact of Halides on the Transformation of Clopidogrel

2.1.1. Influence of Chloride

Our experimental work started with the optimization of conditions for the oxidative chlorination of **CLP** with sodium chloride (NaCl) as a chlorine source, and different oxidant agents (Table 1). We monitored the reaction by NMR and HPLC. In the initial experiments, the treatment of **CLP** with H_2O_2 or *tert*-butyl hydroperoxide (TBHP) in the presence of NaCl did not lead to the formation of products after 24 h at room temperature (Table 1, Entries 1 and 2). Using PMS in D_2O, we observed the complete consumption of **CLP** and the formation of various degradation products after one minute. Inspired by our previous work [31], we decided to investigate the co-solvent effect for this reaction. We performed some experiments with dichloromethane-d$_2$ (CD$_2$Cl$_2$), chloroform-d (CDCl$_3$), dimethyl sulfoxide-d$_6$ ((CD$_3$)$_2$SO), benzene-d$_6$ (C$_6$D$_6$), and toluene-d$_8$ (C$_7$D$_8$), but these deuterated co-solvents did not lead to the formation of any products. However, with acetone-d$_6$, we observed the formation of an N-oxide product, **DP-6** (48%), as well as the chlorinated products **DP-3a** (4%) and **DP-4a** (25%) after 5 h (Entry 4).

Table 1. Screening of conditions for oxidative chlorination reaction [a].

Entry	NaCl (equiv)	Oxidant (equiv)	Deuterated Solvent [b]	Time (h)	DP-3a (%) [d]	DP-4a (%) [d]	DP-6 (%) [d]
1	1.0	H$_2$O$_2$ (30%)	H$_2$O$_2$:CD$_3$OD (2:1)	24	NR	NR	NR
2	1.0	TBHP (70%)	TBHP:CD$_3$OD (2:1)	24	NR	NR	NR
3	1.0	PMS (2.0)	D$_2$O [c]	1 min	degrad.	degrad.	degrad.
4	1.0	PMS (2.0)	D$_2$O:(CD$_3$)$_2$CO (2:1)	5	4	25	48
5	1.0	PMS (2.0)	D$_2$O:CD$_3$OD (2:1)	5	19	32	24
6	1.0	PMS (2.0)	D$_2$O:C$_4$D$_8$O (2:1)	5	1	51	45
7	1.0	PMS (2.0)	D$_2$O:CD$_3$CN (2:1)	4	6	51	36
8	**2.0**	**PMS (2.0)**	**D$_2$O:CD$_3$CN (2:1)**	**4**	**4**	**85**	**10**
9	1.0	PMS (1.5)	D$_2$O:CD$_3$CN (2:1)	4	2	52	42
10	2.0	PMS (1.5)	D$_2$O:CD$_3$CN (2:1)	4	17	59	19
11	1.0	PMS (1.0)	D$_2$O:CD$_3$CN (2:1)	4	5	27	52
12	2.0	PMS (1.0)	D$_2$O:CD$_3$CN (2:1)	4	15	32	33
13	1.0	PMS (0.5)	D$_2$O:CD$_3$CN (2:1)	4	5	7	42

[a] Reaction conditions: **CLP** (1.0 equiv), NaCl, PMS, 0.2 mL of co-solvent, 0.4 mL of D$_2$O. [b] List of abbreviations: (TBHP) tert-butyl hydroperoxide; (CD$_3$OD) methanol-d$_4$; (D$_2$O) water-d$_2$; (CD$_3$CN) acetonitrile-d$_3$; ((CD$_3$)$_2$CO) acetone-d$_6$; (C$_4$D$_8$O) tetrahydrofuran-d$_8$; (NR) no reaction; (degrad.) degradation. [c] Only D$_2$O was used (0.6 mL). [d] Conversion determined by HPLC analysis. Note: Entry 8 (in bold) refers to the best condition for the formation of **DP-4a**.

To investigate the selectivity between chlorinated products **DP-3a** and **DP-4a**, the same reaction was performed with other polar solvents (Entries 5–7): methanol-d$_4$ (32%), THF-d$_8$ (51%) and acetonitrile-d$_3$ (51%). We observed increased formation of **DP-4a**, as well as small amounts of **DP-3a**. With the optimized solvent in hand, we extended the method to optimize **DP-3a** or **DP-4a** using varying amounts of PMS and NaCl (Entries 8–13). We did not obtain **DP-3a** with a conversion higher than 20%. Thus, our best conditions for **DP-4a** formation used a mixture of D$_2$O:CD$_3$CN (2:1), PMS (2.0 equiv), and NaCl (2.0 equiv; Entry 8).

2.1.2. Influence of Bromide

In parallel to the chlorination reaction, we submitted **CLP** to the oxidation process using NaBr as a bromine source (Table 2). We observed a quick conversion of the brominated products **DP-3b** and **DP-5b** with all solvents studied. Then, we extended our study varying the amount of PMS and NaBr. The best condition for the formation of **DP-3b** was with D$_2$O:CD$_3$CN (2:1), PMS (0.5 equiv), and NaBr (1.0 equiv; Table 2, Entry 6). The combination of PMS (1.5 equiv) and NaBr (1.0 equiv; Table 2, Entry 8) provided **DP-5b** in excellent yield.

Table 2. Screening of conditions for oxidative bromination reaction [a].

Entry	NaBr (equiv)	PMS (equiv)	Deut. Solvent	Time (min)	DP-3b (%) [b]	DP-5b (%) [b]
1	NaBr (1.0)	1.0	$D_2O:CD_3OD$ (2:1)	1	62	38
2	NaBr (1.0)	1.0	$D_2O:C_4D_8O$ (2:1)	1	3	97
3	NaBr (1.0)	1.0	$D_2O:(CD_3)_2CO$ (2:1)	1	34	66
4	NaBr (1.0)	1.0	$D_2O:CD_3CN$ (1:1)	1	60	39
5	NaBr (2.0)	1.0	$D_2O:CD_3CN$ (1:1)	1	34	66
6	**NaBr (1.0)**	**0.5**	**$D_2O:CD_3CN$ (1:1)**	**1**	**66**	**2**
7	NaBr (2.0)	0.5	$D_2O:CD_3CN$ (1:1)	1	61	10
8	**NaBr (1.0)**	**1.5**	**$D_2O:CD_3CN$ (1:1)**	**1**	**-**	**99**
9	NaBr (2.0)	1.5	$D_2O:CD_3CN$ (1:1)	1	-	99
10	NaBr (1.0)	2.0	$D_2O:CD_3CN$ (1:1)	1	-	99

[a] Reaction conditions: **CLP** (1.0 equiv), NaBr, PMS, 0.2 mL of co-solvent, 0.4 mL of D_2O. [b] Conversion determined by HPLC analysis. Note: Entries 6 and 8 (in bold) refer to the best conditions for the formation of **DP-3b** and **DP-5b**, respectively.

2.1.3. Influence of Iodide

When we used the iodide reagent NaI and other co-solvents (Entries 1–3), solubility problems were observed after the in situ oxidative conversion of the iodide to the reactive iodine species was generated. With NaCl, NaBr and NaI, acetonitrile-d_3 was the most efficient deuterated co-solvent. Unlike with the halides described above, no substitution at the thiophene ring occurred (no formation of **DP-3c**, entries 4 and 5). Instead, we discovered a fast method (as compared to Padi's) [35] to prepare **DP-2** in excellent yield by using the PMS/NaI system (Table 3, Entry 4).

Table 3. Screening of conditions for oxidative iodination reaction [a].

Entry	NaI (equiv)	PMS (equiv)	Deut. Solvent	Time	DP-2 (%) [b]	DP-3c (%) [b]
1	NaI (1.0)	1.0	$D_2O:CD_3OD$ (2:1)	12 h	<1	-
2	NaI (1.0)	1.0	$D_2O:C_4D_8O$ (1:1)	12 h	NR	NR
3	NaI (1.0)	1.0	$D_2O:(CD_3)_2CO$ (1:1)	12 h	NR	NR
4	**NaI (1.0)**	**1.0**	**$D_2O:CD_3CN$ (1:1)**	**10 min**	**98**	**-**
5	NaI (1.0)	0.5	$D_2O:CD_3CN$ (1:1)	10 min	65	-

[a] Reaction conditions: **CLP** (1.0 equiv), NaI, PMS, 0.2 mL of co-solvent, 0.4 mL of D_2O. [b] Conversion determined by HPLC analysis. Note: Entry 4 (in bold) refers to the best condition for the formation of **DP-2**.

2.2. Determination of Reaction Progress by NMR

2.2.1. With NaCl

In Figure 2, a compilation of the ^{1}H NMR spectra on the main functional transformations that occurred in the oxidative process, which resulted in the formation of the products in high yields, is shown. In addition, the data of the starting materials (as reference) and non-chlorinated intermediates **DP-6** (mixture of diastereomers) are also included in this compilation of spectra, as described in Figure 2A,B [37]. The main observation among the NMR spectra of this work is the disappearance of the doublet (δ_H ~6.7 ppm, CD$_3$CN) of proton H-3 of the thiophene ring and a conversion into a singlet in that same region which results from the insertion of a heteroatom at carbon-2 of this heterocycle (Figure 2C).

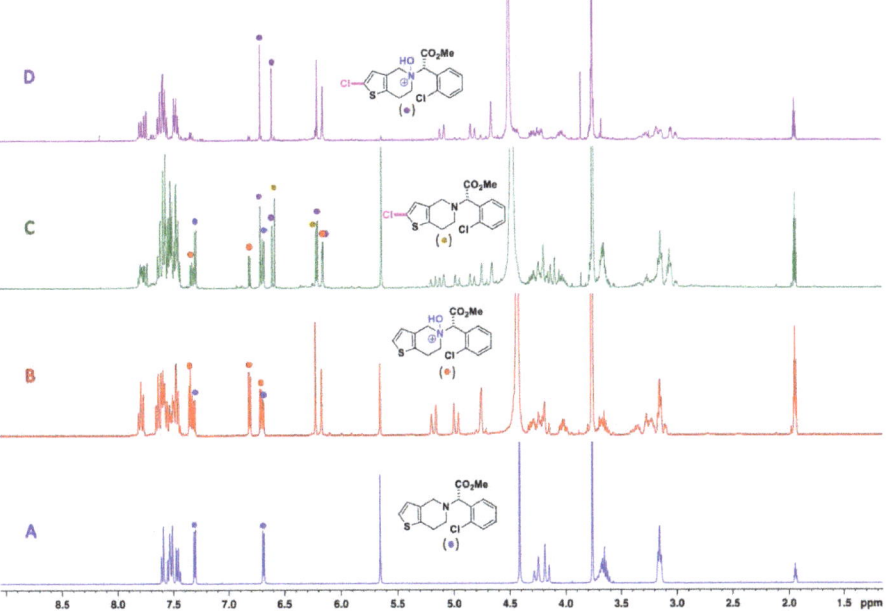

Figure 2. Compilation of ^{1}H NMR spectra (400 MHz, CD$_3$CN) of all compounds involved. Reaction conditions: (**A**) **CLP** (as ^{1}H NMR reference); (**B**) Experiment performed using only **CLP** (1.0) and PMS (1.0); (**C**) Progress of reaction after 2 h (see Table 1, Entry 12); (**D**) Crude ^{1}H NMR spectrum of diastereomers **DP-4a** after 12 h (see Table 1, Entry 8).

Figure 2C represents the progress of oxidative chlorination of **CLP** and NaCl after two hours (Table 1, Entry 12). In the first two hours of reaction, we observed the appearance of non-halogenated diastereomers of intermediate **DP-6** with a new N$^+$-OH bond, as well as the mono-chlorinated compound **DP-3a**. As the reaction progressed towards the production of **DP-4a**, the signals of **DP-6** and **DP-3a** decreased in the spectrum. This led us to conclude that the reaction pathway to obtain **DP-4a** occurred via two simultaneous processes: oxidation-chlorination via **DP-6** and chlorination-oxidation via **DP-3a** (Figure 3). The final product, **DP-4a**, is represented in Figure 4D, and bears a chlorine on the thiophene ring and an *N*-oxide in the piperidine structure. The mechanism of this reaction was shown in our previous work [31].

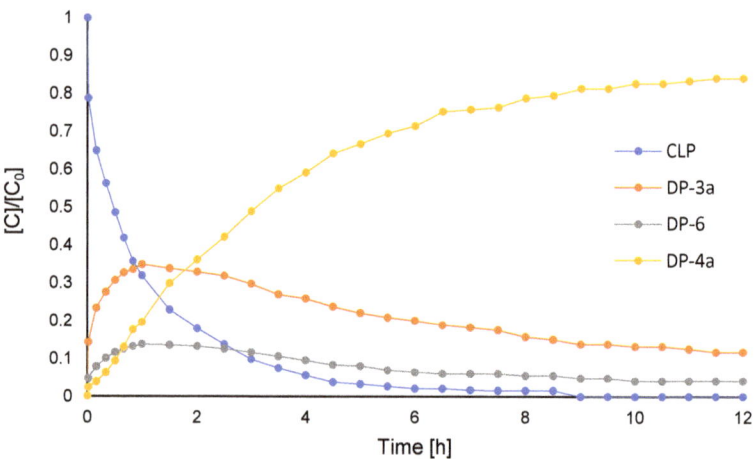

Figure 3. Reaction progress for oxidative chlorination reaction of **CLP** with PMS/NaCl system measured by ^{1}H NMR (see Table 1, Entry 8).

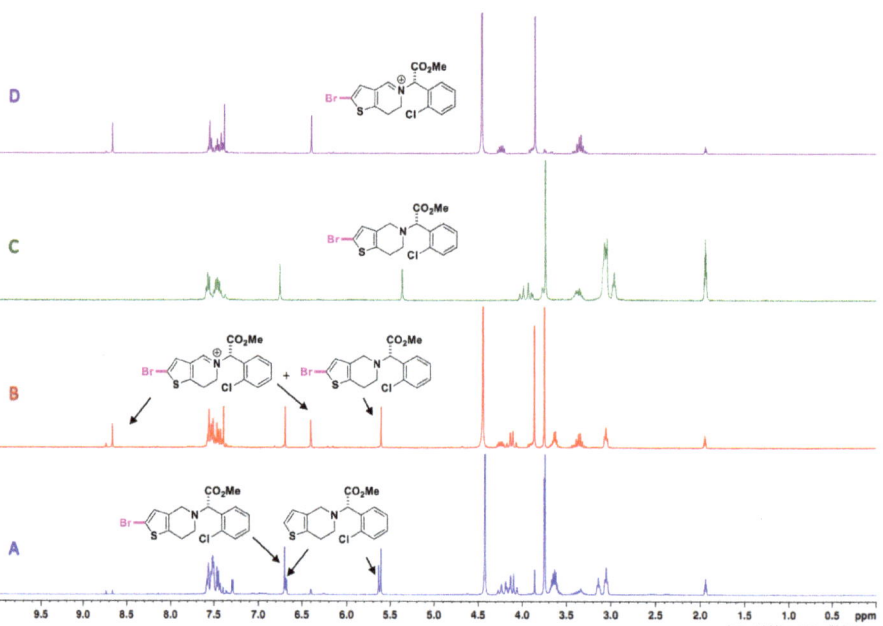

Figure 4. Compilation of crude ^{1}H NMR spectra (400 MHz, CD$_3$CN) of all compounds involved (**CLP**, **DP-3b** and **DP-5b**). Reaction conditions: (**A**) PMS (0.5), NaBr (1.0), see Table 2, Entry 6; (**B**) PMS (1.0), NaBr (1.0), see Table 2, Entry 3; (**C**) ^{1}H NMR spectrum of **DP-3b** isolated; (**D**) PMS (1.5), NaBr (1.0), see Table 2, Entry 8.

2.2.2. With NaBr

Contrary to the PMS/NaCl process, the reaction with NaBr did not present N-oxide products, but a C-2 halogen/endo-iminium **DP-5b** when PMS (1.0 equiv) was used (Figure 4B,D). To understand the development of this reaction, we started with 0.5 equivalent of PMS (Table 2, Entry 6; Figure 4A) and observed a mixture between **CLP** and mono-brominated product **DP-3b**. In addition, we increased the amount of PMS to provide **DP-5b** in a quantitative yield (Table 2, Entry 8; Figure 4D).

2.2.3. With NaI

With NaI, we tried to obtain an iodinated product, **DP-3c**, under our experimental conditions. Instead, we observed the formation of endo-iminium **DP-2** in high yield (Table 3, Entry 4; Figure 5B). Reducing the molar amounts of PMS, we detected a mixture between **CLP** and endo-iminium **DP-2** (Table 3, Entry 5, Figure 5A).

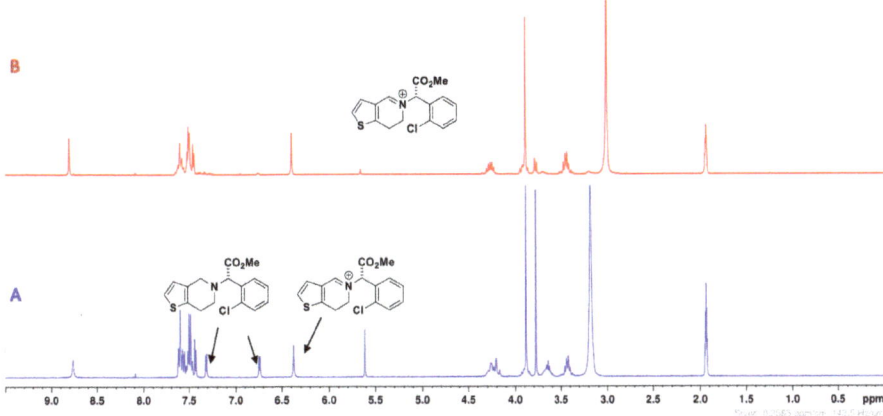

Figure 5. Compilation of crude ^{1}H NMR spectra (400 MHz, CD$_3$CN) of all compounds involved (**CLP** and **DP-2**). Reaction conditions: (**A**): PMS (0.5), NaI (1.0), see Table 3, Entry 5; (**B**) PMS (1.0), NaI (1.0), see Table 3, Entry 4.

2.3. Characterization of the Products

Degradation products **DP-2**, **DP-3a**, **DP-4a**, **DP-3b** and **DP-5b** were characterized directly from the reaction mixtures by HPLC-MS, HRMS, as well as by 1D- and 2D-NMR spectroscopy. Two-dimensional correlations ^{1}H-^{1}H COSY, ^{1}H-^{13}C HSQC and ^{1}H-^{13}C HMBC were used for the elucidation of the structure.

2.3.1. Characterization of DP-2

Using NMR spectrometry (in CD$_3$CN, see Table 4 and SI), fifteen protons were detected, a value consistent with the molecular formula of **DP-2**. In addition, the bidimensional ^{1}H-^{1}H COSY spectrum showed important correlations for structural elucidation of this molecule, including a cross peak between endo-iminium H-4 (δ_H 8.81, δ_C 162.40) and the singlet for H-10, which is part of a CH group (δ_H 6.39, δ_C 72.08). Shift data of the latter are similar to that observed for this position in the other molecules (Table 4). This feature proves that an endo-iminium compound, and not an exo-iminium one, is indeed formed. The edited HSQC exhibited the CH$_2$ groups H-6 (δ_H 4.30–4.23 and 3.95–3.86, δ_C 49.94) and H-7 (δ_H 3.52–3.36, δ_C 24.04). With HRMS/ESI-TOF analysis, the molecular formula of **DP-2** was determined as C$_{16}$H$_{15}$ClNO$_2$S$^+$, with m/z calculated at 320.0517, and m/z observed at 320.0516 (−0.3 ppm error), indicating loss of one hydrogen atom in the molecule and the formation of the endo-iminium product. These analyses do not show a halogenation bond in the molecule and confirm the structure of this degradation product.

Table 4. ¹H and ¹³C assignments for all clopidogrel degradation products.

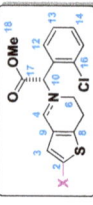

Position	X = H (DP-2) ¹H	X = H (DP-2) ¹³C	X = Cl (DP-3a) ¹H	X = Cl (DP-3a) ¹³C	X = Cl, N-OH (DP-4a) [a] ¹H	X = Cl, N-OH (DP-4a) [a] ¹³C	X = Cl, N-OH (DP-4a') [a] ¹H	X = Cl, N-OH (DP-4a') [a] ¹³C	X = Br (DP-3b) ¹H	X = Br (DP-3b) ¹³C	X = Br (DP-5a) ¹H	X = Br (DP-5a) ¹³C
2 (CX)	-	128.15	-	133.59	-	125.72	-	125.39	-	110.38	-	128.41
3 (CH)	-	129.44	6.50 s	125.54	6.70 s	125.26	6.60 s	125.26	6.70 s	129.36	-	133.01
4 (CH/H₂)	8.81 s	162.40	3.64 d (14.5) 3.54 d (14.5)	51.07	5.09 d (15.5) 4.82 d (15.5)	62.50	4.65 (s)	62.27	3.63 dt (14.4, 1.9) 3.53 dt (14.4, 1.9)	51.10	9.05 s	162.46
6 (CH₂)	4.30–4.23 m 3.95–3.86 m	49.94	2.98–2.84 m	49.94	4.22 dd (12.1, 5.3) 4.03 m	60.68	4.40 m 4.29 dd (9.2, 5.1)	61.99	2.95–2.80 m	49.21	4.44–4.37 m 4.00–3.91 m	49.64
7 (CH₂)	3.52–3.36 m	24.04	2.76–2.70 m	25.78	3.04 m	21.54	3.24 m	22.19	2.77–2.74 m	26.02	3.59–3.40 m	24.38
8 (C)	-	156.50	-	128.22	-	130.56	-	130.48	-	136.39	-	158.09
9 (C)	-	129.02	-	133.28	-	130.39	-	130.18	-	135.32	-	130.19
10 (CH)	6.40 s	72.08	4.97 s	68.55	6.19 (s)	76.95	6.15 (s)	74.93	4.93 s	68.68	6.62 s	72.72
11 (C_ipso)	-	129.35	-	134.13	-	134.58	-	134.55	-	134.76	-	115.56
12 (C_Ar,H)	7.62–7.49 m	133.81	7.44–7.41 m	130.99	7.65–7.53 m	129.20	7.65–7.53 m	129.09	7.48–7.43 m	131.02	7.68–7.56 m	131.29
13 (C_Ar,H)	-	129.16	7.33–7.29 m	128.48	7.65–7.53 m	134.58	7.65–7.53 m	134.54	7.37–7.31 m	128.50		129.61
14 (C_Ar,H)	-	132.86		131.11	7.50–7.43 m	131.91	7.50–7.43 m	131.83		131.05		133.01
15 (C_Ar,H)	7.63–7.60 m	131.83		130.99	7.73 dd (7.9, 1.6)	133.93	7.77 dd (7.9, 1.6)	133.93	7.66–7.61 m	131.02		132.08
16 (CCl)	-	136.18	-	135.80	-	137.59	-	137.17	-	135.91	-	136.54
17 (C=O)	-	167.81	-	172.10	-	166.27	-	166.17	-	172.76	-	168.20
18 (OCH₃)	3.89 s	55.02	3.68 s	52.84	3.75 (s)	55.06	3.74 (s)	55.06	3.70 s	52.71	3.97 s	54.85

Note: Positions show the number on the chemical structure above. Chemical shifts (δ) are reported in ppm relative to TMS (see Section 3.2). J- values are shown in Hz in parentheses. Abbreviations: d, doublet; dd, doublet doublet; dt, doublet triplet; m, multiplet; s, singlet; t, triplet. [a] ¹H NMR (400 MHz, CD₃CN) signals correspond to 56:44 mixture of diastereomers DP-4a and DP-4a' (cf. ref. [37]).

2.3.2. Characterization of DP-3a

The molecular formula of product **DP-3a** was established as $C_{16}H_{11}O_2NSCl_2$ by HRMS/ESI-TOF data in which m/z of 356.0279 was observed and was calculated for [M + H]$^+$ 356.0288 (2.5 ppm error), indicating the insertion of a new chlorine atom into the molecule. In the NMR spectrum (in CD$_3$OD, see Table 4 and SI), the addition of the chlorine atom at carbon 2 resulted in an absence of the H-2 signal and the appearance of the singlet associated with H-3 (δ_H 6.50, δ_C 125.54) consistent with the molecular formula of **DP-3a**. ^{1}H-^{13}C HSQC edited indicated the methylene groups: H-4 (δ_H 3.64 and 3.54, δ_C 51.07), H-6 (δ_H 2.98–2.84, δ_C 49.21) and H-7 (δ_H 2.76–2.70, δ_C 25.78). The homonuclear correlation spectroscopy (^{1}H-^{1}H COSY) showed a correlation with the protons H-3/H-4 and H-3/H-7. This analysis was important to explain the structure elucidation of **DP-3a**. Also, ^{1}H-^{13}C HMBC correlations from proton signals for H-3, H-4, H-6 and H-10 were observed.

2.3.3. Characterization of DP-4a

The oxidative chlorinated product **DP-4a** was found to have the molecular formula $C_{16}H_{15}O_3NSCl_2$ by HRMS/ESI-TOF, in which m/z 372.0228 was observed and was calculated for [M + H]$^+$ 372.0227 (−0.3 ppm error). This mass can also be seen in two retention times in the HPLC-MS chromatogram of Figures S1 and S2 (see SI) indicating the formation of two diastereomers with N$^+$-OH bonds. In comparison to the data of ^{1}H NMR with **CLP**, and based in our previous report about **DP-6** (Figure 2B,D in CD$_3$CN; also see ref. [37]), it is possible to notice a significant change in the chemical shift in H-4 (δ_H 6.62 and 6.52 ppm; δ_C 62.50, 62.27 ppm) due to the asymmetric electric field (AMEF) generated in the molecule. This electrical field is the result of the polarization of the molecule through the dipole N$^+$-OH generated in this oxidative process.

2.3.4. Characterization of DP-3b

Similar to the results seen for the compound **DP-3a**, the product **DP-3b** showed the molecular formula $C_{16}H_{15}O_2NSClBr$ by HRMS/ESI-TOF, with m/z calculated 399.9774 and observed for [M + H]$^+$ 399.9772 (−0.5 ppm error), specifying the insertion of a bromine atom into the molecule. In the NMR spectrum (in CD$_3$OD, see Table 4 and SI), the inclusion of a bromine atom at carbon 2 resulted in an absence of the H-2 signal and the appearance of a signal associated with H-3 (δ_H 6.70, δ_C 129.36), consistent with the molecular formula of **DP-3b**. In addition, other changes in chemical shift in H-4 (δ_H 3.63 and 3.53, δ_C 51.10), H-6 (δ_H 2.95–2.80, δ_C 49.21) and H-7 (δ_H 2.77–2.74, δ_C 26.02) were noted. HMBC correlations from proton signals for H-3, H-4, H-6 and H-10 were observed.

2.3.5. Characterization of DP-5b

Finally, in the NMR spectrum (in CD$_3$OD, see Table 4 and SI), fourteen protons were observed, a value consistent with the molecular formula of **DP-5b**. In addition, HSQC showed the existence of an endo-iminium group in the piperidine moiety as a singlet (δ_H 9.05, δ_C 162.47), two methylene groups in H-6 (δ_H 4.44–4.37 and 4.00 -3.91, δ_C 49.65), H-7 (δ_H 3.59- 3.40, δ_C 24.38) and H-10 protons (δ_H 6.62, δ_C 72.72). HMBC correlations from proton signals for H-3, H-4, H-6 and H-10 were observed. Also, the molecular formula of **DP-5b** was determined to be $C_{16}H_{14}BrClNO_2S^+$ by HRMS/ESI-TOF, with an observed m/z of 397.9622 and a calculated m/z of 397.9623 (0.3 ppm error) indicating the insertion of a bromine atom into the molecule, as well as the formation of the endo-iminium. This analysis corroborated with the data obtained from the NMR spectra.

3. Materials and Methods

3.1. Materials

Clopidogrel hydrogen sulphate (**CLP**) was kindly provided from RD&C Research, Development & Consulting GmbH, Vienna, Austria. Peroxymonosulfate (PMS, Oxone$^\circledR$: $2KHSO_5 \cdot KHSO_4 \cdot K_2SO_4$, MW = 614.74 g·mol^{-1}) was purchased from TCI Deutschland.

NaCl, NaBr and NaI were purchased from Sigma Aldrich, Inc. and used directly without further purification. All deuterated solvents were purchased from Deutero GmbH, Kastellaun, Germany.

3.2. Nuclear Magnetic Resonance Spectroscopy

All NMR spectra were recorded on a Bruker AVANCE III HD 400 MHz spectrometer at 297 K in D_2O/CD_3OD or D_2O/CD_3CN (2:1, v/v) solvents. 19 mg of **CLP** were dissolved in 0.6 mL of deuterated solvent mixture and used for ^1H, ^{13}C NMR, ^1H-^1H COSY, HSQC and HMBC analysis. Chemical shifts are reported in ppm (δ) and residual CD_3OD (δ_H = 3.31 ppm, δ_C = 49.0 ppm) or CD_3CN (δ_H = 1.94 ppm, δ_C = 1.32 ppm). Processing of the raw data was performed using Bruker TOPSPIN software.

3.2.1. Recording of One-Dimensional NMR Spectra

The pulse conditions were as follows: ^1H NMR, spectra (pulse sequence = zg30): number of data points (TD) = 43008, number of scans (NS) = 16, dummy scans (DS) = 2, spectra width (SWH) = 8012.820 Hz, acquisition time (AQ) = 2.6837 sec, spectrometer operating frequency (SFO1) = 400.13 MHz, $\pi/2$ pulse for ^1H (P1) = 14.30 μs, relaxation delay (D1) = 1.27 s, line broadening (LB) = 0.10 Hz. ^{13}C NMR spectra (pulse sequence = zgpg30): TD = 43702, NS = 256, DS = 2, SWH = 29411.766 Hz, AQ = 0.7429 sec, SFO1 = 100.626 MHz, LB = 1.00 Hz, D1 = 2.0 sec, P1 = 10.0 μs.

3.2.2. Recording of Two-Dimensional NMR Spectra

$^1H/^1H$ COSY (pulse sequence = cosygpppqf): TD = 2048 (F2), TD = 256 (F1) NS = 2, DS = 16, SFO1 = 400.132 MHz, D1 = 2.00 sec. ^1H–^{13}C HSQC (pulse sequence = hsqcedetgp): TD = 1024 (F2), TD = 256 (F1) NS = 2, DS = 16, SFO1 = 400.132 (F2) MHz, SFO1 = 100.622 (F1) MHz, D1 = 2.00 sec. ^1H–^{13}C HMBC (pulse sequence = hmbcgpndqf): The parameters were very similar to those used in the HSQC experiment.

3.3. Mass Spectrometry

The compounds were dissolved (about 0.05 mg·mL^{-1}) in acetonitrile, using a solvent system of acetonitrile: formic acid, 0.1% in water [90:10, v/v] at a flow rate of 0.5 mL·min^{-1}. The mass spectrum of the isolated products was acquired on a Xevo G2-XS Tof Mass Spectrometry instrument from Waters (Wilmslow, UK) in positive spray ionization (ES+) mode. The column used was an ACQUITY UPLC BEH C18 1.7 μm and with the following dimensions: 2.1 mm × 50 mm. The ES+ capillary was set at 3.0 kV, the source temperature at 120 °C and the desolvation temperature at 500 °C. Mass range was scanned between 50 and 750 amu.

3.4. General Description of the Experiment

To a glass vial with a solution of **CLP** (19 mg; 45.49 μmol; 1.0 equiv) in D_2O/CD_3CN (0.6 mL, 2:1 v/v) was added the corresponding halide salt (1.0 equiv) and PMS ($2KHSO_5 \cdot KHSO_4 \cdot K_2SO_4$, MW = 614.74 g·mol^{-1})—see Tables 1–3 for specific data. The solution was stirred before being transferred to an NMR tube and continuously monitored by ^1H NMR at room temperature. After complete conversion, the reaction mixture was quenched with sodium thiosulfate and extracted with dichloromethane (3×). The organic layers were combined, dried over anhydrous $MgSO_4$, filtered, and concentrated in vacuo. The desired halogenated products were then characterized without further purification. Characterization of the products by LC-MS is also possible directly from the reaction mixture.

4. Conclusions

In summary, we showed a selective transformation of clopidogrel hydrogen sulfate (**CLP**) by a PMS/halide system in aqueous acetonitrile media without employing a metal catalyst. With this method, we have prepared three major halogenated products using different halide salts. With this condition, a fast preparation of known endo-iminium clopidogrel impurity (**DP-2**, new counterion) was described as well. The new degradation products **DP-3a–b**, **DP-4a** and **DP-5b** were characterized using spectroscopic techniques (namely 1D-NMR, 2D-NMR and HRMS). We believe that this procedure is not only useful for generating clopidogrel derivatives but also very important to the study of drug degradation under hypersaline conditions. We are currently extending this study with other active pharmaceutical ingredients.

Supplementary Materials: The following are available online, Analytical data; Figures S1–S4: HPLC-MS chromatogram: PMS/NaCl; Figure S5: UV Spectrum; Figures S6–S32: HR-MS/NMR spectrum.

Author Contributions: E.F.K.: Conceptualization, Validation, Investigation, Visualization, Writing—Original Draft; W.B. Supervision, Writing—review & editing. All authors have read and agreed to the published version of the manuscript.

Funding: Financial support provided by Leibniz Association, project PHARMSAF (K136/2018).

Institutional Review Board Statement: Not applicable.

Informed Consent Statement: Not applicable.

Data Availability Statement: The data presented in this study are available in Supplementary Material.

Acknowledgments: We thank the analytical staff of the Leibniz-Institute for Catalysis, Rostock, for their excellent service.

Conflicts of Interest: The authors declare no conflict of interest.

Sample Availability: Samples of the compounds are available from the authors.

References

1. Larock, R.C.; Zhang, L. Aromatic Halogenation. In *Comprehensive Organic Transformations: A Guide to Functional Group Preparations*, 3rd ed.; Larock, R.C., Ed.; John Wiley & Sons, Inc.: Chichester, UK, 2018.
2. Kosjek, T.; Heath, E. *Halogenated Heterocycles as Pharmaceuticals*; Springer: Berlin, Heidelberg, 2011; Volume 27.
3. Gramec, D.; Mašič, P.L.; Dolenc, S.M. Bioactivation Potential of Thiophene-Containing Drugs. *Chem. Res. Toxicol.* **2014**, *27*, 1344–1358. [CrossRef]
4. Shah, R.; Verma, P.K. Therapeutic importance of synthetic thiophene. *Chem. Cent. J.* **2018**, *12*, 137. [CrossRef]
5. Podgoršek, A.; Zupan, M.; Iskra, J. Oxidative halogenation with "green" oxidants: Oxygen and hydrogen peroxide. *Angew. Chem. Int. Ed.* **2009**, *48*, 8424–8450. [CrossRef] [PubMed]
6. Hussain, H.; Green, I.R.; Ahmed, I. Journey describing applications of oxone in synthetic chemistry. *Chem. Rev.* **2013**, *113*, 3329–3371. [CrossRef] [PubMed]
7. Baertschi, S.W.; Alsante, K.M.; Reed, R.A. *Pharmaceutical Stress Testing: Predicting Drug Degradation*, 2nd ed.; Informa Healthcare: London, UK, 2011.
8. Sheng, B.; Huang, Y.; Wang, Z.; Yang, F.; Ai, L.; Liu, J. On peroxymonosulfate-based treatment of saline wastewater: When phosphate and chloride co-exist. *RSC Adv.* **2018**, *8*, 13865–13870. [CrossRef]
9. Lefebvre, O.; Moletta, R. Treatment of organic pollution in industrial saline wastewater: A literature review. *Water Res.* **2006**, *40*, 3671–3682. [CrossRef]
10. Woolard, C.R.; Irvine, R.L. Treatment of hypersaline wastewater in the sequencing batch reactor. *Water Res.* **1995**, *29*, 1159–1168. [CrossRef]
11. Kim, E.-H.; Koo, B.-S.; Song, C.-E.; Lee, K.-J. Halogenation of aromatic methyl ketones using Oxone® and sodium halide. *Synth. Commun.* **2001**, *31*, 3627–3632. [CrossRef]
12. Tamhankar, B.V.; Desai, U.V.; Mane, R.B.; Wadgaonkar, P.P.; Bedekar, A.V. A simple and practical halogenation of activated arenes using potassium halide and Oxone® in water-acetonitrile medium. *Synth. Commun.* **2001**, *31*, 2021–2027. [CrossRef]
13. Narender, N.; Srinivasu, P.; Kulkarni, S.J.; Raghavan, K.V. Highly efficient, para-selective oxychlorination of aromatic compounds using potassium chloride and oxone®. *Synth. Commun.* **2002**, *32*, 279–286. [CrossRef]

14. Bovicelli, P.; Bernini, R.; Antoniolettia, R.; Mincione, E. Selective halogenation of flavanones. *Tetrahedron Lett.* **2002**, *43*, 5563–5567. [CrossRef]
15. Desai, U.V.; Pore, D.M.; Tamhankar, B.V.; Jadhavb, S.A.; Wadgaonkar, P.P. An efficient deprotection of dithioacetals to carbonyls using Oxone–KBr in aqueous acetonitrile. *Tetrahedron Lett.* **2006**, *47*, 8559–8561. [CrossRef]
16. Firouzabadi, H.; Iranpoor, N.; Kazemi, S. Direct halogenation of organic compounds with halides using oxone in water—A green protocol. *Can. J. Chem.* **2009**, *87*, 1675–1681. [CrossRef]
17. Takada, Y.; Hanyu, M.; Nagatsu, K.; Fukumura, T. Radiolabeling of aromatic compounds using K[*Cl]Cl and OXONE®. *J. Label. Compd. Radiopharm.* **2012**, *55*, 383–386. [CrossRef]
18. Ren, J.; Tong, R. Convenient in situ generation of various dichlorinating agents from oxone and chloride: Diastereoselective dichlorination of allylic and homoallylic alcohol derivatives. *Org. Biomol. Chem.* **2013**, *11*, 4312–4315. [CrossRef]
19. Brucher, O.; Hartung, J. Oxidative chlorination of 4-pentenols and other functionalized hydrocarbons. *Tetrahedron* **2014**, *70*, 7950–7961. [CrossRef]
20. Lai, L.; Wang, H.; Wu, J. Facile assembly of 1-(4-haloisoquinolin-1-yl) ureas via a reaction of 2-alkynylbenzaldoxime, carbodiimide, and halide in water. *Tetrahedron* **2014**, *70*, 2246–2250. [CrossRef]
21. Wang, Y.; Wang, Y.; Jiang, K.; Zhang, Q.; Li, D. Transition-metal-free oxidative C5 C–H-halogenation of 8-aminoquinoline amides using sodium halides. *Org. Biomol. Chem.* **2016**, *14*, 10180–10184. [CrossRef]
22. Bikshapathi, R.; Parvathaneni, S.P.; Rao, V.J. An atom-economical protocol for direct conversion of Baylis-Hillman alcohols to β-chloro aldehydes in water. *Green Chem.* **2017**, *19*, 4446–4450. [CrossRef]
23. Olsen, K.L.; Jensen, M.R.; MacKay, J.A. A mild halogenation of pyrazoles using sodium halide salts and Oxone. *Tetrahedron Lett.* **2017**, *58*, 4111–4114. [CrossRef]
24. Sriramoju, V.; Kurva, S.; Madabhushi, S. New method for the preparation of N-chloroamines by oxidative N-halogenation of amines using oxone-KCl. *Synth. Commun.* **2018**, *48*, 699–704. [CrossRef]
25. Lakshmireddy, V.M.; Veera, Y.N.; Reddy, T.J.; Rao, V.J.; Raju, B.C. A Green and Sustainable Approach for Selective Halogenation of Anilides, Benzanilides, Sulphonamides and Heterocycles†. *Asian J. Org. Chem.* **2019**, *8*, 1380–1384. [CrossRef]
26. Uyanik, M.; Sahara, N.; Ishihara, K. Regioselective oxidative chlorination of arenols using NaCl and oxone. *Eur. J. Org. Chem.* **2019**, 27–31. [CrossRef]
27. Semwal, R.; Ravi, C.; Kumar, R.; Meena, R.; Adimurthy, S. Sodium salts (NaI/NaBr/NaCl) for the halogenation of imidazo-fused heterocycles. *J. Org. Chem.* **2019**, *84*, 792–805. [CrossRef]
28. Kim, K.-M.; Park, I.-H. A convenient halogenation of α, β-unsaturated carbonyl compounds with OXONE® and hydrohalic acid (HBr, HCl). *Synthesis* **2004**, *16*, 2641–2644. [CrossRef]
29. Lee, H.S.; Lee, H.J.; Lee, K.Y.; Kim, J.N. Controlled C-5 Chlorination and Dichlorohydrin Formation of Uracil Ring with HCl/DMF/Oxone®System. *Bull. Korean Chem. Soc.* **2012**, *33*, 1357–1359. [CrossRef]
30. Qiao, L.; Cao, X.; Chai, K.; Shen, J.; Xu, J.; Zhang, P. Remote radical halogenation of aminoquinolines with aqueous hydrogen halide (HX) and oxone. *Tetrahedron Lett.* **2018**, *59*, 2243–2247. [CrossRef]
31. Krake, E.F.; Baumann, W. Unprecedented Formation of 2-Chloro-5-(2-chlorobenzyl)-4, 5, 6, 7-tetrahydrothieno [3, 2-c] pyridine 5-oxide via Oxidation-Chlorination Reaction Using Oxone: A Combination of Synthesis and 1D-2D NMR Studies. *ChemistrySelect* **2019**, *4*, 13479–13484. [CrossRef]
32. Wu, Y.J. Heterocycles and medicine: A survey of the heterocyclic drugs approved by the US FDA from 2000 to present. *Prog. Heterocycl. Chem.* **2012**, 1–53.
33. Herbert, J.M.; Frehel, D.; Vallée, E.; Kieffer, G.; Gouy, D.; Necciari, J.; Defreyn, G.; Maffrand, J.P. Clopidogrel, A Novel Antiplatelet and Antithrombotic Agent. *Cardiovasc. Drug Rev.* **1993**, *11*, 180–198. [CrossRef]
34. Ferri, N.; Corsini, A.; Bellosta, S. Pharmacology of the new P2Y 12 receptor inhibitors: Insights on pharmacokinetic and pharmacodynamic properties. *Drugs* **2013**, *73*, 1681–1709. [CrossRef]
35. Aalla, S.; Gilla, G.; Anumula, R.R.; Charagondla, K.; Vummenthala, P.R.; Padi, P.R. New and Efficient Synthetic Approaches for the Regioisomeric and Iminium Impurities of Clopidogrel Bisulfate. *Org. Process Res. Dev.* **2012**, *16*, 1523–1526. [CrossRef]
36. Song, S.; Li, X.; Wei, J.; Wang, W.; Zhang, Y.; Ai, L.; Zhu, Y.; Shi, X.; Zhang, X.; Jiao, N. DMSO-catalysed late-stage chlorination of (hetero) arenes. *Nat. Catal.* **2020**, *3*, 107–115. [CrossRef]
37. Krake, E.F.; Jiao, H.; Baumann, W. NMR and DFT analysis of the major diastereomeric degradation product of clopidogrel under oxidative stress conditions. *J. Mol. Struct.* **2022**, *1247*, 131309. [CrossRef]

Article

Selenoxide Elimination Triggers Enamine Hydrolysis to Primary and Secondary Amines: A Combined Experimental and Theoretical Investigation

Giovanni Ribaudo [1], Marco Bortoli [2,3], Erika Oselladore [1], Alberto Ongaro [1], Alessandra Gianoncelli [1], Giuseppe Zagotto [4] and Laura Orian [2,*]

1. Dipartimento di Medicina Molecolare e Traslazionale, Università degli Studi di Brescia, Viale Europa 11, 25123 Brescia, Italy; giovanni.ribaudo@unibs.it (G.R.); e.oselladore@unibs.it (E.O.); a.ongaro005@unibs.it (A.O.); alessandra.gianoncelli@unibs.it (A.G.)
2. Dipartimento di Scienze Chimiche, Università degli Studi di Padova, Via Marzolo 1, 35131 Padova, Italy; marco.bortoli@unipd.it
3. Departament de Química, Institut de Química Computacional i Catàlisi, Universitat de Girona, C/M.A. Capmany 69, 17003 Girona, Spain
4. Dipartimento di Scienze del Farmaco, Università degli Studi di Padova, Via Marzolo 5, 35131 Padova, Italy; giuseppe.zagotto@unipd.it
* Correspondence: laura.orian@unipd.it; Tel.: +39-049-8275140

Abstract: We discuss a novel selenium-based reaction mechanism consisting in a selenoxide elimination-triggered enamine hydrolysis. This one-pot model reaction was studied for a set of substrates. Under oxidative conditions, we observed and characterized the formation of primary and secondary amines as elimination products of such compounds, paving the way for a novel strategy to selectively release bioactive molecules. The underlying mechanism was investigated using NMR, mass spectrometry and density functional theory (DFT).

Keywords: selenoxide elimination; one-pot; imine-enamine; reaction mechanism; DFT calculations; selenium

1. Introduction

In the past few decades, the selenoxide elimination reaction has been used to obtain alkenes and has been largely applied in the synthesis of small molecules such as natural products and bioactive compounds [1–4]. Generally, the organoselenides required for such a reaction can be straightforwardly synthesized from electrophilic phenylselenyl chloride or using diphenyl diselenides as precursors for the nucleophilic selenate. In the presence of oxidants, such as peroxides or other reactive oxygen species (ROS), organoselenium compounds are readily oxidized to selenoxides [5–7]. Then, an intramolecular *syn* elimination occurs in opportune substrates, involving the hydrogen atom in vicinal position with respect to the selenium nucleus. This leads to the formation of the corresponding *trans*-olefine [8–10].

In the context of our studies on selenium-based derivatives of compounds of pharmaceutical interest, we observed a peculiar behavior of amino organoselenides [10]. In particular, under oxidative conditions, we previously detected the formation of primary and secondary amines from mono- or disubstituted 2-phenyl-2-(phenylselanyl)ethan-1-amines (Scheme 1).

Among the functional groups of biological relevance, the amino moiety is one of the most frequently occurring. Nevertheless, drugs containing primary or secondary amines, despite being very common, are endowed with some pharmacokinetic limitations such as poor diffusion through membranes or blood–brain barrier under particular physiologic conditions. Moreover, instability affects such compounds, as amines physiologically undergo metabolic phase 1 transformations such as oxidations or dealkylations or conjugations in phase 2. These limitations can be partially overcome by derivatization to generate prodrugs that can be "unmasked" under certain conditions that can be pH-, redox- or enzyme-dependent [11]. Recent reports showed that selenoxide elimination can be employed by drug-like derivatives responsive to reactive oxygen species (ROS) in innovative targeted therapeutic strategies [12]. Moreover, we previously showed that compounds endowed with bioactivity can be generated after release from such substrates. Molecules selectively providing elimination products in response to oxidative stress represent attractive tools, in particular in the context of central nervous system (CNS) drugs [10].

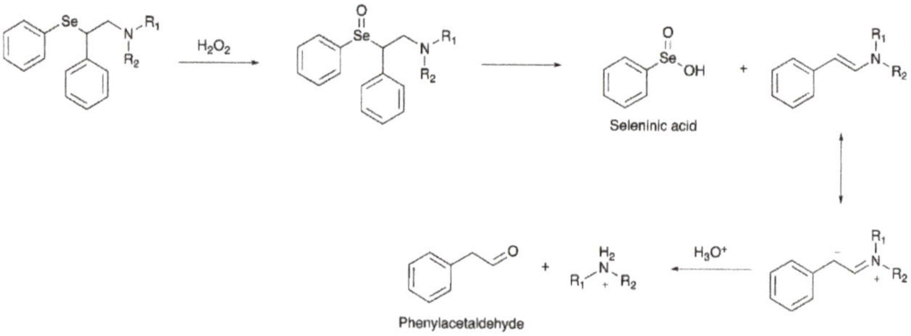

Scheme 1. Proposed mechanism for selenoxide elimination-triggered enamine hydrolysis.

Taking the step from these considerations, we investigated and characterized a one-pot model reaction consisting in a selenoxide elimination-triggered enamine hydrolysis that allows the formation of primary and secondary amines under oxidative conditions.

Additionally, this combined process is endowed with synthetic value, as it represents an innovative approach for obtaining primary and secondary amines as elimination products under these specific conditions. Although selenium-catalyzed organic reactions, and particularly selenoxide elimination, have been extensively studied [13–15], no examples of selenium-based mechanisms for obtaining such amines were previously reported in the literature.

2. Results and Discussion

From the point of view of the molecular mechanism, the considered oxidation-triggered elimination is distinctive of organoselenium compounds having protons in the β-position with respect to the chalcogen atom. This process occurs through a *syn* mechanism and promotes the formation of olefins with high *trans* selectivity [8]. The reaction is initiated by the oxidation of the selenium atom to the corresponding selenoxide, a step which can be induced by different agents such as hydrogen peroxide, meta-chloroperoxybenzoic acid (mCPBA) and ozone [7,16,17]. Then, an intramolecular *syn* elimination takes place: the Se-C bond breaks producing the *trans*-olefin and selenenic acid, that is readily oxidized to seleninic acid [10]. Interestingly, in the case of the studied compounds (Scheme 2), an enamine is produced after the oxidation–elimination step. This can be protonated on both the nitrogen atom and the β-carbon atom in acidic conditions. The latter event is favored and the molecule, rearranging to an iminium ion and undergoing hydrolysis, subsequently produces the corresponding amine (Scheme 1) [18,19].

In the field of synthetic organic chemistry, the preparation of primary aliphatic amines can be achieved by *Gabriel* synthesis [20], by azidation–reduction [21], by *Leuckart* reaction [22] or its variation involving the use of benzylamine and hydrogenolysis [23]. In analogy, secondary aliphatic amines can be obtained through different processes: direct alkylation [24], reductive amination [25] and the *Fukuyama* amine synthesis [26]. In the current study, we aimed at investigating the formation of primary and secondary amines based on the combination of known selenoxide elimination and enamine hydrolysis reactions. In particular, this reaction was studied in aqueous solution and under oxidative conditions. Thus, mono- and disubstituted symmetric, asymmetric and cyclic derivatives reported in Scheme 2 were synthesized and reacted with hydrogen peroxide in the presence of water. The overall reaction was studied in detail by NMR spectroscopy, mass spectrometry and DFT calculations.

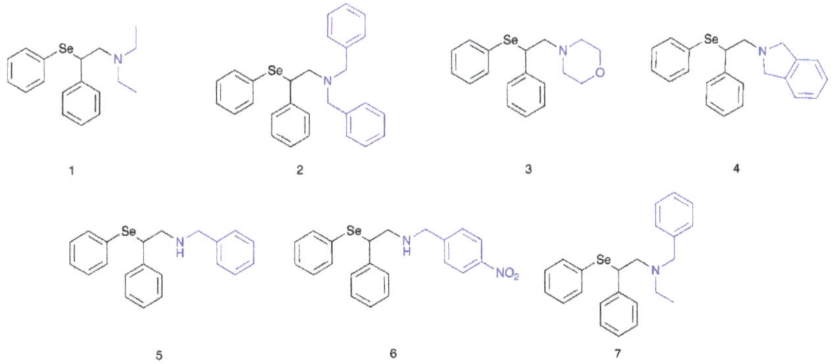

Scheme 2. Chemical structures of the studied compounds. The portion of the molecules providing the resulting amine is highlighted in blue.

Compound **8** was synthesized by styrene azido-phenylselenenylation with the hypervalent iodine reagent (diacetoxyiodo)benzene, sodium azide and diphenyl diselenide in dichloromethane (Scheme 3). The reaction involves the formation of an azido radical that, after the addition to the double bond of the styrene, generates a carbon radical which is then trapped by diphenyldiselenide to provide the product [27]. Racemic compound **9** was obtained by reduction of the azido group of compound **8** with lithium aluminum hydride in THF (Scheme 3).

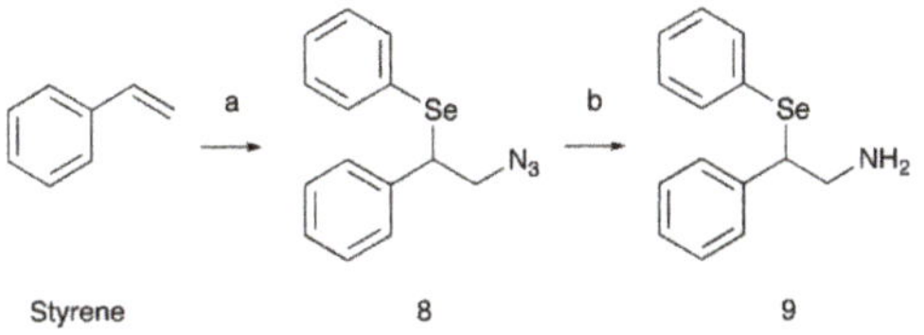

Scheme 3. (**a**) (diacetoxyiodo)benzene/diphenyl diselenide/NaN3/DCM; (**b**) LiAlH4/THF.

Then, **9** was subjected to different alkylation reactions to obtain compounds **1–7** as hydrochlorides (Scheme 2; see Figures S1–S27 in Supplementary Materials for NMR and mass spectra). In particular, compounds **1–4** were obtained through direct alkylation using the opportune halide, such as iodoethane, benzyl bromide, bis(2-bromoethyl) ether and α,α'-dibromo-o-xylene. Compounds **5** and **6** were obtained by reductive amination of compound **9** with benzaldehyde and p-nitrobenzaldehyde, respectively. By reacting compound **5** with iodoethane, the corresponding tertiary amine (compound **7**) was prepared.

First, the combined reaction was studied in detail for compound **3** in aqueous solution by ^{1}H-NMR. This substrate was chosen as a model also because the corresponding elimination product, morpholine, represents an outstanding example of pharmacologically relevant moiety. Interestingly, it can be found in drugs acting on CNS (e.g., in the antidepressants reboxetine and viloxazine and in the withdrawn stimulant phenmetrazine), and which are used to treat pathologies where oxidative stress has been demonstrated to play a role on disease onset and progression [28,29]. More in detail, compound **3** was dissolved in deuterated water and reacted with 1.2 equivalents of hydrogen peroxide at room temperature.

The analysis of this reaction showed the progressive disappearance of the α-hydrogen signal, paralleled by the appearance of the signals corresponding to the two selenoxide diastereoisomers (Figure S28 in Supplementary Materials). Calculations at COSMO-ZORA-OLYP/TZ2P level of theory [30,31] showed that in water, this reaction displays an energy barrier of 14.5 kcal mol^{-1} which separates the reactants from the selenoxide form (**Iox**, see Figure 1a) lying at -39.5 kcal mol^{-1}. As the reaction further proceeded, the signals of such intermediates disappeared and the subsequent formation of the two particular signals corresponding to morpholine hydrochloride (3.25 and 3.90 ppm) was observed. After 70 min, 72% relative abundance of secondary amine was observed in the NMR spectrum. At the same time point, the relative abundance of the starting material was < 10% (Figure 2). The mechanism for the last part of the reaction was computationally modelled in three elementary steps. First, starting from the selenoxide, an intramolecular elimination reaction requiring 7.6 kcal mol^{-1} takes place (**TSelim**, Figure 1a) that results in the cleavage of one C-Se bond and the formation of selenenic acid (the further oxidation of selenenic acid to seleninic acid detected experimentally was not computationally investigated) [9] and the corresponding enamine (**SeOH + En**, Figure 1a) with a ΔG of -76.7 kcal mol^{-1} relative to the starting reactants. Finally, the process that was involved in the cleavage of the C-N bond to give the final morpholine molecule consist in two steps. The first requires the addition of an H_3O^+ ion to the enamine with an activation energy of 51.8 kcal mol^{-1} (**TS1**, Figure 1b) and leads to an intermediate which is destabilized by 25.5 kcal mol^{-1} (**I**, Figure 1b). The final reaction of the C-N bond cleavage leads to the formation of morpholine and phenylacetaldehyde (**P**, Figure 1b) which are found to be 6.4 kcal mol^{-1} more stable than the starting enamine. The reaction involves a proton transfer from the OH moiety of I to the N atom. This step requires a somewhat extended network of explicit water molecules to be successfully modelled [32]. The inclusion of explicit water molecules in mechanisms that involve proton transfers usually helps to recover more favorable energetics [32–35]. Nevertheless, mechanistic features are not modified by this inclusion and conformational studies on the model hydrogen bond network to obtain the most stable configuration are quite tricky. As the detailed mechanistic analysis of this reaction is out of the scope of this work, the effect of explicit water molecules was not addressed in this context. All the solvation contributions were incorporated in the polarizing effect of the dielectric continuum employed in the calculations. On the same basis, the reaction was investigated by ^{1}H-NMR spectroscopy for compounds 1–7 (see Figures S31, S33 and S36 in Supplementary Materials for representative NMR and mass spectra), demonstrating the disappearance of starting material and the formation of the corresponding products (43–93% relative abundance). Electrospray mass spectrometry (ESI-MS) studies, which were performed in parallel, confirmed the identity of involved intermediates, including enamines, and products. More in detail, primary and secondary amines generated after the oxidation of compound **1–7** were characterized using positive ionization ESI-MS on the same samples, while negative ionization ESI-MS analysis confirmed the presence of seleninic acid and benzoic acid as final oxidation products. The latter very likely generates from phenylacetaldehyde oxidation (see Figures S29, S30, S32, S34, S35 and S37 in Supplementary Materials for mass spectra) [36].

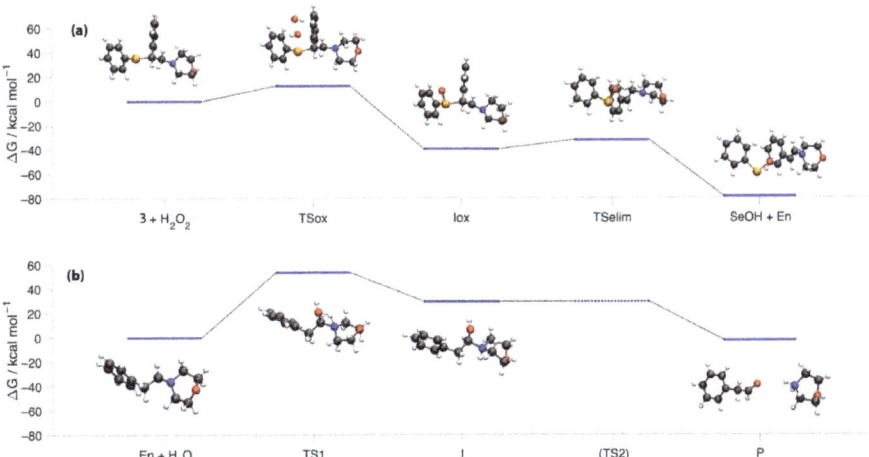

Figure 1. Gibbs free energies of the relevant intermediates and transition states characterized in water for the oxidation and subsequent elimination of compound **3**: (**a**) corresponds to the top reaction of Scheme 1 whereas (**b**) to the bottom one. Water and hydrogen peroxide molecules are not shown in the minimum energy structures for clarity. Level of theory: COSMO- ZORA-OLYP/TZ2P.

Following this NMR-scale mechanistic investigation and to gain further insight on the underlying processes, the reactions involving compounds **1–7** were scaled up and performed on the same substrates, in order to isolate the elimination products, verify the identity and calculate the yield. Compounds were subjected to an overnight oxidation using H_2O_2 and the corresponding amines were isolated with an average yield of 68%, with a best isolated yield of 90%. ^{1}H-NMR, ^{13}C-NMR and ESI-MS analysis confirmed the identity of the products, thus demonstrating the formation of primary and secondary amines in the laboratory scale (see Supplementary Materials for experimental procedures, yields, computational details and Figures S38–S58 for spectra).

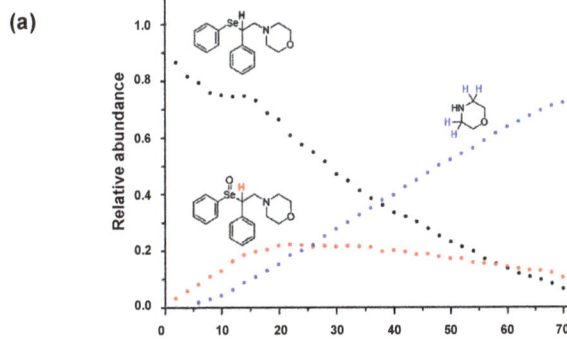

Figure 2. *Cont.*

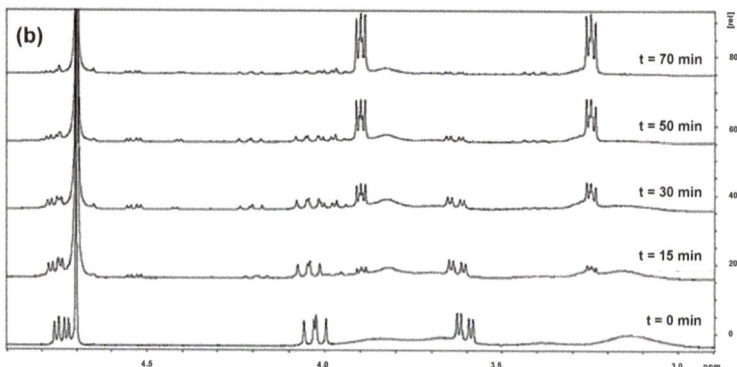

Figure 2. Reaction profiles for compound **3**, intermediate and reaction products. The signals considered for relative quantification are depicted as in the following: • starting material, • intermediate, • secondary amine. Integrals were measured using acetonitrile as internal standard (**a**). Detailed views of representative NMR spectra showing the considered signals (**b**).

3. Materials and Methods

3.1. Chemistry

Commercially available chemicals were purchased from Sigma-Aldrich and used without any further purification if not specified elsewhere. NMR experiments were performed on an Avance III 400 (Bruker, Billerica, MA, USA) spectrometer (frequencies: 400.13, 100.62 MHz for ^{1}H and ^{13}C nuclei, respectively) equipped with a multinuclear inverse z-field gradient probe head (5 mm). For data processing, TopSpin 4.0.8 software was used, and the spectra were calibrated using solvent signal (^{1}H-NMR, δ_H = 7.26 ppm for CDCl$_3$, δ_H = 3.31 ppm for CD$_3$OD, δ_H = 2.50 ppm for DMSO-$d6$; ^{13}C-NMR, δ_C = 77.16 ppm for CDCl$_3$, 39.52 ppm for DMSO-$d6$, δ_C = 49.00 for CD$_3$OD). Multiplicities are reported as follows: s, singlet; d, doublet; t, triplet; q, quartet; m, multiplet; b, broad; dd, doublet of doublets. Mass spectra were recorded by direct infusion ESI on a LCQ Fleet ion trap (Thermo Fisher Scientific, Waltham, MA, USA) mass spectrometer. For data processing, Qual Browser Thermo Xcalibur 4.0.27.13 software was used. ESI parameters for samples acquired in positive ionization mode: spray voltage 3.2 kV, capillary temperature 160 °C, capillary voltage 43 V. ESI parameters for samples acquired in negative ionization mode: spray voltage 5.0 kV, capillary temperature 160 °C, capillary voltage −8 V.

3.2. Synthesis of Compound **1**

Compound **9** (100 mg, 0.36 mmol) was dissolved in iodoethane (2 mL) and DIPEA (190 µL, 1.08 mmol) was added to the solution that was then deaerated, purged with nitrogen and refluxed under stirring at 90 °C for 2 h. After TLC showed the disappearance of the starting material (DCM/MeOH/NH$_3$, 97.5/2.0/0.5), the mixture was cooled to r.t. and the solvent was evaporated under reduced pressure. DCM (30 mL) was added, and the solution was washed with 1 M KOH (3 × 10 mL), dried over magnesium sulphate and evaporated at reduced pressure obtaining a viscous brownish liquid which was purified by chromatographic column (silica gel, DCM/MeOH/NH$_3$, 97.5/2.0/0.5). The hydrochloride salt was then prepared, dissolving the product in anhydrous diethyl ether and adding a 2 M solution of HCl in diethyl ether dropwise.

White solid, 119 mg, 0.32 mmol, 89%. ^{1}H-NMR (CD$_3$OD, 400 MHz): 7.63–7.61 (m, 2H), 7.46–7.34 (m, 8H), 4.82 (dd, J = 10.7, 4.5 Hz, 1H), 4.06 (dd, J = 14.0, 10.7 Hz, 1H), 3.52 (dd, J = 14.0, 4.5 Hz, 1H), 3.20–3.06 (m, 4H), 1.19–1.10 (m, 6H); ^{13}C{^{1}H} NMR (CD$_3$OD, 101 MHz): 138.8, 137.2, 130.6, 130.4, 130.3, 129.7, 129.2, 129.0, 56.3, 41.8, 9.0, 8.5; (ESI+) m/z calcd for C$_{18}$H$_{24}$NSe$^+$ [M + H]$^+$: 334.11, found: 333.95.

3.3. Synthesis of Compound 2

Compound **9** (100 mg, 0.36 mmol) was dissolved in EtOH (5 mL). Benzyl bromide (87 µL, 124 mg, 0.73 mmol) and DIPEA (190 µL, 1.08 mmol) were added to the solution that was deaerated, purged with nitrogen and refluxed under stirring at 90 °C for 2 h. After TLC showed the disappearance of the starting material (hexane/diethyl ether, 6/1), the mixture was cooled to r.t. and the solvent was evaporated under reduced pressure. DCM (30 mL) was added, and the obtained solution was washed with 1 M KOH (3 × 10 mL), dried over magnesium sulphate and evaporated under reduced pressure. The crude product was purified by column chromatography (silica gel, hexane/DCM, 9/1) obtaining a viscous liquid. The hydrochloride salt was then prepared dissolving the product in anhydrous diethyl ether and adding a 2 M solution of HCl in diethyl ether dropwise.

White solid, 93 mg, 0.19, 52%. ^{1}H-NMR (CD$_3$OD, 400 MHz): δ_H (ppm) 7.57–7.20 (m, 18H), 6.80–6.78 (m, 2H), 4.56–4.43 (m, 3H), 4.31 (d, J = 12.7 Hz, 1H), 4.17 (d, J = 12.6 Hz, 1H), 3.87 (dd, J = 13.9, 10.9 Hz, 1H), 3.61 (dd, J = 13.9, 5.0 Hz, 1H); ^{13}C{^{1}H} NMR (CD$_3$OD, 101 MHz): 138.0, 137.8, 132.9, 132.6, 131.6, 130.7, 130.5, 130.5, 130.3, 129.6, 128.7, 127.9, 60.4, 55.5, 41.6; (ESI+) m/z calcd for C$_{28}$H$_{28}$NSe$^+$ [M + H]$^+$: 458.14, found: 458.05.

3.4. Synthesis of Compound 3

Compound **9** (100 mg, 0.36 mmol) was dissolved in EtOH (5 mL). Bis(2-bromoethyl) ether (84 mg, 0.36 mmol) and DIPEA (190 µL, 1.08 mmol) were added to the solution that was deaerated, purged with nitrogen and refluxed under stirring at 90 °C for 2 h. After TLC showed disappearance of starting material (DCM/MeOH/TEA, 95/4.5/0.5), the mixture was cooled to r.t. and the solvent was evaporated under reduced pressure, DCM (30 mL) was added and the so obtained solution was washed with 1 M KOH (3 × 10 mL), dried over magnesium sulphate and evaporated at reduced pressure. The crude product was purified by column chromatography (silica gel, DCM/MeOH/TEA, 95/4.5/0.5), obtaining a viscous liquid. The hydrochloride salt was then prepared, dissolving the product in anhydrous diethyl ether and adding a 2 M solution of HCl in diethyl ether dropwise.

White solid, 70 mg, 0.18 mmol, 50%. ^{1}H-NMR (CD$_3$OD, 400 MHz): 7.52–7.50 (m, 2H), 7.38–7.28 (m, 8H), 4.92 (dd, J = 10.4, 4.7 Hz, 1H), 4.13–4.07 (m, 1H), 4.00–3.69 (m, 4H), 3.64 (dd, J = 13.6, 4.7 Hz, 1H), 3.24–3.20 (m, 4H); ^{13}C{^{1}H} NMR (CD$_3$OD, 101 MHz): 139.2, 136.9, 130.5, 130.2, 130.0, 129.5, 129.1, 129.1, 64.6, 62.1, 53.7, 41.4; (ESI+) m/z calcd for C$_{18}$H$_{22}$NOSe$^+$ [M + H]$^+$: 347.95, found: 347.98.

3.5. Synthesis of Compound 4

Compound **9** (150 mg, 0.54 mmol) was dissolved in EtOH (6 mL). α,α'-dibromo-o-xylene (154 mg, 0.54 mmol) and K$_2$CO$_3$ (225 mg, 1.62 mmol) were added to the solution that was deareated, purged with nitrogen and refluxed under stirring at 90 °C for 3 h. After TLC showed the disappearance of the starting material (hexane/diethyl ether, 6/1), the mixture was cooled to r.t. and the solvent was evaporated under reduced pressure, DCM (30 mL) was added and the obtained solution was washed with 1 M KOH (3 × 10 mL), dried over magnesium sulphate and evaporated under reduced pressure. The crude product was purified by column chromatography (silica gel, hexane/diethyl ether, 89.75/9.75/0.5) obtaining a viscous liquid. The hydrochloride salt was then prepared dissolving the product in anhydrous diethyl ether and adding a 2 M solution of HCl in diethyl ether dropwise.

Green solid, 45 mg, 0.11 mmol, 20 %. ^{1}H NMR (CD$_3$OD, 400 MHz): 7.53–7.51 (m, 2H), 7.40–7.30 (m,12H), 4.82–4.79 (m, 1H), 4.67–4.58 (m, 4H), 4.29 (dd, J = 13.2, 10.2 Hz, 1H) 3.96 (dd, J = 13.2, 5.7 Hz, 2H); ^{13}C{^{1}H} NMR (CD$_3$OD, 101 MHz): 138.9, 137.1, 134.8, 130.5, 130.2, 130.1, 130.1, 129.6, 129.1, 128.7, 123.8, 60.4, 60.0, 43.2; (ESI+) m/z calcd for C$_{22}$H$_{21}$NSe$^+$ [M + H]$^+$: 379.08, found: 380.00.

3.6. Synthesis of Compound 5

Compound **9** (100 mg, 0.36 mmol) was dissolved in dry MeOH (5 mL). Benzaldehyde (37 µL, 0.36 mmol) was added to the solution that was deaerated, purged with nitrogen and stirred at r.t. for 2 h. After TLC showed disappearance of starting materials (hexane/diethyl ether 6/1), the mixture was cooled in an ice bath and NaBH$_4$ (16 mg, 0.43 mmol) was added in portions. The ice bath was removed, and the mixture was stirred at r.t. for other 2 h. After TLC showed the disappearance of the starting material (hexane/diethyl ether 6/1), the solution was cooled in an ice bath and first acidified to pH = 2 by the addition of 1 M HCl and the basified to pH = 14 by the addition of 8 M KOH. The solvent was evaporated under reduced pressure. DCM (30 mL) was added, and the solution was washed with 1 M KOH (3 × 10 mL), dried over magnesium sulphate and evaporated under reduced pressure, obtaining a viscous liquid. The crude product was purified by column chromatography (silica gel hexane/EtOAc, 9/1 with 0.5% TEA). The hydrochloride salt was then prepared, dissolving the product in anhydrous diethyl ether and adding a 2 M solution of HCl in diethyl ether dropwise.

White solid, 125 mg, 0.31 mmol, 86%. ^{1}H-NMR (CDCl$_3$, 400 MHz): 7.42–7.19 (m, 15H), 4.49 (dd, J = 7.9, 7.0 Hz, 1H), 3.0 (d, J = 3.4 Hz, 2H), 3.26 (dd, J = 12.6, 8.2 Hz, 1H), 3.16 (dd, J = 12.6, 7.0 Hz, 1H); ^{13}C{^{1}H} NMR (CDCl$_3$, 101 MHz): 140.7, 140.0, 135.4, 128.8, 128.6, 128.5, 128.4, 128.1, 127.9, 127.8, 127.2, 127.0, 53.4, 53.3, 48.2; (ESI+) m/z calcd for C$_{21}$H$_{22}$NSe$^+$ [M + H]$^+$: 368.08, found: 367.93.

3.7. Synthesis of Compound 6

Compound 9 (150 mg, 0.36 mmol) was dissolved in dry MeOH (6 mL). p-Nitrobenzaldehyde (82 mg µL, 0.36 mmol) was added to the solution that was deareated, purged with nitrogen and stirred overnight at r.t. After TLC showed the disappearance of the starting material (6/1 hexane/diethyl ether), the mixture was cooled in an ice bath and NaBH$_4$ (16 mg, 0.43 mmol) was added portion wise. The ice bath was removed, and the mixture was stirred at r.t. for other 2 h. After TLC showed the disappearance of the starting material (6/1 hexane/diethyl ether), the solution was cooled in an ice bath and first acidified to pH = 2 by the addition of 1 M HCl and the basified to pH = 14 by the addition of 8 M KOH. The solvent was evaporated under reduced pressure. DCM (30 mL) was added, and the solution was washed with 1 M KOH (3 × 10 mL), dried over magnesium sulphate and evaporated under reduced pressure, obtaining a viscous liquid. The crude product was purified by column chromatography (silica gel hexane/EtOAc, 9/1 with 0.5% TEA). The hydrochloride salt was then prepared, dissolving the product in anhydrous diethyl ether and adding a 2 M solution of HCl in diethyl ether dropwise.

White solid, 135 mg, 0.31 mmol, 87%. ^{1}H-NMR (DMSO-d_6, 400 MHz): δ_H (ppm) 9.56 (br, 1H), 9.25 (br, 1H), 8.24 (d, J = 8.9 Hz, 2H), 7.75 (d, J = 8.9 Hz, 2H), 7.44–7.42 (m, 2H), 7.36–7.25 (m, 8H), 4.88 (dd, J = 9.7, 5.6 Hz, 1H), 4.26 (s, 2H), 3.74–3.72 (m, 2H); ^{13}C{^{1}H} NMR (DMSO-d_6, 101 MHz): 148.2, 139.5, 138.2, 134.8, 134.0, 131.9, 129.8, 129.2, 128.7, 128.5, 128.4, 124.0, 50.5, 49.8, 42.2; (ESI+) m/z calcd for C$_{21}$H$_{21}$N$_2$O$_2$Se$^+$ [M + H]$^+$: 413.08, found: 412.89.

3.8. Synthesis of Compound 7

Compound 5 (100 mg, 0.27 mmol) was dissolved in iodoethane (5 mL). The solution was deaerated, purged with nitrogen and refluxed under stirring at 90 °C for 2 h. After TLC showed the disappearance of the starting material (hexane/diethyl ether, 6/1), the mixture was cooled to r.t. and the solvent was evaporated under reduced pressure. DCM (30 mL) was added, and the solution was washed with 1 M KOH (3 × 10 mL), dried over magnesium sulphate and evaporated under reduced pressure obtaining a viscous colorless liquid. The crude product was purified by column chromatography (silica gel, hexane/EtOAc/TEA, 97/2.5/0.5) obtaining a viscous liquid. The hydrochloride salt was then prepared dissolving the product in anhydrous diethyl ether and adding a 2 M solution of HCl in diethyl ether dropwise.

White solid, 69 mg, 0.15 mmol, 57 %. ^{1}H-NMR (CD$_3$OD, 400 MHz): 7.51–7.06 (m, 15H), 4.77 (dd, J = 10.8, 4.7 Hz, 0.5H), 4.60 (dd, J = 10.5, 4.6 Hz, 0.5H), 4.39–4.22 (m, 2H), 4.06 (dd, J = 13.9, 10.6 Hz, 0.5H), 3.88 (dd, J = 13.9, 10.7 Hz, 0.5H), 3.63–3.48 (m, 1H), 3.30–3.13 (m, 2H), 1.27–1.22 (m, 3H); ^{13}C{^{1}H} NMR (CD$_3$OD, 101 MHz): 138.8, 138.3, 137.4, 137.3, 132.3,

132.2, 131.4, 130.6, 130.5, 130.5, 130.4, 130.3, 130.3, 130.1, 129.7, 129.6, 129.0, 128.9, 128.8, 128.6, 58.8, 56.3, 56.1, 50.9, 50.2, 42.1, 41.4, 28.0, 8.9, 8.5; (ESI+) m/z calcd for $C_{23}H_{26}NSe^+$ [M + H]$^+$: 396.12, found: 395.94.

3.9. Synthesis of Compound **8**

Diphenyldiselenide (414 mg, 1.33 mmol), (diacetoxyiodo)benzene (1.000 g, 3.10 mmol) and sodium azide (346 mg, 5.32 mmol) were added to a solution of styrene (231 mg, 2.22 mmol) in DCM (20 mL). The mixture was deaerated and purged with nitrogen. After overnight stirring at r.t., a saturated solution of $NaHCO_3$ (10 mL) was added, the organic phase was separated, and the aqueous phase was extracted with DCM (2 × 20 mL). The organic phases were reunited and evaporated at reduced pressure. The obtained yellow liquid was purified by column chromatography (silica gel, pure hexane then 90:10 hexane/diethylether), obtaining a colorless liquid.

Colorless liquid, 257 mg, 0.85 mmol, 38%. ^{1}H NMR (CDCl$_3$, 400 MHz,): δ$_H$ 7.54–7.51 (m, 2H), 7.37–7.27 (m, 8H), 4.44 (dd, J = 9.7, 5.7 Hz, 1H), 3.91 (dd, J = 12.5, 9.8 Hz, 1H), 3.75 (dd, J = 12.5, 5.8 Hz, 1H); ^{13}C{^{1}H} NMR (CDCl$_3$, 101 MHz): 139.0, 135.7, 129.3, 128.9, 128.6, 128.5, 128.0, 127.9, 55.6, 46.5; (ESI+) m/z calcd for $C_{14}H_{14}N_3Se^+$ [M + H]$^+$: 304.19, found: 304.15.

3.10. Synthesis of Compound **9**

Compound **8** (1.010 g, 3.33 mmol) was dissolved in anhydrous THF (20 mL) and the solution was deaerated and purged with nitrogen. The system was equipped with a reflux apparatus and a solution of 1 M LiAlH$_4$ (5 mL, 5.01 mmol) was added portion wise to the mixture; after the addition, the mixture was stirred for 2 h. After TLC showed disappearance of starting material (DCM/MeOH/NH$_3$, 97.5/2.0/0.5), wet diethyl ether was slowly added to the solution and distilled water was then added. The resulting biphasic mixture was filtered, and the aqueous phase was separated and extracted with diethyl ether (2 × 20 mL). The organic phase was dried over magnesium sulphate and evaporated at reduced pressure obtaining a viscous colorless liquid that was purified by column chromatography (silica gel, DCM/MeOH/NH$_3$, 97.5/2.0/0.5).

Colorless liquid, 503 mg, 1.82 mmol, 55%. ^{1}H NMR (CDCl$_3$, 400 MHz): 7.53–7.51 (m, 2H), 7.37–7.27 (m, 8H), 4.43 (dd, J = 9.6, 5.7 Hz, 1H), 3.91 (dd, J = 12.6, 9.6 Hz, 1H), 3.74 (dd, J = 12.6, 5.7 Hz, 1H); ^{13}C{^{1}H} NMR (CDCl$_3$, 101 MHz): 140.2, 135.1, 128.8, 128.7, 128.3, 127.8, 127.6, 127.0, 51.8, 46.8; (ESI+) m/z calcd for $C_{14}H_{16}NSe^+$ [M + H]$^+$: 278.04, found: 277.80.

3.11. Laboratory Scale Oxidation Procedures

3.11.1. Oxidation of Compound **3**

Compound **3** (1.50 g, 3.92 mmol, 1 eq) was dissolved in 120 ml of distilled H_2O in a round-bottom flask and H_2O_2 (530 μL, 4.70 mmol, 1.2 eq) was then added to the mixture. The solution was stirred overnight and checked by TLC. Further 1.2 eq of H_2O_2 were then added in order to consume the remained reactant. After TLC showed the disappearance of the starting material, the mixture was cooled in an ice bath, filtered and the filtrate was evaporated under reduced pressure, adding ethanol to facilitate solvent removal. The resulting yellowish precipitate was washed with small aliquots of acetone, obtaining the desired product as a white solid.

Morpholine·HCl: white solid, 435 mg, 3.52 mmol, 90%; ^{1}H-NMR (CD$_3$OD, 400 MHz): 3.91–3.88 (m, 4H), 3.25–3.22 (m, 4H); ^{13}C{^{1}H} NMR (CD$_3$OD, 101 MHz): 64.9, 44.6; (ESI+) m/z calcd for $C_4H_{10}NO^+$ [M + H]$^+$: 88.08, found: 87.93.

3.11.2. Oxidation of Compound **1, 4, 5** and **7**

The selenoamine (1 eq) was dissolved in a MeOH/H$_2$O mixture (85/15) in a round-bottom flask. 1.2 eq of H_2O_2 were then added to the mixture and the solution was stirred overnight. The solution was stirred overnight and checked by TLC. Further 1.2 eq of H_2O_2 were then added in order to consume the remained reactant. After TLC showed the disappearance of the starting material, the solvent was evaporated under reduced pressure. Water was added to the reaction flask in order to dissolve the desired product and eliminate most of the insoluble byproducts. The final compounds were isolated following the procedure described above.

Diethylamine·HCl: white solid; 55 mg, 0.50 mmol, 82%; ^{1}H-NMR (CD$_3$OD, 400 MHz): 3.06 (q, J = 7.3 Hz, 4H), 1.32 (t, J = 7.3 Hz, 6H); ^{13}C{^{1}H} NMR (CD$_3$OD, 101 MHz): 43.5, 11.6; (ESI+) m/z calcd for C$_4$H$_{12}$N$^+$ [M + H]$^+$: 74.10, found: 73.9.

Isoindoline·HCl: brown solid, 69 mg, 0.44 mmol, 69%; ^{1}H NMR (DMSO-d_6, 400 MHz): 10.21 (br, 2H), 7.52–7.36 (m, 4H), 4.48 (s, 4H); ^{13}C{^{1}H} NMR (DMSO-d_6, 101 MHz): 134.4, 127.6, 122.4, 49.2; (ESI+) m/z calcd for C$_8$H$_{10}$N$^+$ [M + H]$^+$: 120.08, found: 119.98.

Benzylamine·HCl: white solid, 183 mg, 1.27 mmol, 85%; ^{1}H NMR (CD$_3$OD, 400 MHz): 7.49–7.40 (m, 5H), 4.12 (s, 2H); ^{13}C{^{1}H} NMR (CD$_3$OD, 101 MHz): 134.4, 130.2, 130.0, 129.9, 44.4; (ESI+) m/z calcd for C$_7$H$_{10}$N$^+$ [M + H]$^+$: 108.08, found: 107.88.

N-ethylbenzylamine·HCl: white solid, 114 mg, 0.66 mmol, 75%; ^{1}H-NMR (CD$_3$OD, 400 MHz): 7.52–7.44 (m, 5H), 4.19 (s, 2H), 3.12 (q, J = 7.3 Hz, 2H), 1.34 (t, J = 7.3 Hz, 3H); ^{13}C{^{1}H} NMR (CD$_3$OD, 101 MHz): 132.7, 130.9, 130.6, 130.3, 52.0, 43.8, 11.5; (ESI+) m/z calcd for C$_9$H$_{14}$N$^+$ [M + H]$^+$: 136.11, found: 135.97.

3.11.3. Oxidation of Compound **2** and **6**

The compounds were obtained following the procedure reported above. Although, after the evaporation of the MeOH/H$_2$O mixture and the addition of water, the mixture was acidified with 0.2 M HCl to pH 2–3 and heated for a couple of hours at 40 °C in order to facilitate the hydrolysis of the enamine.

Dibenzylamine·HCl: white solid, 47 mg, 0.22 mmol, 33%; ^{1}H NMR (CD$_3$OD, 400 MHz): 7.52–7.45 (m, 10H), 4.25 (s, 4H); ^{13}C{^{1}H} NMR (CD$_3$OD, 101 MHz): 131.8, 130.1, 128.8, 128.5, 49.5; (ESI+) m/z calcd for C$_{14}$H$_{16}$N$^+$ [M + H]$^+$: 198.13, found: 198.06.

p-Nitrobenzylamine: white solid, 15 mg, 0.08 mmol, 45%; ^{1}H NMR (DMSO-d_6, 400 MHz): 8.78 (br, 3H), 8.26 (d, J = 8.2 Hz, 2H), 7.80 (d, J = 8.2 Hz, 2H), 4.17 (s, 2H); ^{13}C{^{1}H} NMR (DMSO-d_6, 101 MHz): 147.4, 141.7, 130.2, 123.5, 41.3; (ESI+) m/z calcd for C$_7$H$_9$N$_2$O$_2$$^+$ [M + H]$^+$: 153.07, found: 152.93.

3.12. Computational Details

The oxidation–elimination–hydrolysis mechanism of compound **3** was modeled with an in silico approach based on DFT. The oxidation and elimination initial reactions (Scheme 1, top line) were modelled on the basis of a recent in-depth study on analogous reactions by some of the authors of this work [10]. The only difference with Scheme 1 is that the in the computational model, the direct product of the elimination process is the selenenic acid analogue of the seleninic acid. The oxidation from selenenic to seleninic acid was not investigated in silico. The oxidation reaction keeps the stereochemistry of the C carbon bonded to Se fixed and can take two pathways, leading to diastereomeric geometries of oxidized products (R–R and R–S). In light of the conclusions drawn in a recent work [10], only the pathway leading to the R–R diastereomer was investigated due to its more favorable reaction energies. The quantum chemistry calculations for the oxidation mechanism were performed using the Amsterdam Density Functional (ADF) [37–39]. The energy profiles were obtained from geometries and energies computed by using the OLYP functional [40,41], which is known to perform well for reactivity studies on organic compounds, and it has been recently benchmarked [30] and applied [31] to organic dichalcogenides and amines [42]. OLYP was combined with the TZ2P basis set for all the atoms [43]. The TZ2P basis set is of triple- quality and has been augmented with two sets of polarization functions. Core shells of the atoms (1s for C, F, N and O and up to 3p for Se) were treated by using the frozen-core approximation. Scalar relativistic effects were accounted for using the Zeroth Order Regular Approximation (ZORA) [44–46]. The numerical integration was performed by using the fuzzy-cell integration scheme developed by Becke [47,48]. Energy minima and transition states have been verified through vibrational analysis. All minima were found to have zero imaginary frequencies and all transition states have one that correspond to the mode of the reaction under consideration. For single point calculations in water, the conductor-like screening model was employed (COSMO), as implemented in ADF [49–51]. Water was parameterized using the default values in ADF, i.e., a dielectric constant of 78.39 and a solvent radius of 1.93 Å. The empirical parameter in the scaling function in the COSMO equation was set to 0.0. The radii of the atoms were taken to be MM3 radii divided by 1.2, giving 1.350 Å for H, 1.700 Å for C, 1.608 for N, 1.517 for O and 1.908 for Se [49,52]. This level of theory is referred to as (COSMO)-ZORA-OLYP/TZ2P. Geometry optimizations were conducted in the gas phase and frequency calculations were employed to establish the nature of the stationary points found (all real frequencies for minima and one imaginary frequency for transition states). Thereafter, single point calculations in COSMO water were used to obtain the ΔG values reported in the main text. Relative free Gibbs energies are shown for the gas phase calculations and for the water single points in Table S1.

4. Conclusions

Amines are present in a vast portion of biologically active compounds, and their relevance in medicinal chemistry is crucial. More specifically, such moieties are often present in CNS-targeting molecules. In several drugs or drug-like compounds, variously substituted amines represent a part of the pharmacophore or are employed to enhance water solubility. We here reported the experimental and in silico characterization of an innovative combined reaction consisting in a selenoxide elimination-triggered hydrolysis for the preparation and/or selective release under oxidative conditions of several model primary and secondary amines. This may allow an effective in situ release of high-value pharmaceutical compounds that overcomes multiple limitations to which amines are often subjected to in the organism. Moreover, the ability of this reaction to be triggered in an oxidizing environment is very promising for the development of amino-containing molecules that can be selectively activated only in case of excessive oxidative stress levels.

Supplementary Materials: The following are available online, Figures S1–S27: NMR and ESI-MS characterization of compounds 1–9; Figures S28–S37: representative ^{1}H-NMR and ESI-MS spectra of the oxidation study with H_2O_2; Figures S38–S58: ^{1}H-NMR, ^{13}C-NMR and ESI-MS spectra of the isolated compounds obtained by oxidation with H_2O_2; Table S1: Gibbs free energies relative to free reactants for the selenoxide-triggered amine formation; Table S2: geometries of the optimized structures of the selenoxide-triggered amine formation reaction.

Author Contributions: Conceptualization, G.R., M.B.; data curation, G.R., M.B., E.O., A.O.; funding acquisition, A.G., L.O.; investigation, G.R., M.B., E.O., A.O.; methodology, G.R., M.B., L.O.; project administration, L.O.; supervision, G.Z., L.O.; validation, G.Z., L.O.; visualization, G.R., M.B.; writing—original draft preparation, G.R., M.B.; writing—review and editing, G.R., M.B., A.G., G.Z., L.O. All authors have read and agreed to the published version of the manuscript.

Funding: This research was funded by Università degli Studi di Padova, thanks to the P-DiSC (BIRD2018-UNIPD) project MAD³S (Modeling Antioxidant Drugs: Design and Development of computer-aided molecular Systems), P.I. L.O., and by Università degli Studi di Brescia. Calculations were carried out on the Bastion system of the CNAF institute (Bologna, Italy) thanks to the INCIpit grant (Insights on Chalcogen Nitrogen Interaction, ISCRA C HP10C15ZCK).

Institutional Review Board Statement: Not applicable.

Informed Consent Statement: Not applicable.

Data Availability Statement: The data presented in this study are available in the supplementary materials.

Conflicts of Interest: The authors declare no conflict of interest.

Sample Availability: Not available.

References

1. Jones, D.N.; Mundy, D.; Whitehouse, R.D. Steroidal Selenoxides Diastereoisomeric at Selenium; Syn-Elimination, Absolute Configuration, and Optical Rotatory Dispersion Characteristics. *J. Chem. Soc. D Chem. Commun.* **1970**, *2*, 86–87. [CrossRef]
2. Sharpless, K.B.; Lauer, R.F.; Teranishi, A.Y. Electrophilic and Nucleophilic Organoselenium Reagents. New Routes to Alpha, Beta-Unsaturated Carbonyl Compounds. *J. Am. Chem. Soc.* **1973**, *95*, 6137–6139. [CrossRef]
3. Heffner, R.J.; Jiang, J.; Joullie, M.M. Total synthesis of (−)-nummularine F. *J. Am. Chem. Soc.* **1992**, *114*, 10181–10189. [CrossRef]
4. Nicolaou, K.C.; Reddy, K.R.; Skokotas, G.; Sato, F.; Xiao, X.Y.; Hwang, C.K. Total Synthesis of Hemibrevetoxin B and (7aα)-Epi-hemibrevetoxin B. *J. Am. Chem. Soc.* **1993**, *115*, 3558–3575. [CrossRef]
5. Vargas, D.; Fronczek, F.R.; Fischer, N.H.; Hostettmann, K. The Chemistry of Confertiflorin and the Molecular Structure of Confertiflorin and Allodesacetylconfertiflorin, Two Molluscicidal Sesquiterpene Lactones. *J. Nat. Prod.* **1986**, *49*, 133–138. [CrossRef]
6. Callant, P.; Ongena, R.; Vandewalle, M. Iridoids: Novel Total Synthesis of (±)- Isoiridomyrmecin and of (±)-Verbenalol. *Tetrahedron* **1981**, *37*, 2085–2089. [CrossRef]
7. Waring, A.J.; Zaidi, J.H. Synthesis of a 4-acylcyclohexa-2,5-dienone: 3,4-dihydro-3,3,8a-trimethyl-naphthalene-1,6(2H,8aH)-dione. *J. Chem. Soc. Perkin. Trans.* **1985**, *1*, 631–639. [CrossRef]
8. Sharpless, K.B.; Young, M.W.; Lauer, R.F. Reactions of Selenoxides: Thermal Syn-elimination and H218O Exchange. *Tetrahedron Lett.* **1973**, *14*, 1979–1982. [CrossRef]
9. Ribaudo, G.; Bellanda, M.; Menegazzo, I.; Wolters, L.P.; Bortoli, M.; Ferrer-Sueta, G.; Zagotto, G.; Orian, L. Mechanistic Insight into the Oxidation of Organic Phenylselenides by H_2O_2. *Chem. Eur. J.* **2017**, *23*, 2405–2422. [CrossRef] [PubMed]
10. Ribaudo, G.; Bortoli, M.; Ongaro, A.; Oselladore, E.; Gianoncelli, A.; Zagotto, G.; Orian, L. Fluoxetine Scaffold to Design Tandem Molecular Antioxidants and Green Catalysts. *RSC Adv.* **2020**, *10*, 18583–18593. [CrossRef]
11. Simplício, A.L.; Clancy, J.M.; Gilmer, J.F. Prodrugs for Amines. *Molecules* **2008**, *13*, 519–547. [CrossRef]
12. Sun, C.; Wang, L.; Xianyu, B.; Li, T.; Gao, S.; Xu, H. Selenoxide Elimination Manipulate the Oxidative Stress to Improve the Antitumor Efficacy. *Biomaterials* **2019**, *225*, 119514. [CrossRef]
13. Guo, R.; Huang, J.; Huang, H.; Zhao, X. Organoselenium-Catalyzed Synthesis of Oxygen- and Nitrogen-Containing Heterocycles. *Org. Lett.* **2016**, *18*, 504–507. [CrossRef]
14. Santi, C.; Bagnoli, L. Celebrating Two Centuries of Research in Selenium Chemistry: State of the Art and New Prospective. *Molecules* **2017**, *22*, 2124. [CrossRef] [PubMed]
15. Shao, L.; Li, Y.; Lu, J.; Jiang, X. Recent Progress in Selenium-Catalyzed Organic Reactions. *Org. Chem. Front.* **2019**, *6*, 2999–3041. [CrossRef]
16. Krief, A.; Dumont, W.; De Mahieu, A.F. Novel Synthesis of Selenones: Application to the Synthesis of Alkyl Cyclopropanecarboxylates. *Tetrahedron Lett.* **1988**, *29*, 3269–3272. [CrossRef]
17. Sevrin, M.; Dumont, W.; Krief, A. Synthetic Route to Vinyl Selenides and Vinyl Selenoxides. *Tetrahedron Lett.* **1977**, *18*, 3835–3838. [CrossRef]
18. Stamhuis, E.J.; Maas, W. Mechanism of Enamine Reactions. II. 1 The Hydrolysis of Tertiary Enamines. *J. Org. Chem.* **1965**, *30*, 2156–2160. [CrossRef]
19. Capon, B.; Wu, Z.P. Comparison of the Tautomerization and Hydrolysis of Some Secondary and Tertiary Enamines. *J. Org. Chem.* **1990**, *55*, 2317–2324. [CrossRef]
20. Gabriel, S. Über eine Darstellungsweise primärer Amine aus den entsprechenden Halogenverbindungen. *Ber. Dtsch. Chem. Ges.* **1887**, *20*, 2224–2236. [CrossRef]
21. Corey, E.J.; Link, J.O. A General, Catalytic, and Enantioselective Synthesis of Alpha.-Amino Acids. *J. Am. Chem. Soc.* **1992**, *114*, 1906–1908. [CrossRef]
22. Leuckart, R. Über eine neue Bildungsweise von Tribenzylamin. *Ber. Dtsch. Chem. Ges.* **1885**, *18*, 2341–2344. [CrossRef]
23. Doi, H.; Sakai, T.; Iguchi, M.; Yamada, K.; Tomioka, K. Chiral Ligand-Controlled Asymmetric Conjugate Addition of Lithium Amides to Enoates. *J. Am. Chem. Soc.* **2003**, *125*, 2886–2887. [CrossRef]
24. Von Hofmann, A.W.V. Researches Regarding the Molecular Constitution of the Volatile Organic Bases. *Philos. Trans. R. Soc.* **1850**, *140*, 93–131. [CrossRef]
25. Russell Bowman, W.; Coghlan, D.R. A Facile Method for the N-alkylation of α-Amino Esters. *Tetrahedron* **1997**, *53*, 15787–15798. [CrossRef]
26. Fukuyama, T.; Jow, C.-K.; Cheung, M. 2- and 4-Nitrobenzenesulfonamides: Exceptionally Versatile Means for Preparation of Secondary Amines and Protection of Amines. *Tetrahedron Lett.* **1995**, *36*, 6373–6374. [CrossRef]
27. Tingoli, M.; Tiecco, M.; Chianelli, D.; Balducci, R.; Temperini, A. Novel Azido-phenylselenenylation of Double Bonds. *Evidence for a Free-radical Process*. *J. Org. Chem.* **1991**, *56*, 6809–6813. [CrossRef]
28. Bortoli, M.; Dalla Tiezza, M.; Muraro, C.; Pavan, C.; Ribaudo, G.; Rodighiero, A.; Tubaro, C.; Zagotto, G.; Orian, L. Psychiatric Disorders and Oxidative Injury: Antioxidant Effects of Zolpidem Therapy disclosed In Silico. *Comput. Struct. Biotechnol. J.* **2019**, *17*, 311–318. [CrossRef]
29. Ribaudo, G.; Bortoli, M.; Pavan, C.; Zagotto, G.; Orian, L. Antioxidant Potential of Psychotropic Drugs: From Clinical Evidence to In Vitro and In Vivo Assessment and toward a New Challenge for In Silico Molecular Design. *Antioxidants* **2020**, *9*, 714. [CrossRef]

30. Bortoli, M.; Zaccaria, F.; Dalla Tiezza, M.; Bruschi, M.; Fonseca Guerra, C.; Bickelhaupt, F.M.; Orian, L. Oxidation of organic diselenides and ditellurides by H2O2 for bioinspired catalyst design. *Phys. Chem. Chem. Phys.* **2018**, *20*, 20874–20885. [CrossRef]
31. Bortoli, M.; Bruschi, M.; Swart, M.; Orian, L. Sequential oxidations of phenylchalcogenides by H2O2: Insights into the redox behavior of selenium via DFT analysis. *New J. Chem.* **2020**, *44*, 6724–6731. [CrossRef]
32. Yildiz, I. A DFT Approach to the Mechanistic Study of Hydrozone Hydrolysis. *J. Phys. Chem. A* **2016**, *120*, 3683–3692. [CrossRef]
33. Bayse, C.A. DFT Study Of The Glutathione Peroxidase-Like Activity of Phenylselenol Incorporating Solvent-Assisted Proton Exchange. *J. Phys. Chem. A* **2007**, *111*, 9070–9075. [CrossRef]
34. Bayse, C.A. Transition States for Cysteine Redox Processes Modeled by DFT and Solvent-Assisted Proton Exchange. *Org. Biomol. Chem.* **2011**, *9*, 4748–4751. [CrossRef] [PubMed]
35. Antony, S.; Bayse, C.A. Modeling the Mechanism of the Glutathione Peroxidase Mimic Ebselen. *Inorg. Chem.* **2011**, *50*, 12075–12084. [CrossRef] [PubMed]
36. Chen, H.; Wang, W.; Yang, Y.; Jiang, P.; Gao, W.; Cong, R.; Yang, T. Solvent Effect on the Formation of Active Free Radicals From H_2O_2 Catalyzed by Cr-Substituted PKU-1 Aluminoborate: Spectroscopic Investigation and Reaction Mechanism. *Appl. Catal. A Gen.* **2019**, *588*, 117283. [CrossRef]
37. te Velde, G.; Bickelhaupt, F.M.; Baerends, E.J.; Fonseca Guerra, C.; van Gisbergen, S.J.A.; Snijders, J.G.; Ziegler, T. Chemistry with ADF. *J. Comput. Chem.* **2001**, *22*, 931–967. [CrossRef]
38. Fonseca Guerra, C.; Snijders, J.G.; te Velde, G.; Baerends, E.J. Towards an Order- N DFT Method. *Theor. Chem. Acc.* **1998**, *99*, 391–403. [CrossRef]
39. Baerends, E.J.; Ziegler, T.; Atkins, A.J.; Autschbach, J.; Bashford, D.; Baseggio, O.; Bérces, A.; Bickelhaupt, F.M.; Bo, C.; Boerritger, P.M.; et al. *ADF2018; SCM, Theoretical Chemistry*; Vrije Universiteit: Amsterdam, The Netherlands, 2018.
40. Handy, N.C.; Cohen, A.J. Left-right Correlation Energy. *Mol. Phys.* **2001**, *99*, 403–412. [CrossRef]
41. Handy, N.C.; Cohen, A.J. A Dynamical Correlation Functional. *J. Chem. Phys.* **2002**, *116*, 5411–5418. [CrossRef]
42. Savoo, N.; Laloo, J.Z.A.; Rhyman, L.; Ramasami, P.; Bickelhaupt, F.M.; Poater, J. Activation Strain Analyses of Counterion and Solvent Effects on the Ion-Pair S_N2 Reaction of and CH_3Cl. *J. Comput. Chem.* **2020**, *41*, 317–327. [CrossRef] [PubMed]
43. Van Lenthe, E.; Baerends, E.J. Optimized Slater-Type Basis Sets For the Elements 1-118. *J. Comput. Chem.* **2003**, *24*, 1142–1156. [CrossRef]
44. van Lenthe, E.; Baerends, E.J.; Snijders, J.G. Relativistic Regular Two-Component Hamiltonians. *J. Chem. Phys.* **1993**, *99*, 4597–4610. [CrossRef]
45. van Lenthe, E.; Baerends, E.J.; Snijders, J.G. Relativistic Total Energy Using Regular Approximations. *J. Chem. Phys.* **1994**, *101*, 9783–9792. [CrossRef]
46. van Lenthe, E.; Snijders, J.G.; Baerends, E.J. The Zero-Order Regular Approximation for Relativistic Effects: The Effect of Spin–Orbit Coupling in Closed Shell Molecules. *J. Chem. Phys.* **1996**, *105*, 6505–6516. [CrossRef]
47. Becke, A.D. A Multicenter Numerical Integration Scheme for Polyatomic Molecules. *J. Chem. Phys.* **1988**, *88*, 2547–2553. [CrossRef]
48. Franchini, M.; Philipsen, P.H.T.; Visscher, L. The Becke Fuzzy Cells Integration Scheme in the Amsterdam Density Functional Program Suite. *J. Comput. Chem.* **2013**, *34*, 1819–1827. [CrossRef] [PubMed]
49. Pye, C.C.; Ziegler, T. An Implementation of the Conductor-Like Screening Model of Solvation within the {Amsterdam} Density Functional Package. *Theor. Chem. Acc.* **1999**, *101*, 396–408. [CrossRef]
50. Klamt, A.; Schüürmann, G. COSMO: A New Approach to Dielectric Screening in Solvents with Explicit Expressions for the Screening Energy and its Gradient. *J. Chem. Soc. Perkin Trans.* **1993**, *2*, 799–805. [CrossRef]
51. Klamt, A.; Jonas, V. Treatment of the Outlying Charge in Continuum Solvation Models. *J. Chem. Phys.* **1996**, *105*, 9972–9981. [CrossRef]
52. Allinger, N.L.; Zhou, X.; Bergsma, J. Molecular Mechanics Parameters. *J. Mol. Struct. THEOCHEM* **1994**, *312*, 69–83. [CrossRef]

Article

Molecular Docking Study on Several Benzoic Acid Derivatives against SARS-CoV-2

Amalia Stefaniu *, Lucia Pirvu *, Bujor Albu and Lucia Pintilie

National Institute for Chemical-Pharmaceutical Research and Development, 112 Vitan Av., 031299 Bucharest, Romania; abujor@gmail.com (B.A.); lucia.pintilie@gmail.com (L.P.)
* Correspondence: astefaniu@gmail.com (A.S.); lucia.pirvu@yahoo.com (L.P.)

Academic Editors: Giovanni Ribaudo and Laura Orian
Received: 15 November 2020; Accepted: 1 December 2020; Published: 10 December 2020

Abstract: Several derivatives of benzoic acid and semisynthetic alkyl gallates were investigated by an in silico approach to evaluate their potential antiviral activity against SARS-CoV-2 main protease. Molecular docking studies were used to predict their binding affinity and interactions with amino acids residues from the active binding site of SARS-CoV-2 main protease, compared to boceprevir. Deep structural insights and quantum chemical reactivity analysis according to Koopmans' theorem, as a result of density functional theory (DFT) computations, are reported. Additionally, drug-likeness assessment in terms of Lipinski's and Weber's rules for pharmaceutical candidates, is provided. The outcomes of docking and key molecular descriptors and properties were forward analyzed by the statistical approach of principal component analysis (PCA) to identify the degree of their correlation. The obtained results suggest two promising candidates for future drug development to fight against the coronavirus infection.

Keywords: SARS-CoV-2; benzoic acid derivatives; gallic acid; molecular docking; reactivity parameters

1. Introduction

Severe acute respiratory syndrome coronavirus 2 is an international health matter. Previously unheard research efforts to discover specific treatments are in progress worldwide. Virtual screening of existing compound databases against three protein targets (main protease, RNA dependent RNA polymerase and spike protein) to inhibit coronavirus replication is one of the actual approaches that allows researchers to quickly select best drug candidates for further in vitro assays [1–5].

Boceprevir is a direct-acting antiviral agent (DAA), acting as an inhibitor of NS3/4A, a serine protease enzyme encoded by hepatitis C virus (HCV) [6]. The serine protease enzyme plays a vital role in the replication and cleavage of viral proteins. We found boceprevir as a cocrystallized ligand in a complex with a protein named 3C-like proteinase, the main protease found in coronaviruses, characterized by X-ray diffraction, introduced in the protein data bank with the entry ID: 6WNP [7], at 1.45 Å resolution. The main protease in coronaviruses contains a cysteine-histidine dyad able to achieve catalytic cleavage of the coronavirus polyprotein [8]. Cysteine acts as a nucleophile due to its free electron pair on the sulfur atom, donated to form intramolecular bonds and histidine, respectively, that acts as a general base by its imidazole heterocycle [9]. The main protease of SARS-CoV-2 consists of three domains and a characteristic CYS145-HIS41 dyad in the active site [10].

Due to their low toxicity and antioxidant activity, phenolic acids and flavonoids appear as the most feasible and secure natural antiviral compounds. The potential antiviral activity of vegetal polyphenols is generally based on their capacity to alter virus replication and functional protein synthesis [11]. Another aspect to be mentioned is the capacity of volatile oils, saponins and triterpenic acids, to act as effective solvents and detergents, therefore to solubilize and destroy the lipid layer of the enveloped viruses. Regarding the general chemical aspects associated with antiviral activity, the presence of phenyl ring(s), vinyl and carboxyl moieties and ester, hydroxyl and methoxy groups appear to be the basis of plant phenolics antiviral efficacy [11–13]. For example, it is considered that phenolics with free hydroxyl groups interfere with viral adsorption and further cell penetration [11], which is sustained by the fact that high polar phenolics create a protective coating on the cell's surface. On the other hand, the plant compounds' bioavailability allows them to reach the circulation and be able to manifest antiviral activity. Accordingly, clinical studies have revealed that gallic acid, catechins, flavones and quercetin glucosides have the best bioavailability in humans [14]; also, data suggests that anthocyanins are fully absorbed in humans [15]. The most potent antiviral compounds proved to interfere with virus replication and/or viral essential protein synthesis are some of the most common vegetal polyphenols, namely quercetin, kaempferol and apigenin, ellagic, rosmarinic and gallic acids, catechin and epicatechin, and various alkyl gallates [11]. Similary, chrysin, acacetin and apigenin inhibited viral transcription of the human rhinovirus 14 [16], while proanthocyanidins from *Myrothamnusf labellifolia* have proved antiviral activity against herpes simplex virus type-1 by viral adsorption and cell penetration inhibition [17].

Furthermore, molecular docking studies in the last two years, made on dozens of natural and synthesized antiviral compounds, associate their antiviral activity with the capacity of their active groups (phenyl groups and phenyl moieties such as hydroxyl, carbonyl, amino, azo, nitrile, sulfonyl) to bind the active groups of several amino acids found in the active site of SARS-CoV-2 main protease. GLU166, HIS41, ASN142, GLY143, HIS163 and THR 190, are on the top of the most frequent amino acids bound [10,18–20].

Recent virtual screening of 22 U.S.-FDA approved antiviral drugs in the parallel with 24 natural plant-based molecules, indicated theaflavin digallate (from green tea) with the best docking score (−10.574), a result obtained using the Glide (Schrödinger) module on the COVID-19 main protease (structure 6LU7) [20]. It must be noted that excepting HIS41, which bound Osp^2 from the carboxyl moiety of gallic acid, all other interactions occurred at hydroxyl groups from the flavan ring. Therefore, it is confirmed that the high degree of hydroxylation is associated with a good antiviral effect.

Another aspect to be mentioned is the conclusion of a computational survey to a drug repurposing study claiming that five neutral antiviral drugs have inhibitory activities against SARS-CoV-2 main protease [21]. As it is well known, the logP value describes lipophilicity for neutral compounds. Therefore, the compounds with an octanol-water partition coefficient (logP) value between one to three have good passive absorption across lipid membranes (bioavailability), while those with a logP value greater than three or less than one usually have lesser bioavailability [22]. In this context, further calculations can be made in connection with the formulation of antiviral products to obtain targeted bioavailability in the human intestine.

Gallic acid, one of the most abundant phenolic acids in plants, has various health benefits including anti-inflammatory, antioxidant and antimicrobial activities [23] and is viewed as the lead compound with promising pharmacological properties to design and develop new drugs [24]. Regarding semisynthetic derivatives of gallic acid, the alkyl gallates with antimicrobial activity were reported too [25,26]. Their biological activity is associated with the length with their alkyl side chain, which affects membrane binding capability [25]. Generally, membrane binding of the alkyl gallates increases with increasing alkyl chain length and is correlated with their antiviral activity.

Starting from such intriguing premises, in molecular docking approach we conducted in silico screening on several benzoic acid derivatives and on a homologue series of alkyl gallates, starting from the lead compound gallic acid, against SARS-CoV-2 main protease.

2. Results and Discussion

We selected the following compounds: benzoic acid, 4-aminobenzoic acid, 4-hydroxybenzoic acid, 3,4,5-trimethoxybenzoic acid (eudesmic acid), 3,4-dihydroxybenzoic acid (protocatehuic acid), 2,5-dihydroxybenzoic acid (gentisic acid), 4-hydroxy-3-methoxybenzaldehyde (vanillin), 4-hydroxy-3,5-dimethoxybenzoic acid (syringic acid), 4,5-dihydroxy-3-oxocyclohex-1-ene-1-carboxylic acid, epicatechin, 3,4,5- trihydroxybenzoic acid (gallic acid), methyl 3,4,5-trihydroxybenzoate, ethyl 3,4,5-trihydroxybenzoate, propyl 3,4,5-trihydroxybenzoate, isopropyl 3,4,5-trihydroxybenzoate, butyl 3,4,5-trihydroxybenzoate, isobutyl 3,4,5-trihydroxybenzoate, pentyl 3,4,5-trihydroxybenzoate, isopentyl 3,4,5-trihydroxybenzoate, octyl 3,4,5-trihydroxybenzoate. Their 3D optimized structures obtained with the Spartan program, along with the atomic numbering scheme, arbitrary chosen by the software, are given in Figure 1.

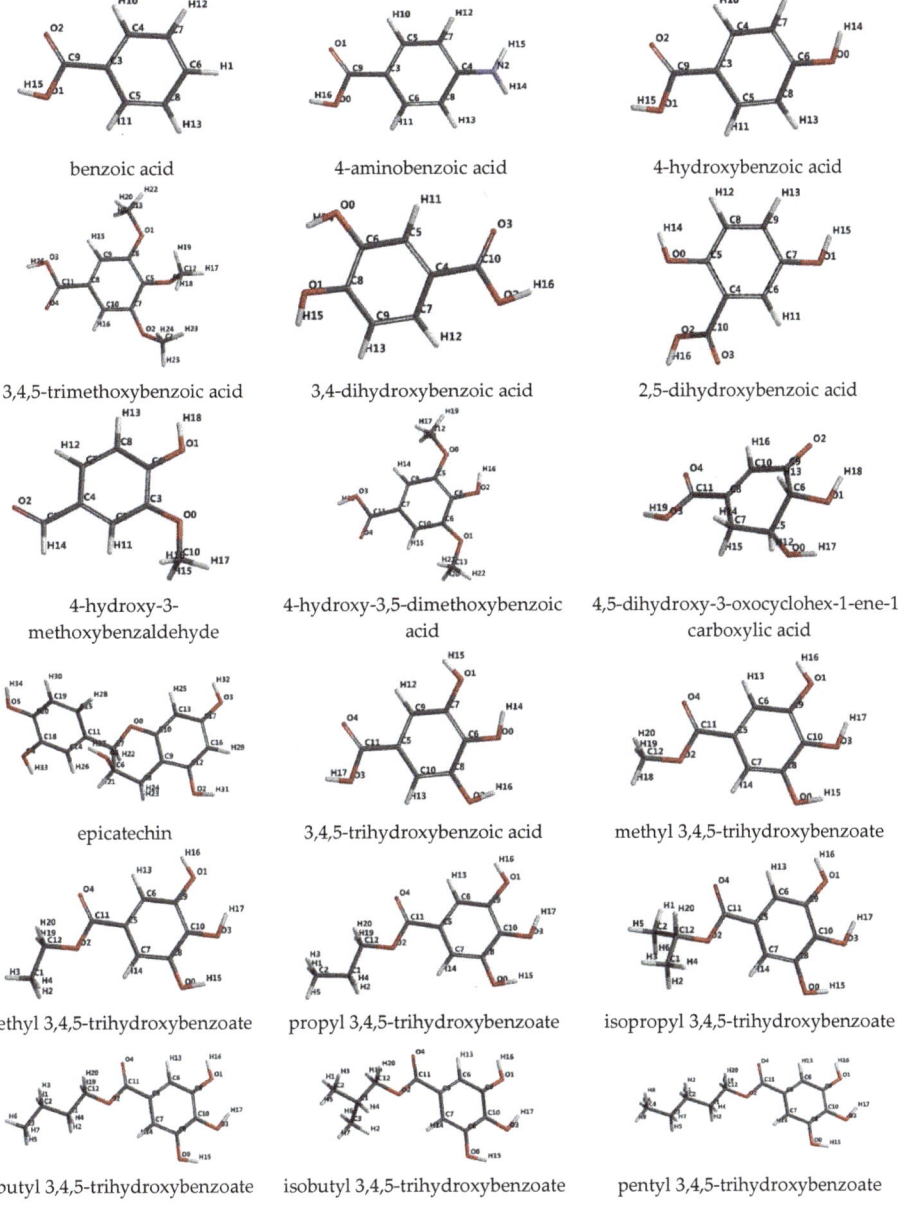

Figure 1. *Cont.*

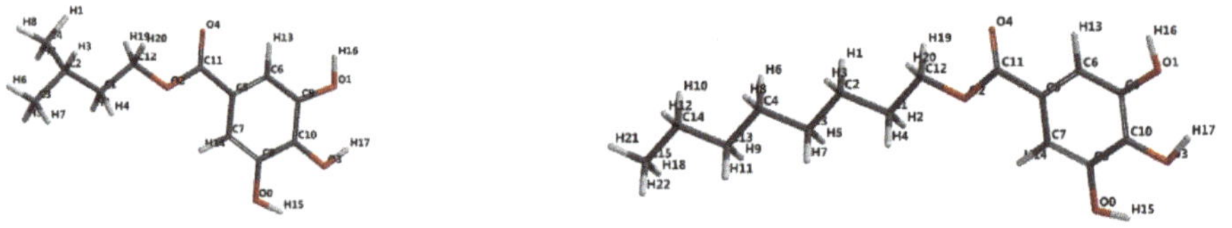

Molecules **2020**, *25*, 5828

Isopentyl 3,4,5-trihydroxybenzoate

octyl 3,4,5-trihydroxybenzoate

Figure 1. 3D optimized structures of investigated ligands and their numbering atomic labels.

2.1. Results of Molecular Docking Simulations

Intermolecular interactions occurring in boceprevir, benzoic acid and alkyl gallate derivatives in complex with the 6WNP protein fragment were identified. The lengths of hydrogen-bonding interactions were measured. The results are given in terms of docking score function and RMSD (root mean square deviation).

To validate the molecular docking protocol, boceprevir was initially docked into the crystal structure of the main protease fragment and its interactions with the target 3C-like proteinase were analyzed. As can be seen Figure 2a, the native ligand forms eight hydrogen bonding interactions with residues CYS145, SER144, GLY143, HIS41, HIS164, GLU166, GLU166 and GLU166. In Figure 2b, the superposition of the binding pose of boceprevir, obtained by redocking, is displayed. As illustrated in Figure 2c, all docked ligands were found to have similar binding poses to the native ligand, thus validating the chosen docking approach.

Figure 3 reveals the obtained docking scores for the cocrystallized and selected ligands. Boceprevir exhibits the greatest score (−63.95), due to its numerous interactions with the amino acid residues from the protein's active binding site, i.e., eight hydrogen bonding interactions with nitrogen or oxygen atoms of amino acids residues N sp^2 CYS145 (2.900 Å), N sp^2 SER144 (3.053 Å), N sp^2 GLY143 (2.783 Å), N sp^2 HIS41 (2.604 Å), O sp^2 HIS164 (3.103 Å), N sp^2 GLU166 (3.118 Å), O sp^2 GLU166 (2.908 Å and O sp^2 GLU166 (3.229 Å). Remarkable is the docking score for octyl-gallate (−60.22, RMSD 1.12), close to boceprevir, and for epicatechin (−49.57). Benzoic acid derivatives, in the first half of the graph exhibit moderate scores ranging between −29.59 (benzoic acid) and −37.25 (syringic acid). We observed an increasing trend of score with the increasing number of hydroxyl groups. For instance, gallic acid docking simulations resulted in a −38.31 score. Regarding the gallates, the increasing length of alkyl chain led to docking scores in the order gallic acid < methyl gallate < ethyl gallate < isopropyl gallate < propyl gallate < butyl gallate < isobutyl gallate < isopentyl gallate < pentyl gallate < octyl gallate. The ramification of lateral side chain led to a slight decrease of binding affinity, noticeable with an increasing number of -CH_2 (e.g., isopentyl: −46.17 versus pentyl gallate: −48.77).

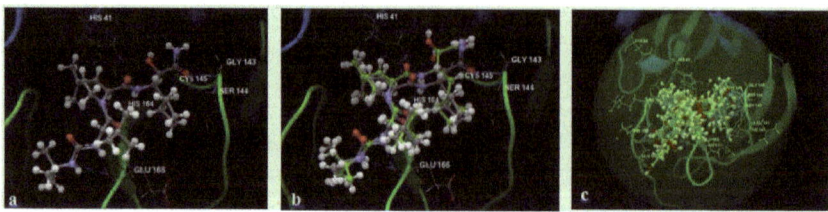

Figure 2. Hydrogen bonding interactions of the native ligand (boceprevir); (**a**) superposition of the native ligand; (**b**) superposition of all docked ligands in the active binding site of 3C-like proteinase (6WNP) (**c**).

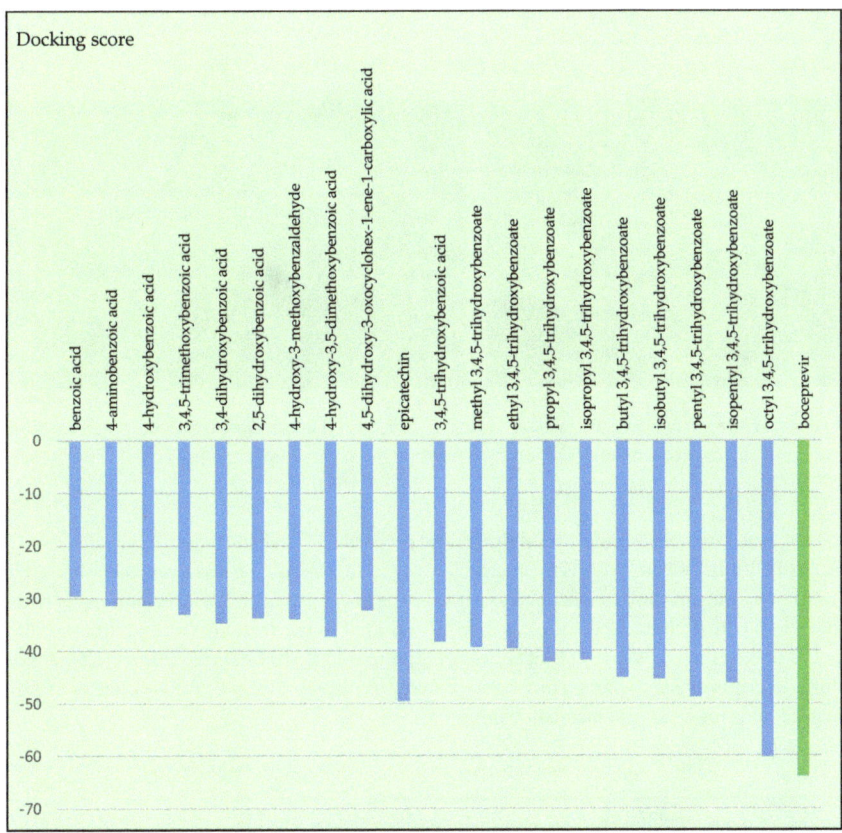

Figure 3. Docking scores for investigated ligands against SARS-CoV-2 main protease (6WNP).

In Table S1 from the Supplementary files available online are listed the obtained values for docking score and RMSD (root mean square deviation), the amino acids group interactions and type and length of hydrogen-bonding interactions formed by each ligand in complex with SARS-CoV-2 main protease.

Figure 4 illustrates the intramolecular interactions by H-bonding (a), and amino acid group interactions occurring in the complex formed by octyl gallate and the 6WNP main protease fragment.

Amino acid residues CYS145 and SER144 interacted by hydrogen-bonding both with boceprevir and with octyl-gallate (see Figure 4a); similar interactions resulted in similar docking results. Propyl gallate and pentyl gallate revealed the same ten intramolecular interactions with N sp^2 GLU166, N sp^2 HIS163, O sp^3 SER144, O sp^3 SER144, O sp^2 LEU141, N sp^2 SER144, N sp^2 GLY143, N sp^2 CYS145, N sp^2 GLY143 and O sp^2 ASN142. Their obtained docking scores were −42.13 and −48.77, respectively; lower than for octyl-gallate but greater than the results for gallic acid (−38.31). These observations are in good agreement with experimental findings of Takai E. et al., 2019 [25], anticipating increasing antiviral effects of alkyl gallates with increasing alkyl chain length, except for cetyl and stearyl gallate (which are not included in this study), a fact experimentally validated by fluorescence analysis of the binding of alkyl gallates to phospholipid membranes. Beyond a certain alkyl chain length (8–11), a reduction of antibacterial and antiviral activities of the alkyl gallates was observed, probably due to a self-association process [25–27]. This was the reason for choosing to break off the gallates screening at octyl. Increasing docking results were observed with increasing length of the *n*-alkyl side chain. The good docking result for octyl gallate recommends it as good alternative for developing new therapeutic antiviral agents.

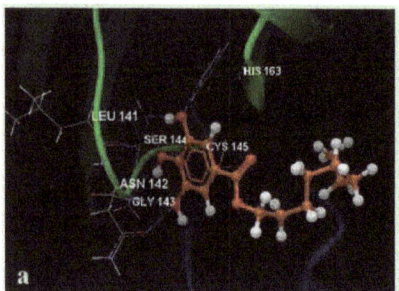
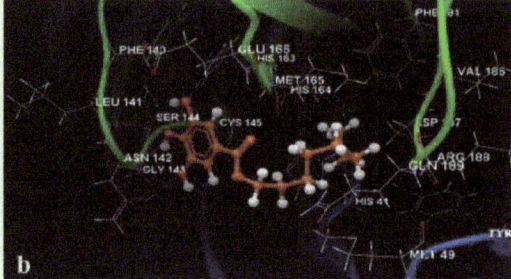

Figure 4. Hydrogen bonding interactions of octyl-gallate (**a**) and interactions with amino acid residues from the active binding site of 6WNP protein fragment (**b**).

2.2. Results of Oral Bioavailability Evaluation

In Table 1 are listed key molecular descriptors and properties to evaluate the oral bioavailability [28,29] and Veber's [30] rules, where: MW is the molecular weight that should be less than 500 Daltons, HBD is the number of hydrogen bond donors (recommended to be lower than 5), HBD is the number of hydrogen bond acceptors with acceptable values less than 10 and log P is the water-octanol partition coefficient, that should be less than 5. Veber D.F. et al., 2002 [30] imposed additional restrictions related with the molecular descriptor PSA (polar surface area), namely, no larger than 140 Å^2 and with a maximum 10 rotatable bonds for good oral bioavailability.

Table 1. Lipinski and Veber's parameters for drugability assessment.

Ligand	MW	PSA	HBD	HBA	LogP	rb	LV
Benzoic acid	122.123	33.690	1	2	0.79	1	0
4-Aminobenzoic acid	137.138	58.471	3	3	−0.93	1	0
4-Hydroxybenzoic acid	138.122	53.444	2	3	−0.29	1	0
3,4,5-Trimethoxybenzoic acid	212.201	53.223	1	5	−2.14	4	0
3,4-Dihydroxybenzoic acid	154.121	71.217	3	4	−1.37	1	0
2,5-Dihydroxybenzoic acid	154.121	71.262	3	4	0.81	1	0
4-Hydroxy-3-methoxybenzaldehyde	152.149	41.012	1	3	−1.53	2	0
4-Hydroxy-3,5-dimethoxybenzoic acid	198.174	64.706	2	5	−2.24	3	0
4,5-Dihydroxy-3-oxocyclohex-1-ene-1-carboxylic acid	172.136	83.671	3	5	−0.92	1	0
Epicatechin	290.271	101.294	5	6	−3.72	1	0
3,4,5-Trihydroxybenzoic acid	170.12	89.408	4	5	−2.46	1	0
Methyl 3,4,5-trihydroxybenzoate	184.147	75.752	3	5	−2.19	2	0
Ethyl 3,4,5-trihydroxybenzoate	198.174	75.425	3	5	−1.86	3	0
Propyl 3,4,5-trihydroxybenzoate	212.201	75.433	3	5	−1.37	4	0
i-Propyl 3,4,5-trihydroxybenzoate	212.201	75.068	3	5	−1.54	3	0
Butyl 3,4,5-trihydroxybenzoate	226.228	75.433	3	5	−0.95	5	0
i-Butyl 3,4,5-trihydroxybenzoate	226.228	75.149	3	5	−0.97	4	0
Pentyl 3,4,5-trihydroxybenzoate	240.255	75.426	3	5	−0.54	6	0
i-Pentyl 3,4,5-trihydroxybenzoate	240.255	75.415	3	5	−0.62	5	0
Octyl 3,4,5-trihydroxybenzoate	282.336	75.390	3	5	0.72	9	0

MW—molecular weight (g mol^{-1}); PSA—polar surface area (Å^2) HBD—hydrogen bond donor; HBA—hydrogen bond acceptor; rb—rotatable bonds count; LV—Lipinski's violations.

Therefore, the boceprevir antiviral exhibited two violations of Lipinski's criteria, namely molecular mass (521.69 g mol^{-1}) and six hydrogen bond donors. The structure of boceprevir presents the maximum allowed number of flexible bonds (10) and maximum hydrogen bond acceptors (10).

Although it had these exceptions, the docking score was the highest among the investigated ligands, suggesting strong interactions and stability of the complex formed with the SARS-CoV-2 main protease. The results indicated that all proposed ligands met the restrictive criteria for good oral bioavailability. Increased hydrophilicity was observed for all compounds due to their hydroxyl, carboxyl and/or methoxy groups on their skeleton. These functional groups offer good absorption and permeation properties. Thus, by means of NH/OH/N/O groups, hydrophilic interactions were favored and further reflected in good and high docking scores. Concerning logP values, there were observed positive values for benzoic acid (0.79), 2,5-dihydroxybenzoic acid (0.81) and 0.72 for octyl 3,4,5-trihydroxy benzoate. A combination of molecular factors and properties, mainly due to the increased hydrophobicity of the lateral *n*-octyl chain of octyl gallate, was also reflected by its best docking score function, indicating this compound as the best antiviral candidate among all screened compounds in the study.

2.3. Results of Quantum Reactivty Analysis

Frontier molecular orbitals, the highest occupied molecular orbital (HOMO) and the lowest unoccupied molecular orbital (LUMO) localization and energy levels for octyl-gallate, are illustrated in Figure 5.

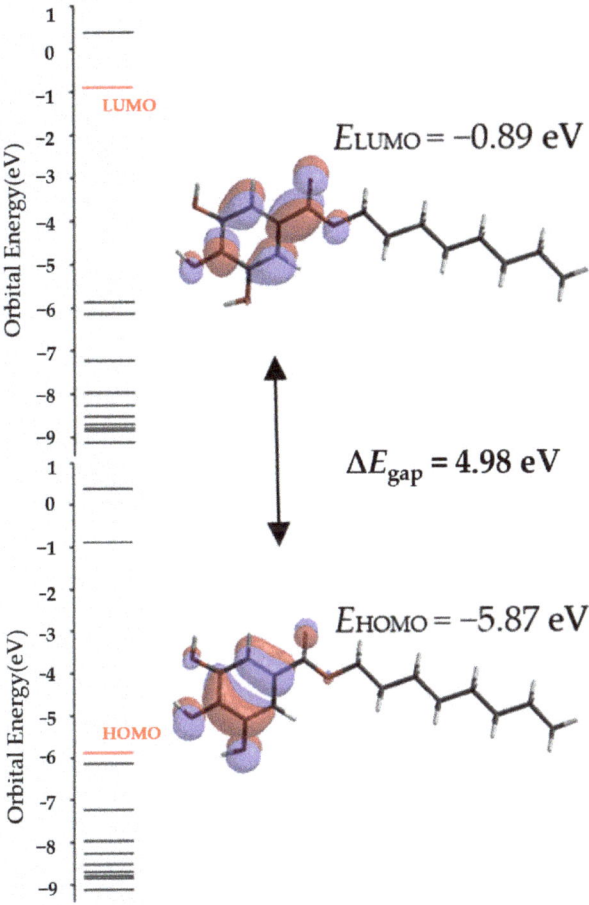

Figure 5. Highest occupied molecular orbital (HOMO) and lowest unoccupied molecular orbital (LUMO) molecular frontier orbitals and their energy gap for octyl-gallate.4

The resulting band gap (ΔE) can provide useful information on the chemical reactivity and kinetic stability of each ligand. The same values for energy gap were given for alkyl gallates, starting at ethyl to octyl-gallate, suggesting the same stability. Slight differences were found for the values of ionization potential (I = −E_{HOMO}) and electron affinity (A = −E_{LUMO}), according Koopmans' theorem [31]. The theorem allows estimation of quantum global reactivity parameters, starting from calculated energies of frontier molecular orbitals, and describes the molecules in terms of chemical hardness (η), global softness (σ), ionization (I), electron affinity (A), electronegativity (χ) and electrophyliciy index (ω) [32,33]. Obtained quantum reactivity parameters for all investigated ligands are given in Table S2 of the Supplementary Materials.

The global reactivity parameters analysis provides deep structural insights, a holistic characterization for revealing the properties of interest leading to strong binding affinity related to the protease target. The quantum reactivity parameters are related to relative nucleophilicity and electrophilicity. Ionization potential (I), refers to the energy needed to remove an electron from a molecule, and electron affinity (A) measures the ability of a molecule to accept electrons and form anions species [34,35]. Such parameters are useful to estimate further chemical reactivity behavior. Some of the investigated molecules can also be seen as lead compounds for a new series of (semi)synthetic molecules. Therefore the data on their reactivity are useful.

2.4. Results of Principal Component Analysis (PCA)

Principal component analysis (PCA) is a statistical tool for the identification of linear combinations of the variables which account for certain proportions of the variance of a set of variables. The selection is based on the eigenvalues of the dispersion matrix of the variables. The principal components are associated with decreasing eigenvalues and, therefore, share the amount of the variance. Typically, the first few principal components account for virtually all the variance. PCA also represents the pattern of similarity of the observations and the variables by displaying them as points in maps [36–39]. PCA analysis of all properties was calculated with Spartan software, along with docking scores, and data are listed in Table S4 of the Supplementary Materials.

PCA analysis was employed to find the degree of correlation between molecular descriptors and properties and their involvement in the resulting docking score.

The PCA correlation matrix (Table S4 in Supplementary Materials) revealed fairly good correlations between area and mass (r = 0.95), area and ovality (r = 0.96), docking score and polarizability (r = 0.97), volume and area (r = 0.97), volume and ovality (r = 0.98) and a moderate correlation (r = 0.55–0.66) between the dipole moment and docking score, mass, area, volume, and ovality, and between PSA and polarizability and mass, respectively.

Table 2 and Figure 6 are related to the eigenvalues which reflect the quality of the projection from the N–dimensional initial (n = 6) to a lower number of dimensions.

Table 2. Eigenvalues from the Principal Component Analysis (PCA) analysis.

	F1	F2	F3	F4	F5	F6
Eigenvalue	6.529	1.440	1.087	0.480	0.291	0.084
Variability, %	65.29%	14.40%	10.87%	4.80%	2.91%	0.84%
Cumulative, %	65.29%	79.69%	90.56%	95.36%	98.27%	99.11%

From Table 2 it can be seen that the first eigenvalue equals 6.529 and represents 65.29 % of the total variability. Each eigenvalue corresponds to a factor, and each factor to a one dimension. A factor is a linear combination of the initial variables and all the factors are uncorrelated (r = 0). The eigenvalues and the corresponding factors are sorted by descending order of how much of the initial variability they represent (converted to %). Therefore, the first two factors allow us to represent 70.69% of the initial variability of the data.

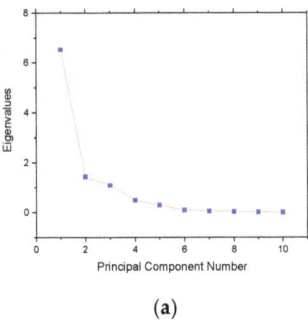 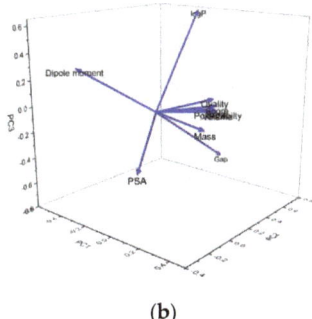

(a) (b)

Figure 6. Results of PCA analysis: scree plot of the eigenvalues and cumulative variability versus the F1-F6 components (**a**) and PC1 PC2 PC3 3D Loading plot (**b**). The scree plot (Figure 6a) is a useful visual aid for determining the number of the principal components, which depends on the elbow point at which the remaining eigenvalues are relatively small and all about the same size. This point is not so evident in the scree plot, but we may say that our elbow point is the third point. In conclusion Plot 4 and the Table 2 indicate that the first three PCs are sufficient to explain most of the variance (more than 90.56%) of the data set without overfitting the model. Detailed information about the best principal component locations of the extracted eigenvectors PC1-PC2-PC3 3D loading plots are also presented (Figure 6b). It is worth mentioning that the dipole moment is located in the PC2 × PC3 space, PSA in the PC1 × PC2 space and the other variables in the PC1 × PC3 space.

3. Methods

3.1. Methods for Molecular Docking Simulations

The docking simulation was carried out using CLC Drug, Discovery Work Bench (QIAGEN, Aarhus, Denmark,). SARS-CoV-2 main protease bound to boceprevir at 1.45 Å (PDB ID: 6WNP) [7], which was imported from the Protein Data Bank. Ligands were constructed using the Spartan'16 program [40,41] and their geometries were optimized to obtain the lowest energy conformers. The viral protein fragment contains three binding pockets: 48.13 Å^3, 40.45 Å^3, 36.62 Å^3. Redocking of the cocrystallized ligand (boceprevir) was realized to validate the docking protocol. The amino acid residues forming the binding site were protonated and water molecules were removed.

3.2. Methods for Energy Minimization and Computation of Molecular Properties

The examined ligands were first generated in 3D by importing their corresponding files from the Pubchem database (National Center for Biotechnology Information, U.S. National Library of Medicine Rockville Pike, Bethesda, MD, USA, https://pubchem.ncbi.nlm.nih.gov) or directly constructed (long chain alkyl gallates) in the Spartan'18 program [40,41]. Their geometry was optimized in a multistep procedure by molecular mechanics force fields, developed at Merck Pharmaceuticals (Merck Research Labs, Kenilworth, NJ, USA), to realize energy minimization [42]. Molecular descriptor and properties were calculated using the DFT, B3LYP algorithm [43] and the 6-31G (d, p) polarization basis set [40]. Calculations were performed for equilibrium geometry at ground state in gas for neutral molecules. No solvent corrections were done.

3.3. Methods for Principal Component Analysis (PCA)

The predicted ligand data were processed to compute principal component analysis using the free Excel add-in Real Statistics Resource Pack software (Release 7.2, www.real-statistics.com), Copyright (2013–2020) Charles Zaiontz (www.real-statistics.com).

4. Conclusions

In this work we analyzed and predicted the antiprotease activity of natural derivatives of benzoic and semisynthetic alkyl gallate acids on SARS-CoV-2 main protease. The investigation was corroborated with drugability and quantum reactivity evaluations. The docking results of the two studied series (benzoic acid series versus gallic acid series) suggested 2,5-dihydroxybenzoic acid (gentisic acid) and octyl gallate as the best potential candidates among the investigated structures. The two compounds had similar logP values (0.81 and 0.72). Octyl gallate had the best docking score (−60.22), but decreased dipole moment (1.31). 2,5-dihydroxybenzoic acid (gentisic acid) had a lower score (−33.84) but an increased dipole moment (4.62) which means higher polarity and also higher reactivity. As is known, the dipole moment represents a measure of net molecular polarity and, therefore, the larger the difference in electronegativities of bonded atoms the larger the reactivity of the molecule. Accordingly, a combination of the two compounds can be considered. On the other hand, if the benzoic acid series indicated that the position of hydroxyl groups on the ring was more important than the ester, hydroxyl and methoxy groups' number or type, the galllic acid series clearly indicated similar results for all studied compounds, apart from octyl gallate which was proved to be the best potential candidate among investigated structures, exhibiting antiviral activity against the SARS-CoV-2 protease and, therefore, could be studied further for drug development.

Supplementary Materials: The following are available online. Table S1: The list of intermolecular interactions between the compounds docked with 6WNP, Table S2: Quantum chemical reactivity parameters calculated with Koopman's relationships, Table S3: Molecular properties for the investigated ligands, calculated with Spartan Software, Table S4: Correlation matrix of PCA.

Author Contributions: Conceptualization, A.S. and L.P. (Lucia Pirvu); methodology, A.S.; software, L.P. (Lucia Pintilie) and B.A.; validation, A.S., B.A. and L.P. (Lucia Pirvu); formal analysis, L.P. (Lucia Pintilie); investigation, L.P. (Lucia Pintilie) and B.A.; resources, L.P. (Lucia Pintilie) and B.A.; data curation, L.P. (Lucia Pirvu); writing—original draft preparation, A.S.; writing—review and editing, L.P. (Lucia Pirvu); visualization, A.S.; supervision, L.P. (Lucia Pirvu); project administration, L.P. (Lucia Pintilie). All authors have read and agreed to the published version of the manuscript.

Funding: This research work was funded by the Romanian National Authority for Scientific Research ANCS, Competitiveness Operational Programme COP-A1-A1.2.3-G-2015, Project title "Innovative technologies for new, natural health products", ID P_40_406, SMIS 105542.

Conflicts of Interest: The authors declare no conflict of interest.

Abbreviations

A	A-electron affinity
B3LYP	Becke, 3-parameter, Lee-Yang-Parr
DAA	direct-acting antiviral agent
DFT	Density Functional Theory
E_{HOMO}	energy of the highest occupied molecular orbital
E_{LUMO}	energy of the lowest unoccupied molecular orbital
FDA	Food and Drug Administration (U.S.)
HBA	hydrogen bond acceptor
HBD	hydrogen bond donor
HCV	hepatitis C virus
HOMO	the highest occupied molecular orbital
I	ionization potential
logP	octanol-water partition coefficient
LUMO	the highest occupied molecular orbital
LV	Lipinski's violations
MW	Molecular weight
PCA	Principal component analysis
PSA	polar surface area
r	Pearson correlation coefficient
rb	rotatable bond

RMSD	Root mean square deviation
RNA	ribonucleic acid
SARS	CoV-2-Severe acute respiratory syndrome coronavirus 2
ΔE_{gap}	energy gap between frontier molecular orbitals
μ	chemical potential
η	global chemical hardness
σ	global softness
χ	electronegativity
ω	global electrophilicity index

References

1. Vijayakumar, B.G.; Ramesh, D.; Joji, A.; Prakasan, J.J.; Kannan, T. In silico pharmacokinetic and molecular docking studies of natural flavonoids and synthetic indole chalcones against essential proteins of SARS-CoV-2 proteins. *Eur. J. Pharmacol.* **2020**, *886*, 173448. [CrossRef] [PubMed]
2. Cherrak, S.A.; Merzouk, H.; Mokhtari-Soulimane, N. Potential bioactive glycosylated flavonoids as SARS-CoV-2 main protease inhibitors: A molecular docking and simulation studies. *PLoS ONE* **2020**, *15*, e0240653. [CrossRef] [PubMed]
3. Kumar, C.S.; Ali, D.; Alarifi, S.; Radhakrishnan, S.; Akbar, I. In silico molecular docking: Evaluation of coumarin based derivatives against SARS-CoV-2. *J. Infect. Public Health* **2020**, *13*, 1671–1677.
4. El-hoshoudy, A.N. Investigating the potential antiviral activity drugs against SARS-CoV-2 by molecular docking simulation. *J. Mol. Liq.* **2020**, *318*, 113968. [CrossRef] [PubMed]
5. Narkhede, R.R.; Pise, A.V.; Cheke, R.S.; Shinde, S.D. Recognition of Natural Products as Potential Inhibitors of COVID-19 Main Protease (Mpro): In-Silico Evidences. *Nat. Prod. Bioprospect.* **2020**, *10*, 297–306. [CrossRef] [PubMed]
6. Kiser, J.J.; Flexner, C. Direct-acting antiviral agents for hepatitis C virus infection. *Annu. Rev. Pharmacol. Toxicol.* **2013**, *53*, 427–449. [CrossRef]
7. Anson, B.; Mesecar, A. 6WNP X-ray Structure of SARS-CoV-2 Main Protease Bound to Boceprevir at 1.45 A. Available online: https://www.rcsb.org/structure/6WNP (accessed on 3 December 2020).
8. Fan, K.; Wei, P.; Feng, Q.; Chen, S.; Huang, C.; Ma, L.; Lai, B.; Pei, J.; Liu, Y.; Chen, J.; et al. Biosynthesis, purification, and substrate specificity of severe acute respiratory syndrome coronavirus 3C-like proteinase. *J. Biol. Chem.* **2004**, *279*, 1637–1642. [CrossRef]
9. Ryu, Y.B.; Park, S.J.; Kim, Y.M.; Lee, J.Y.; Seo, W.D.; Chang, J.S.; Park, K.H.; Rho, M.C.; Lee, W.S. SARS-CoV 3CLpro inhibitory effects of quinone-methide triterpenes from Tripterygium regelii. *Bioorg. Med. Chem. Lett.* **2010**, *20*, 1873–1876. [CrossRef]
10. Kumar, Y.; Singh, H.; Patel, C.N. In silico prediction of potential inhibitors for the Main protease of SARS-CoV-2 using molecular docking and dynamics simulation-based drug-repurposing. *J. Infect. Public Health* **2020**. [CrossRef]
11. Kamboj, A.; Saluja, A.K.; Kumar, M.; Atri, P. Antiviral activity of plant polyphenols. *J. Pharm. Res.* **2012**, *5*, 2402–2412.
12. Park, E.S.; Moon, W.S.; Song, M.J.; Kim, M.N.; Chung, K.H.; Yoon, J.S. Antimicrobial activity of phenol and benzoic acid derivatives. *Int. Biodeterior. Biodegrad.* **2001**, *47*, 209–214. [CrossRef]
13. Kratky, M.; Konecna, K.; Janousek, J.; Brablikova, M.; Jand'ourek, O.; Trejtnar, F.; Stolarikova, J.; Vinsova, J. 4-Aminobenzoic Acid Derivatives: Converting Folate Precursor to Antimicrobial and Cytotoxic Agents. *Biomolecules* **2019**, *10*, 9. [CrossRef] [PubMed]
14. Manach, C.; Scalbert, A.; Morand, C.; Rémésy, C.; Jimenez, L. Polyphenols: Food sources and bioavailability. *Am. J. Clin. Nutr.* **2004**, *79*, 727–747. [CrossRef] [PubMed]
15. Passamonti, S.; Terdoslavich, M.; Franca, R.; Vanzo, A.; Tramer, F.; Braidot, E.; Petrussa, E.; Vianello, A. Bioavailability of flavonoids: A review of their membrane transport and the function of bilitranslocase in animal and plant organisms. *Curr. Drug Metab.* **2009**, *10*, 369–394. [CrossRef]
16. Smith, T.J.; Kremer, M.J.; Luo, M.; Vriend, G.; Arnold, E.; Kamer, G.; Rossman, M.G.; McKinlay, M.A.; Diana, G.D.; Otto, M.J. The site of attachment in human rhinovirus 14 for antiviral agents that inhibit uncoating. *Science* **1986**, *233*, 1286–1293. [CrossRef]

17. Kirsten, G.; Joachim, K.; Eva, L.; Wali, H.; Derksena, A.; Detersa, A.; Andreas, H. Proanthocyanidin-enriched extract from Myrothamnus flabellifolia Welw. exerts antiviral activity against herpes simplex virus type-1 by inhibition of viral adsorption and penetration. *J. Ethnopharmacol.* **2011**, *134*, 468–474.
18. Shah, B.; Modi, P.; Sagar, S.R. In silico studies on therapeutic agents for COVID-19: Drug repurposing approach. *Life Sci.* **2020**, *252*, 117652. [CrossRef]
19. Cardoso, W.B.; Mendanha, S.A. Molecular dynamics simulation of docking structures of SARS-CoV-2 main protease and HIV protease inhibitors. *J. Mol. Struct.* **2021**, *1225*, 129143. [CrossRef]
20. Peele, K.A.; Durthi, C.P.; Srihansa, T.; Krupanidhi, S.; Sai, A.V.; Babu, D.J.; Indira, M.; Reddy, A.R.; Venkateswarulu, T.C. Molecular docking and dynamic simulations for antiviral compounds against SARS-CoV-2: A computational study. *Inform. Med. Unlocked.* **2020**, *19*, 100345. [CrossRef]
21. Wang, J. Fast identification of possible drug treatment of coronavirus disease-19 (COVID-19) through computational drug repurposing study. *J. Chem. Inf. Model.* **2020**, *60*, 3277–3286. [CrossRef]
22. Hetal, T.; Bindesh, P.; Sneha, T. A review on techniques for oral bioavailability enhancement of drugs. *Int. J. Pharm. Sci. Rev. Res.* **2010**, *4*, 203–223.
23. Kahkeshani, N.; Farzaei, F.; Fotouhi, M.; Alavi, S.S.; Bahramsoltani, R.; Naseri, R.; Momtaz, S.; Abbasabadi, Z.; Rahimi, R.; Farzaei, M.H.; et al. Pharmacological effects of gallic acid in health and diseases: A mechanistic review. *Iran J. Basic Med. Sci.* **2019**, *22*, 225–237. [CrossRef] [PubMed]
24. Nayeem, N.; Asdaq, S.M.B.; Salem, H.; Alfqy, S.A. Gallic Acid: A Promising Lead Molecule for Drug Development. *J. Appl. Pharm.* **2016**, *8*, 1000213. [CrossRef]
25. Takai, E.; Hirano, A.; Shirak, K. Effects of alkyl chain length of gallate on self-association and membrane binding. *J. Biochem.* **2011**, *150*, 165–171. [CrossRef] [PubMed]
26. Król, E.; de Sousa Borges, A.; da Silva, I.; Polaquini, C.R.; Regasini, L.O.; Ferreira, H.; Scheffers, D.J. Antibacterial activity of alkyl gallates is a combination of direct targeting of FtsZ and permeabilization of bacterial membranes. *Front. Microbiol.* **2015**. [CrossRef]
27. Fujita, K.; Kubo, I. Plasma membrane injury induced by nonyl gallate in Saccharomyces cerevisiae. *J. Appl. Microbiol.* **2002**, *92*, 1035–1042. [CrossRef] [PubMed]
28. Lipinski, C.A. Lead-and drug-like compounds: The rule-of-five revolution. *Drug Discov. Today Technol.* **2004**, *1*, 337–341. [CrossRef] [PubMed]
29. Lipinski, C.A.; Lombardo, F.; Dominy, B.W.; Feeney, P.J. Experimental and computational approaches to estimate solubility and permeability in drug discovery and development settings. *Adv. Drug Deliv. Rev.* **2001**, *46*, 3–26. [CrossRef]
30. Veber, D.F.; Johnson, S.R.; Cheng, H.Y.; Smith, B.R.; Ward, K.W.; Kopple, K.D. Molecular properties that influence the oral bioavailability of drug candidates. *J. Med. Chem.* **2002**, *45*, 2615–2623. [CrossRef]
31. Koopmans, T. Über die Zuordnung von Wellenfunktionen und Eigenwerten zu den Einzelnen Elektronen Eines Atoms. *Physica* **1934**, *1*, 104–113. [CrossRef]
32. Sastri, V.S.; Perumareddi, J.R. Molecular orbital theoretical studies of some organic corrosion inhibitors. *Corrosion* **1997**, *53*, 617–622. [CrossRef]
33. Yankova, R.; Genieva, S.; Halachev, N.; Dimitrova, G. Molecular structure, vibrational spectra, MEP, HOMO-LUMO and NBO analysis of $Hf(SeO_3)(SeO_4)(H_2O)_4$. *J. Mol. Struct.* **2016**, *1106*, 82–88. [CrossRef]
34. Landeros-Martinez, L.L.; Orrantia-Borunda, E.; Flores-Holguin, N. DFT Chemical Reactivity analysis of biological molecules in the presence of silver ion. *Org. Chem. Curr. Res.* **2015**, *4*, 153. [CrossRef]
35. Pirvu, L.; Neagu, G.; Terchescu, I.; Albu, B.; Stefaniu, A. Comparative studies of two vegetal extracts from Stokesia laevis and Geranium pratense: Polyphenol profile, cytotoxic effect and antiproliferative activity. *Open Chem.* **2020**, *18*, 488–502. [CrossRef]
36. Jolliffe, I.T. *Principal Component Analysis*; Springer: New York, NY, USA, 2002; ISBN 978-0-387-22440-4.
37. Jackson, J.E. *A Use's Guide to Principal Components*; John Wiley & Sons: New York, NY, USA, 1991; ISBN 9780471725336. [CrossRef]
38. Saporta, G.; Niang, N. Chapter 1: Principal component analysis: Application to statistical process control. In *Data Analysis*; Govaert, G., Ed.; John Wiley & Sons: London, UK, 2009; pp. 1–23.
39. Abdi, H.; Williams, L.J. Principal component analysis. *WIREs Comput. Stat.* **2010**, *2*, 433–459. [CrossRef]
40. Shao, Y.; Molnar, L.F.; Jung, Y.; Kussmann, J.; Ochsenfeld, C.; Brown, S.T.; Gilbert, A.T.B.; Slipchenko, L.V.; Levchenko, S.V.; O'Neill, D.P.; et al. Advances in methods and algorithms in a modern quantum chemistry program package. *Phys. Chem. Chem. Phys.* **2006**, *8*, 3172–3191. [CrossRef]

41. Hehre, W.J. *A Guide to Molecular Mechanics and Quantum Chemical Calculations*; Wavefunction, Inc.: Irvine, CA, USA, 2003.
42. Halgren, A.T. Merck molecular force field. I. Basis, form, scope, parameterization, and performance of MMFF94. *J. Comput. Chem.* **1996**, *17*, 490–519. [CrossRef]
43. Lee, C.; Yang, W.; Parr, R.G. Development of the Colle-Salvetti correlation-energy formula into a functional of the electron density. *Phys. Rev. B* **1988**, *37*, 785–789. [CrossRef]

Sample Availability: Samples of the compounds are not available from the authors.

Publisher's Note: MDPI stays neutral with regard to jurisdictional claims in published maps and institutional affiliations.

© 2020 by the authors. Licensee MDPI, Basel, Switzerland. This article is an open access article distributed under the terms and conditions of the Creative Commons Attribution (CC BY) license (http://creativecommons.org/licenses/by/4.0/).

Review

The Role of Organic Small Molecules in Pain Management

Sebastián A. Cuesta and Lorena Meneses *

Laboratorio de Química Computacional, Facultad de Ciencias Exactas y Naturales, Escuela de Ciencias Químicas, Pontificia Universidad Católica del Ecuador, Av. 12 de Octubre 1076 Apartado, Quito 17-01-2184, Ecuador; sebastian_cuesta@yahoo.com
* Correspondence: lmmeneses@puce.edu.ec; Tel.: +593-2-2991700 (ext. 1854)

Abstract: In this review, a timeline starting at the willow bark and ending in the latest discoveries of analgesic and anti-inflammatory drugs will be discussed. Furthermore, the chemical features of the different small organic molecules that have been used in pain management will be studied. Then, the mechanism of different types of pain will be assessed, including neuropathic pain, inflammatory pain, and the relationship found between oxidative stress and pain. This will include obtaining insights into the cyclooxygenase action mechanism of nonsteroidal anti-inflammatory drugs (NSAID) such as ibuprofen and etoricoxib and the structural difference between the two cyclooxygenase isoforms leading to a selective inhibition, the action mechanism of pregabalin and its use in chronic neuropathic pain, new theories and studies on the analgesic action mechanism of paracetamol and how changes in its structure can lead to better characteristics of this drug, and cannabinoid action mechanism in managing pain through a cannabinoid receptor mechanism. Finally, an overview of the different approaches science is taking to develop more efficient molecules for pain treatment will be presented.

Keywords: anti-inflammatory drugs; QSAR; pain management; cyclooxygenase; multitarget drug; cannabinoid; neuropathic pain

1. Introduction

Inflammation is a very complex self-defense biological process to protect the body against harmful stimuli including pathogens, physical injury, or contact with irritant substances [1]. There are five classical signs of inflammation, i.e., pain, redness, swelling, heat, and loss of function. Although the function of an inflammatory process is to eliminate the cause of injury and heal damaged tissue by clearing dead cells, sometimes this response is too aggressive, causing excessive pain and incapacity [2]. In these cases, an anti-inflammatory drug is needed to ameliorate symptoms and allow the person to continue a normal life. Inflammation is involved in many disorders and complex diseases, including metabolic syndrome, autoimmune, depression, and neurodegenerative diseases [1,3,4]. In first-world societies, the excessive nutrient storage caused by food security and lack of physical activity have stressed humans' metabolic pathways, causing diseases. The metabolic syndrome is composed of a group of conditions including dyslipidemia, hypertension, obesity, and elevated glucose levels, causing diabetes and atherosclerotic cardiovascular disease [5]. It is known that inflammation plays a pivotal role in the pathogenesis of the metabolic syndrome. Although the mechanism is not yet fully understood, it is believed that reactive oxygen species, free fatty acid intermediates, and adipose tissue dysregulation promote inflammation through high levels of proinflammatory adipokines and low levels of anti-inflammatory adiponectines [4,6–8]. The inflammatory process is so complex that whole animal models are needed. In this sense, the zebrafish has emerged as a key tool to study inflammatory diseases. Zebrafish present receptors, mediators, and inflammatory cells similar to mammals and humans making them suitable animal models to study new anti-inflammatory agents and their mechanisms [9]. There are different types of anti-inflammatory agents, including small molecules, peptides, and antibodies. In this review, there will be a focus on small molecules for anti-inflammatory treatments as they have been the center of traditional medicine. Small molecule drugs are compounds with low molecular weight that can easily enter the body and modulate biochemical processes to treat medical conditions [10]. Cyclic small molecules including naturally occurring and synthetic heterocyclic and polycyclic compounds are key to produce new drugs to target the inflammation process [11].

Nature represents one of the greatest sources of anti-inflammatory agents. Recent investigations have found novel anti-inflammatory agents from natural sources, including cyanobacterial extract [12], a norditerpenoid from the Hainan soft coral *Sinularia siaesensis* [13], and peptides extracted from the adzuki bean [14]. Traditional Chinese medicine has used herbs such as *Andrographis paniculata* to treat fever, coughs, and other cold symptoms. The active ingredient andrographolide extracted from this herb has been tested as a multi-target inflammatory drug, showing good results against multiple sclerosis, some respiratory diseases, and osteoarthritis [15]. Complex ancient preparations used for centuries, such as the Shiyifang Vinum containing 13 herbs, have been employed to treat pain and inflammation [16]. One of the first anti-inflammatory treatments described by ancient civilizations, such as Egyptians, Greeks, Sumerians and Chinese, was the use of the willow bark [17–19]. Preparations of Salix species have been used for centuries to alleviate pain and to treat fever and rheumatic conditions [17–20]. Its principal active ingredient is called salicin, which is used as a prodrug of salicylate [21]. Willow bark is considered one of the first examples of modern drug development from herbal plants. Salicylic acid was obtained from hydrolysis followed by oxidation of salicin. An acetylated derivative of salicylic acid became one of the most important drugs in the world, aspirin (Figure 1) [17]. The effectiveness and safety profile of herbal medicines such as willow bark extract is of great interest. In this sense, the ethanolic extract showed to be effective and safe to treat low back pain [21,22], and it showed a moderate analgesic effect against osteoarthritis [17]. The action of willow bark extract as an anti-inflammatory agent was compared to celecoxib and acetylsalicylic acid, and it was found to be as effective as these drugs [23]. The extract contains only 24% salicin, which suggests other components of the extract such as flavonoids are helping to increase its effectiveness [20].

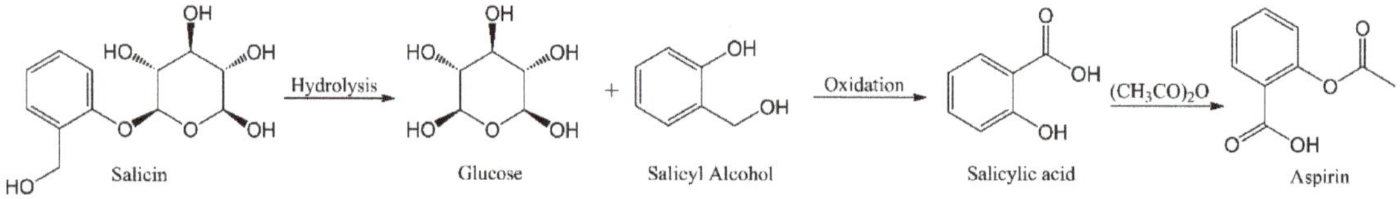

Figure 1. Aspirin synthesis from salicin.

Prostaglandins and the Cyclooxygenase Anti-Inflammatory Pathway

The main target of anti-inflammatory drugs (NSAIDs) is the cyclooxygenase (COX) pathway composed of two isoforms of the enzyme, i.e., cyclooxygenase-1 (COX-2) and cyclooxygenase-2 (COX-2). Both isoforms have an alpha carbon RMSD of only 0.9 Å. Cyclooxygenases are key in the lipid signaling pathway, being the first and rate-dependent step in prostaglandin and thromboxane synthesis [24,25]. Therefore, NSAIDs work by inhibiting the COX pathway and preventing the synthesis of prostaglandins [1]. In the body, arachidonic acid (AA) is transformed by COX-1 and COX-2 in lipid signaling to prostaglandin and thromboxane to mediate inflammatory processes [26]. COX-1 is expressed in all the body, including the kidney and stomach [27], while COX-2 is only expressed at the site of inflammation [28,29].

Functions of prostaglandin other than mediating the inflammation processes are to protect the gastric mucosa and stimulate platelet aggregation. Therefore, reversible COX-1 inhibition may be used as an antiplatelet agent helping patients with cardiovascular conditions when aspiring is not sufficient. Selective COX-1 inhibitors may be safer than nonselective inhibitors associated with higher risks of upper gastrointestinal bleeding [30] and selective COX-2 inhibitors that are linked to cardiovascular effects [31–33].

The active site of COX-1 enzyme is formed by Val116, Arg120, Tyr348, Val349, Leu352, Tyr355, Leu359, Phe381, Leu384, Tyr385, Trp387, Phe518, Ile523, Ala527, Ser530 and Leu531 [24]. Its difference with the COX-2 active site is the presence in COX-2 of Ile523 instead of Val509 and Arg499 instead of His513, which makes the COX-2 active site 20% bigger and with a hydrophilic side chamber [34–36]. COX-2-selective inhibitors take advantage of the side pocket and bind in a different mode compared to nonselective compounds, making an additional salt bridge and three extra hydrogen bonds [24]. NSAIDs inhibit COX enzymes in a reversible competitive manner or in a slow tight-binding way depending on the speed and efficiency of each molecule in displacing water molecules inside the pocket and forming hydrogen bonding [37].

Structural and functional information about COXs has been widely studied, identifying structural features key for binding [25] and using different techniques such as saturation transfer difference NMR spectroscopy (STD) [38]. To fit the active site, the basis of an NSAID chemical structure is generally composed of an aromatic ring and an acidic group. While the aromatic ring makes hydrophobic interactions with the pocket, the acidic group forms hydrogen bonds with Arg120 and Tyr355 [24]. In the study of Viegas et al., known COX-1 and COX-2 inhibitors were evaluated, finding a good correlation between experimental crystallographic structure and STD signal [38]. Ibuprofen, diclofenac, and ketorolac bind in a similar way to COX-2, where the ligand moieties form tighter interactions towards Arg-120 and Tyr-355 than towards Ser-520 and Tyr-385 [38].

The importance of the COX enzymes in cancer-related inflammation has been widely studied [26]. Some types of cancer such as epithelial ovarian cancer are reported to overexpress COX-1. In this scenario, selective COX-1 inhibitors may be useful as clinically proven to detect cancer in an early state (imaging agents) but also as therapeutic agents [39]. The similarity between COX-1 and COX-2 makes it a challenge to synthesize selective inhibitors [40]. The link between inflammation and cancer occurs in the inflammatory pathway and includes proinflammatory agents such as cytokines. Studies have also shown upregulation of COX-2 during cancer, and COX-2 has been associated with some neurodegenerative diseases [40,41].

2. Nonsteroidal Anti-Inflammatory Drugs (NSAIDs)

The most prescribed family of anti-inflammatory drugs are NSAIDs, accounting for 5–10% of total prescriptions [42–45]. Some of them are over-the-counter drugs, which increases their usage [46]. Ibuprofen market alone was valued at USD 294.4 million in 2020 and is expected to reach USD 447.6 million by the end of 2026 [47,48]. NSAIDs are the most cost-effective initial therapy for inflammation and pain relief, including sports injuries, arthritis and headaches [49–51]. Although NSAIDs are somewhat effective for spinal pain and other acute painful conditions, there is an urgent need for new therapies to treat these medical conditions [52,53].

NSAIDs' most common side effects include renal, hepatic, gastrointestinal, and cardiovascular reactions [46,53–56]. These side effects are known to be enhanced when the patient presents medical conditions such as diabetes, obesity and hypertension [53]. The most common and severe one is gastrointestinal (GI) bleeding and ulceration [57–59]. This side effect is mostly attributed to inhibition of COX-1, although there are several other factors involved such as the interaction of NSAID with phospholipids [46]. This is the major impediment to the use of this type of medication in chronic patients [60]. To ameliorate the GI effect, gastroprotective drugs like proton pump inhibitors are used together with NSAIDs [53]. In this sense, more than improving the potency of NSAIDs, what is important regarding this type of drug is to enhance gastrointestinal safety [34,45].

Ibuprofen is considered one of the first and best options when talking about NSAIDs. It was created by a group of scientists from Boots company in 1960 [61], proving to be more effective than its predecessors and, in turn, causing fewer side effects. Ibuprofen belongs to the family of propionic acid derivatives. This family is characterized by moderate efficacy, having analgesic, antipyretic and anti-inflammatory action. Its main difference from other propionic acid derivatives lies in its pharmacokinetic characteristics [62]. It is used in antirheumatic treatments, sports injuries, and menstrual cramps. Although it has been on the market for more than 50 years, there are still ways to change this molecule to enhance its properties and activity. One way is by increasing the number of rotatable bonds to less than or equal to ten, which will increase bioavailability [63]. Another approach reported by Kleemiss et al. is to replace one carbon atom with a silicon one, creating the so-called sila-ibuprofen (**1**; Figure 2) [64,65]. The silicon atom is considered a carbon bioisostere that could enhance biological activity while reducing toxicity. Due to its bigger atomic radius, adding a silicon atom lengthens a single bond by around 20% [65]. Silicon atoms change lipophilicity values, which can improve solubility in some cases and enhance membrane penetration in others, altering potency and selectivity. As silicon is more electropositive than carbon, hydrogen bonding can be enhanced [65].

Some of the drawbacks of ibuprofen's properties are its low solubility in physiological media, high melting point, and high melting enthalpy, which make the production of intravenous formulations challenging [64]. By changing the tertbutyl carbon for silicon, its melting point is reduced by 30 °C (from 75 to 45 °C) and its melting enthalpy is reduced by around 10 kJ/mol. The solubility of sila-ibuprofen is enhanced, going from 21 mg/L for ibuprofen to 83 mg/L. Obtaining insights into the difference in their chemical structure, it was noted that the C–Si change produces an increase in electron density of 7%, a longer distance to the hydrogen atom (0.374 Å), and a change of 0.68° in the Si–C–H angle. A difference in the electrostatic potential is also noticed, as it changes from positive to negative around the silicon atom [64]. Free energy perturbation and experimental results indicate COX-1 and COX-2 inhibition, and a low toxicity profile of the drug was maintained [64].

Another change that was shown to enhance the pharmacological profile of a drug is to form a conjugate with a saccharide. Sodano and coworkers synthesized and evaluated a paracetamol–galactose conjugate as a prodrug [66]. Their results revealed the conjugate improved the pharmacodynamic and toxicological profile of the drug. In this regard, the conjugate presented higher stability in human serum and reduced in vitro metabolism, as the conjugate after was able to be found after 2 h, which does not happen when using paracetamol alone. As paracetamol is slowly released from the conjugate, a longer analgesic effect was found after oral administration lasting up to 12 h being able to treat neuropathic pain such as hyperalgesia. Moreover, the hepatotoxicity was significantly reduced compared to paracetamol [66].

Changing functional groups is also an interesting approach to increase the safety profile of a drug. Some studies suggest that carboxylic acid is key for binding, but it is also linked to gastrotoxicity. In this sense, oxetan-3-ol and thietan-3-ol appear to be interesting bioisosteres in COX inhibitors. As oxetane is a carbonyl isostere, carboxylic acid may be exchanged for oxetan-3-ol. Results showed this isostere improves brain penetration and can be used for CNS drug design [67]. Furthermore, linking furoxan and furazan groups with ibuprofen produces compounds **2** and **3** (Figure 2) with better gastrotoxicity properties without changing the anti-inflammatory effect [27].

A lot of work has been conducted in designing new COX inhibitors [68]. QSAR modeling is the most used and powerful approach that allows correlating chemical modifications in a molecule with its biological activity [69]. This approach has been applied to find new treatments for Alzheimer's disease [70], malaria [71,72], diabetes [73,74], cancer [75–77], and HIV [78]. Based on the anti-inflammatory activity of pyrazolo[1,5-a]pyrimidine, 2,5-diarylpyrazolo[1,5-a]pyrimidin-7-amines, new compounds were synthesized. Eleven compounds (**4–14**; Figure 2) showed interesting anti-inflammatory properties compared to indomethacin (Table 1) [2]. As shown in Table 1, all compounds except for **7** achieve more than 50% inhibition after 4 h. Furthermore, compound **9** presents only 3.5% less inhibition than indomethacin. An enhanced activity is achieved when R is a chlorine atom. For R', it was observed that H and F diminished activity while Cl, CH3, and Br increased it.

Table 1. Anti-inflammatory activity after 4 h of compounds **4** to **14** compared to indomethacin.

Compound	Inhibition (%)
4	54.76
5	51.16
6	66.66
7	34.52
8	64.28
9	80.95
10	67.85
11	76.19
12	65.47
13	66.66
14	64.28
Indomethacin	84.52

In another study, Harrak et al. designed 4-(aryloyl)phenyl methyl sulfones as anti-inflammatory compounds against COX-1 and COX-2 [49]. Molecular modeling results showed how the methylsulfone group in the studied compounds fit the COX-2 pocket, making hydrogen bonds with Arg120, Ser353 and Tyr355. N-Arylindole (**15**; Figure 2) was found to be the most potent and selective COX-2 inhibitor (COX-1/COX-2 ratio of 262), having greater anti-inflammatory activity than ibuprofen in vivo [49]. Computational studies at the B3LYP/6-31G(d) level revealed the optimized dipole moment of the structures related to COX-2 binding ($R^2 = 0.81$). This happens as the dipole moment performs a pivotal role in aligning to the receptor's binding pocket. Looking at the binding mode in COX-2, the methyl sulfone group of the studied molecules aligns to the carboxylic group of the crystallographic structure of flurbiprofen. The pose of the ligands allows forming hydrogen bonds with Arg120, Ser353 and Tyr355 and van de Waals contacts with Val349, Phe518 and Leu352 [49]. Differences were found between the position of the indole ring compared to the pyrrole and pyridine groups of similar compounds, which led to a more efficient π–π stacking of the indole ring interacting with Phe518. Therefore, the conjunction of a methylsulfone and an aryl group is enough to produce interesting anti-inflammatory activity [49].

Yamakawa et al. proposed that gastric lesions are related to membrane permeabilization. Therefore, loxoprofen derivatives with lower membrane permeabilization should produce fewer gastric lesions. Loxoprofen is used in Japan and is considered safer than indomethacin [79,80]. Synthesized compounds **16** and **17** (Figure 2) have lower membrane permeabilization and indeed produced fewer gastric lesions compared to loxoprofen but with equivalent activity [42]. In a subsequent study, Yamakawa studied the properties and designed analogs of 2-fluoroloxoprofen, where **18** (Figure 2) presented an equivalent ulcerogenic effect with an enhanced potency [81]. Other types of compounds that possess low ulcerogenicity compared to indomethacin are 5-substituted-1-(phenylsulfonyl)-2-methylbenzimidazole derivatives, where compounds **19**, **20** and **21** (Figure 2) also presented good anti-inflammatory properties in in vivo assays, making them good anti-inflammatory candidates [60].

Uddin and coworkers created selective COX-1 inhibitors by taking out the SO_2CH_3 group from known COX-2 inhibitor rofecoxib. Starting from the 3,4-diarylfuran-2(5H)-one structure, several fluorinated compounds were designed, synthesized, and tested as COX-1 inhibitors using a structure-activity relationship [39]. It was found that different positions of different functional groups such as trifluoromethyl, halogens, phenoxy, alkyl, alkoxy, and thioalkyl influence COX inhibition and potency. As observed in compound **22** (Figure 2), adding a methoxy group on carbon 4 of one phenyl ring and a fluorine substituent on position 3 or 4 of the other phenyl ring gave the most active compounds with a COX-1 IC$_{50}$ of 0.36 and 0.48 uM, respectively, while their estimated COX-2 IC$_{50}$ is >25 uM. The advantage of having a fluorinated compound is the possibility of using 18F during the synthesis, which will allow the development of prototypes of PET imaging agents [39]. Another approach to find COX-1 inhibitors is the use of protein affinity fingerprints. Hsu and coworkers identified the fingerprint of 19 COX-1 inhibitors and developed a model to screen 62 compounds for new possible COX-1 molecules [82]. Interestingly, although a carboxylate group is present in known COX inhibitors such as ibuprofen or naproxen, that functional group was not found in newly identified COX-1 inhibitors. The structure of **23** (Figure 2) is similar to ketoprofen, suggesting the carboxylic functional group is not needed for activity while the diaryl ketone is [82,83]. Moreover, it has been shown experimentally that flurbiprofen ethyl ester inhibits COX-1 in a similar pose to flurbiprofen with no need for the carboxylic group [37,82].

Vitale et al. found that 3-(5-chlorofuran-2-yl)-5-methyl-4-phenylisoxazole (**24**; Figure 2) is a potent COX-1 inhibitor, where the isoxazole ring and the furanyl group are key for selectivity [30]. Furthermore, a SAR series based on this compound was performed, showing that the incorporation of 5-chlorofuran-2-yl, 4-phenyl and 5-methyl groups on the isoxazole core enhances selective inhibition of COX-1. Two extra changes were shown to increase potency with a slow reversible process: exchanging chlorine with bromine or a methyl group in the furyl core and placing a CF$_3$ group instead of the 5-methyl group (**25**, **26**, **27**; Figure 2). For these compounds, COX-1 IC$_{50}$ values of 0.32, 0.33 and 0.18 uM were found. Furthermore, compound **25** was shown to be >1000-fold more potent for COX-1 than COX-2, having a slow reversible interaction with the first one and a fast one with the second [30].

Selective COX-2 inhibitors such as coxibs allow treating inflammation without the gastric effect [53,57,84]. Starting from known selective COX-2 inhibitors and after elucidating their mode of binding by X-ray crystallography or NMR, a pharmacophore model can be built in order to identify the key features for binding and design new molecules that can inhibit COX-2 selectively. The basis of selective inhibition lies in the large Ile523 on the entrance of the side pocket which prevents bulk and rigid functional groups such as the sulfamoyl or sulfonyl from interacting with the COX-1 side pocket [34]. Compounds such as MK-2894 (**28**; Figure 2) [43] have been discovered to be potent COX-2 inhibitors when looking for novel and effective treatments. Several moieties have been tested, including carbocycle [85], imidazoles [86], thiophenes [87], oxazoles [88], pyrazoles [89], furanones [90], pyridazinone [91], N-benzoyl-5-sulfonylindole [92], and others. Furthermore, synthesizing amides and esters of known inhibitors such as meclofenamic acid [93] and indomethacin [94] resulted in compounds with interesting COX-2 inhibition properties. In this sense, it has been shown that indomethacin esters and amides are slow, tight-binding COX-2-selective inhibitors eliminating gastric side effects. Furthermore, primary and secondary amides are found to be more potent than tertiary ones. The 4-chlorobenzyl group is very important for potency as the change of this group by 4-bromobenzyl or hydrogen produced inactive compounds. The same occurred when replacing the 2-methyl group on the indole ring with hydrogen [94]. Although compounds **29** and **30** (Figure 2) are very potent, compound **31** (Figure 2) is not very potent; this is due to the lack of the carbonyl group which is suggested to be key for interaction. The presence of the methoxy (**29**) instead of the amino group (**30**) sightly improves the activity (0.02 uM) and has no effect on COX-1 IC$_{50}$ (>66 uM).

Using a 4,9-dihydro-3H-pyrido[3,4-b]indole core, several compounds were designed, with **32** (Figure 2) being a promising substrate-selective inhibitor. Structural studies show a movement of Leu531, which can be key for the substrate-specific inhibition [95]. Harmaline (1,2-benzisothiazol-3(2H)-one-1,1-dioxide) derivatives were also studied as COX-2-selective inhibitors. Compound **33** (Figure 2) presents an activity of 0.09 (SI 1/4 135.9); exchanging the benzene ring with a methyl group to form **34** (Figure 2) increased activity and selectivity to 0.06 mM (SI 1/4 154), and the formation of **35** (Figure 2) increased activity and selectivity to 0.05 mM (SI 1/4 236), which is comparable to celecoxib (IC$_{50}$: 0.05 mM, SI 1/4 296.00) [54].

When exchanging a five-member triazole with a pyrazole, enhanced activity was found. Adding an extra pyrazole produces an active and selective inhibitor comparable with celecoxib. On the other hand, exchanging the dipyrazole with dihydrazone produces an important reduction in potency and selectivity. Therefore, pyrazoles present better selectivity and activity than hydrazone. The most active benzenesulfonamides are the ones bearing dipyrazole, followed by those with pyrazole, triazole and oxadiazole, with activities similar to celecoxib and presenting a low ulcer index [50].

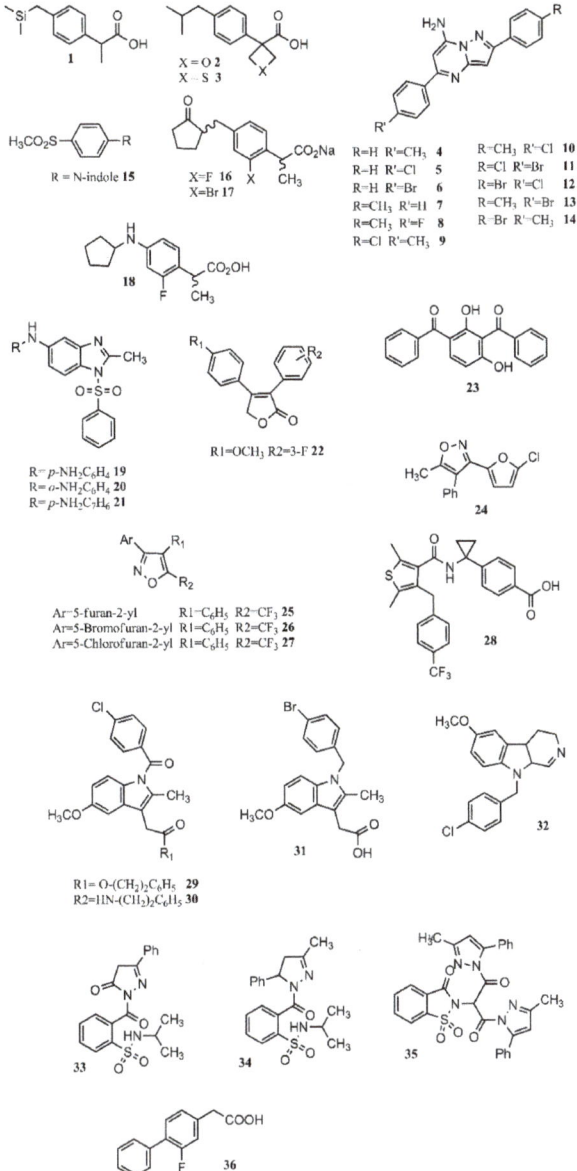

Figure 2. COX inhibitors.

One of the main functions of COX-2 is to oxygenate arachidonic acid, 2-arachidonoy-lglycerol (2-AG), and arachidonoylethanolamide (AEA). Studies have shown that ibuprofen and mefenamic acid are more potent in inhibiting 2-AG and AEA than AA oxygenation [96]. The mechanism proposed involves the interaction of the drugs with one active site of the homodimer which alters the structure of the second one, impeding the oxygenation of 2-AG and AEA but not of AA. Inhibiting the second active site requires higher concentration than inhibiting only the first one, allowing substrate-selective inhibition [97]. It is reported that COX inactive (R)-profens can inhibit endocannabinoid oxygenation but not arachidonic acid oxygenation in a substrate-selective manner. Results show that smaller substituents are more potent but less selective. In this sense, desmethylflurbiprofen (**36**; Figure 2) exhibits an IC_{50} of 0.11 µM. Aryl groups are better than other groups, and fluriprofen is more potent than ibuprofen [97].

In a novel approach, Bhardwaj et al. used the COX-2 enzyme to produce selective inhibitors in a process called in situ click chemistry. Lead compounds are produced through a [2,3]-cycloaddition inside the active site of COX-2. As shown in Figure 3, 1,4-regioisomers were produced. The SO_2Me functional group, the orientation of the azide group inside the pocket, the size, and several interactions are key for the synthesis and later inhibition of this selective compound. When adding a series of precursors, COX-2 was able to choose the one that produced the best fit, producing better inhibitors than celecoxib [40].

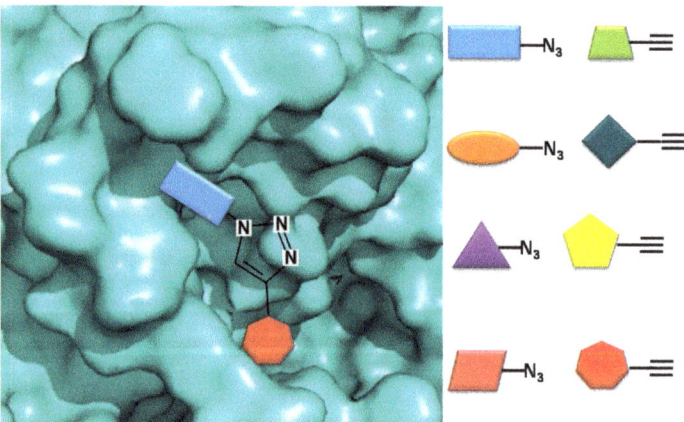

Figure 3. COX-2 inhibitor design using in situ click chemistry.

3. Multitarget Drugs

Sometimes, monotherapy is not considered an effective treatment in complex disorders such as cancer, diabetes, infection, or inflammatory conditions [98]. Fixed-dose drug combination is an alternative, although this type of therapy may increase the risk of adverse drug–drug interactions and produce changes in the pharmacokinetic profile of one of the components [99]. Cocrystals have emerged as compounds with improved properties and performance including solubility, incompatibility, and stability. One example is the tramadol-celecoxib cocrystal, which successfully passed phase II clinical trial for the treatment of acute pain [98,100].

The dual inhibition of COX with other enzymes such as 5-lipoxygenase (5-LOX) [33,101], soluble epoxide hydrolase [102], phosphoinositide-3-kinase delta [103], or fatty acid amide hydrolase (FAAH) [31] enhances therapeutic effect. Advantages of this dual inhibition include greater anti-inflammatory effect with reduced side effects [31]. 5-LOX is involved in the metabolism of arachidonic acid. The leukotrienes produced by 5-LOX are important inflammatory mediators and may be related to cancer, cardiovascular diseases, and gastrointestinal reactions. Still, 5-LOX inhibitors do not show anti-inflammatory effects [33].

Hybrid multiligand molecules containing NSAID structures can lead to enhance anti-inflammatory activity by inhibiting different targets inside the inflammation pathway [104,105]. COX-2/5-LOX inhibitors are compounds that offer more efficacy, fewer side effects, and broader anti-inflammatory properties. Hybridization techniques using COX-2 inhibitors and 5-LOX inhibitors present a good strategy to find dual leads. In this sense, celecoxib analogs were designed by replacing the tolyl ring with an N-difluoromethyl-1,2-dihydropyrid-2-one, which is a 5-LOX pharmacophore [106]. Similarly, by joining β-boswellic acid with different NSAID molecules via a Steglich esterification, compounds with interesting anti-inflammatory and antiarthritic properties can be obtained [3]. All hybrid molecules have synergistic effects in inflammation and arthritis, with **37** and **38** being the best (Figure 4). The compound **37** is more effective than ibuprofen in inhibiting COX-2 in vivo while achieving the same effect with only a third of the concentration. The antiarthritic activity of the hybrid is better than β-boswellic acids or ibuprofen individually [3].

Gedawy et al. synthesized a series of pyrazole sulfonamide as anti-inflammatory agents where benzothiophen-2-yl pyrazole carboxylic acid derivative **39** (Figure 4) was the best lead found, with an IC_{50} of 5.40 nM for COX-1, 0.01 nM for COX-2, and 1.78 nM for 5-LOX. Its selectivity index towards COX-2 was 344.56 [34].

The phenoxy acetamide, indole, chalcone thiosemicarbazone, and quinoline are important groups in anti-inflammatory agents [47,107,108]. Huang, Qian, and coworkers designed indole-2-amides as anti-inflammatory drugs [33,109]. Results showed that **40**, **41**, **42** and **43** (Figure 4) present interesting anti-inflammatory properties, with **40** being the most promising compound with COX-2 selectivity and dual COX-2/5-LOX activity binding in the same way as coxibs into the COX-2 active site [33].

Dual inhibitors can also be obtained from natural sources. Primin, a quinone extracted from Primula obconica, presents good anti-inflammatory activity. Therefore, related compounds including hydroquinone, benzoquinone, and resorcinol groups were designed and tested for activity against COX-1, COX-2 and 5-LOX. The compounds **44** (2-methoxy-6-undecyl-1,4-benzoquinone) and **45** (2-methoxy-6-undecyl-1,4-hydroquinone) (Figure 4) were found to be dual COX/5-LOX inhibitors. A key structural modification to achieve a dual inhibition is to have a longer alkyl chain in position 6 from 5 to 11 carbons. Adding an acetyl group in the ortho position of **46** (Figure 4) produces compound **47** (Figure 4), enhancing 5-LOX inhibition. Although acetylation negatively affects COX inhibition, it enhances 5-LOX inhibition [57]. In a recent approach, 15-LOX has also been targeted to reduce inflammation. 1,2,4-Triazine-quinoline and benzimidazole–thiazole hybrids have shown a dual COX-2/15-LOX inhibition being better than celecoxib and quercetin in acting on COX-2 and 15-LOX, respectively. The best compound found for 1,2,4-triazine–quinoline was **48** (Figure 4), presenting IC_{50} values of 0.047 μM (COX-2) and 1.81 μM (15-LOX) [110]; the best benzimidazole–thiazole hybrid was **49** (Figure 4), with an IC_{50} of 0.045 μM for COX-2 and 1.67 μM for 15-LOX [111].

In the design of multitarget compounds, **50** ((±)-2-[3-fluoro-4-[3-(hexylcarbamoyloxy)-phenyl]phenyl]propanoic acid, ARN2508; Figure 4) was found to be a potent FAAH and COX inhibitor without the gastrotoxic effect. Other potent dual-target compounds found were **58–62** (Figure 4). The **50** enantiomer ((S)-(+)) can be considered a FAAH–COX inhibitor with in vivo potency [31]. The compound **51** (Figure 4) presents the same FAAH potency as c-pentyl (IC_{50} 4.8uM), while **52** (Figure 4), c-butyl, presents a loss in potency (48.7 uM), and **53** (Figure 4), c-propyl, showed no potency at all. Regarding COX activity, **51** is equivalent to **54** (Figure 4), **52** is more potent, and **53** is similar to **54**. It seems that the N-terminal region of the carbamate group may increase the interaction with FAAH. Adding a methylene group next to the c-hexyl ring (**55**; Figure 4) increases FAAH and COX-1 potency but not COX-2 potency. The compound **56** (Figure 4) inhibits the three enzymes. The compound **57** (Figure 4) is the most active compound, showing that small and branched alkyl groups substantially affect inhibitory activity [31].

Other types of dual inhibitors are the ones inhibiting COX and carbonic anhydrase (CAI) [112]. COX–CAI inhibitors are more effective in treating rheumatoid arthritis than common NSAIDs as they do not increase the risk of oxidative stress in patients [4,113]. Rheumatoid arthritis is an autoimmune inflammatory disease affecting joints, cartilage, and bones [114,115]. Its pharmacological treatment includes the use of common NSAIDs; glucocorticoids (prednisolone); and antirheumatic agents, including aminosalicylates (sulfasalazine), antimalarial drugs (hydroxychloroquine), immunosuppressants/cytostatic drugs (methotrexate), and antirheumatic drugs (aurothiomalate) [116]. Hybrid compounds containing NSAID drugs and sulfonamides and carboxylates moieties can be synthesized as dual COX–CAI inhibitors [117]. In this sense, two molecules (**63** and **64**; Figure 4) were designed and presented interesting properties [113,114]. Furthermore, the 7-coumarine ibuprofen (**65**; Figure 4) hybrid showed a high antihyperalgesic effect and activity against human hCAs, although it was limited to the isoforms IV and XII and did not include IX [114].

Finally, by a covalent link between a sulohydroxamic acid moiety and known NSAIDs via a two-carbon ethyl spacer, prodrugs with anti-inflammatory activity and the ability to release nitric oxide and nitroxyl were synthesized [32]. The esters produced with naproxen and ibuprofen showed greater anti-inflammatory activity than their parent compounds, while indomethacin ester was shown to be a selective COX-2 inhibitor with no ulcerogenic effect [32]. The multitarget drugs described in this section are shown in Figure 4.

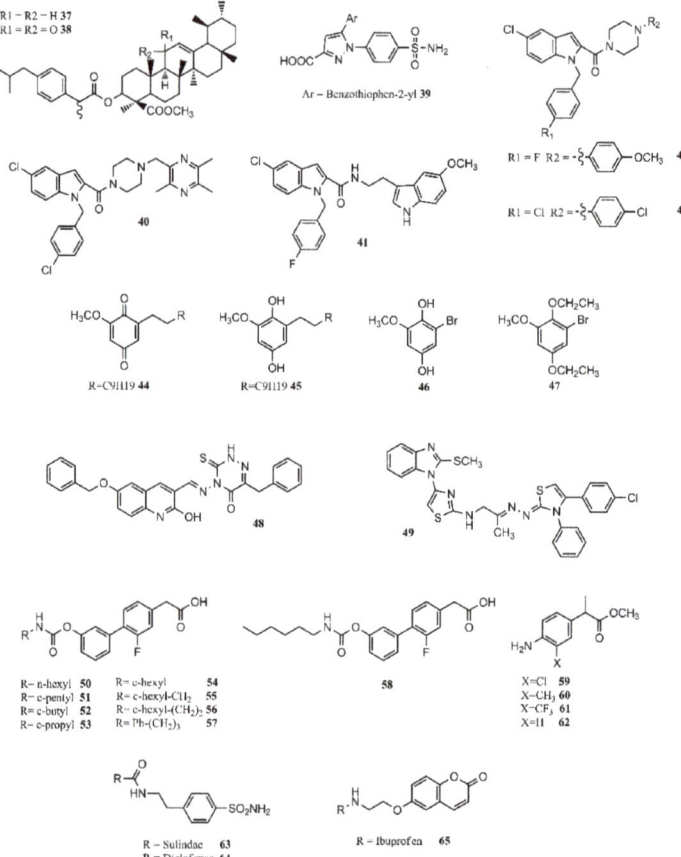

Figure 4. Multitarget drug leads.

4. Cannabis

Opioids and NSAIDs are the first options to treat acute pain [118]. Still, those drugs have shown to be not as effective in treating chronic pain, which may be disabling in some cases [119]. The endocannabinoid system involved in pain and inflammation processes has emerged as a promising approach to treat chronic pain [119–121]. Cannabinoids acting as positive allosteric CB1 receptor modulators, CB1 agonists, CB2 agonists, and mixed CB1/CB2 agonists have been reported to help in inflammatory and neuropathic pain [122–124].

The cannabis plant has been cultivated for hundreds of years and used for medicinal, spiritual, and recreational purposes. The main components in the plant are alkaloids which have been demonstrated to present anti-inflammatory activity [125]. In the cannabis plant, there are more than 540 phytochemicals from 18 different classes. Of those, 100 are phytocannabinoids [120]. Preparations with this plant have been reported to produce anti-inflammatory, anticonvulsant, antianxiety, analgesic, and muscle relaxant effects [126]. Some countries have approved the use of cannabis to treat pain, multiple sclerosis, epilepsy, sleep disturbance, and neurodegenerative diseases. Furthermore, antiseizure effects in Lennox–Gastaut syndrome and Dravet syndrome have been observed in randomized controlled trials [120,127].

The term medical cannabis has arisen in the last years and refers to the use of the cannabis plant to treat different medical conditions as prescribed by a doctor [128,129]. The main use of medical cannabis is for pain management [86]. In this sense, cannabinoids act mainly through cannabinoid receptors, although it has been reported they can also modulate ion channels and enzymes [119]. The most accepted mechanism for the analgesic effect of cannabis is through the reduction of neural inflammation and descending inhibitory pain and the modulation of postsynaptic neuron excitability pathways. Although more than 10,000 scientific articles present "conclusive or substantial evidence" and support the use of cannabis for neuropathic pain, there is a need to carry out long clinical trials to establish its safety, dosage, and indications for medical conditions [119].

Studies have shown cannabinoids act on different targets in the peripheral and central nervous system [130,131]. Their targets include CB1/CB2 receptors, GPR55, GPR18, N-arachidonoyl glycine (NAGly) receptor, nuclear receptors, ion channels, and other potential targets in the CNS [130,132]. Furthermore, cannabinoids may also act on γ-aminobutyric acid, serotonergic, adrenergic and opioid receptors that are part of the analgesic pathway [121].

Although cannabinoids are considered the main active ingredients in cannabis, other compounds inside the plant, including terpenoids and flavonoids, may also be involved in the anti-inflammatory and analgesic effect of cannabis [126,133]. The main components of cannabis are cannabidiol (CBD) and tetrahydrocannabinol (THC) [134]. THC was first isolated in the 1960s, and it is responsible not only for the psychoactive effect but also for some of the analgesic and anti-inflammatory properties. Cannabis prohibition started in the 20th century due to its psychoactive properties (ElSohly et al., 2017). Psychotropic effects include euphoria and paranoia that occur due to the activation of CB1 receptors [119].

THC is a partial CB1 and CB2 receptor agonist, while CBD, a nonpsychoactive cannabinoid, has little affinity to both receptors. Studies have shown it is able to antagonize those receptors in presence of THC acting synergically [124]. Obtaining insight into the action mechanism, CBD is a noncompetitive negative allosteric modulator of the CB1 receptor [135]. CBD modulates non-cannabinoid G protein-coupled receptors (GPCRs), ion channels, and peroxisome proliferator-activated receptor (PPARs), affecting the perception of pain [131,136].

Inhaled cannabinoids are rapidly absorbed in the bloodstream. Gastrointestinal absorption is irregular, causing low bioavailability and poor pharmacokinetic profile, although it can be increased with food [137]. This is a limitation for using cannabis in oral formulations [119]. Nabiximols, a mixture of CBD and THC approved in Spain, Switzerland,

Australia, Canada, Brazil and other countries, is prescribed to treat spasticity in multiple sclerosis [127]. Currently, there are only two synthetic cannabinoids on the market. Dronabinol and nabilone were accepted to treat chemotherapy-associated nausea and vomiting, but they may also be helpful in treating pain [138].

The lack of interest in cannabinoids was caused by the psychoactive side effect that some of these compounds have. Therefore, scientific efforts should focus on preserving and increasing the analgesic effect of these compounds while reducing their psychoactive effect [119]. In this regard, the medicinal chemistry approach towards the use of cannabinoids to treat medical conditions has focused on the modulation of the cannabinoid receptors CB1/CB2 and two endocannabinoid deactivating enzymes, monoacylglycerol lipase (MGL) and enzymes fatty acid amide hydrolase (FAAH), including allosteric inhibition [139]. Aminoalkylindole is the first different chemotype to be introduced in cannabinoid medicinal chemistry. Compounds such as AM2201 (**66**; Figure 5) and AM678 (**67**; Figure 5) present potent antinociceptive properties, but due to their structure, some of them have been classified as controlled substances. BAY 38-7271 (**68**; Figure 5) is a potent CB1/CB2 receptor agonist with clinical efficacy against severe traumatic brain injury [139].

One approach to eliminate undesirable side effects is to design peripherally restricted agonists. In this sense, SAB378 (**69**; Figure 5) and AZD1940 (**70**; Figure 5) are good candidates due to their interesting properties. The discovery of selective CB2 receptors was important because their presence is peripheral, exhibiting antinociceptive and anti-inflammatory action without the CB1 side effects. AM1241 (**71**; Figure 5) and AM1710 (**72**; Figure 5) were tested in rodent models and found to be successful for treating inflammatory and neuropathic pain [139].

In 29 trials performed between 2003 and 2014, more than 75% of the studies showed a significant analgesic effect in chronic noncancer pain, including neuropathic pain, rheumatoid arthritis, and fibromyalgia [140]. The compounds targeting CB1 and/or CB2 receptors are shown in Figure 5.

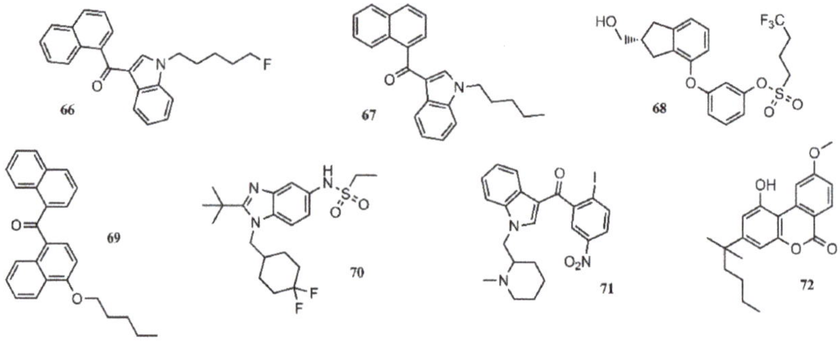

Figure 5. Compounds targeting CB1 and/or CB2 receptors.

5. Calcium Channel

Neuropathic pain is a chronic condition affecting 7 to 8% of the population [141,142]. Although NSAIDs and cannabinoids have been used, their effectiveness has not been as expected [121,143]. Neuropathic pain is characterized by discomfort and soreness that is stimulus-independent, affecting the quality of life of the patients [144]. Anticonvulsant drugs including pregabalin and gabapentin have emerged as a new therapy to treat neuropathic pain, showing efficacy in more than ten double-blind clinical trials [142,145]. The analgesic effect of these drugs comes through a different pathway [146]. In 1996, the pregabalin and gabapentin target was found to be the $\alpha 2\text{-}\delta$ subunit of the P/Q type voltage-gated calcium channel [147,148]. Its main mechanism involves returning hyperexcited neurons to a normal state through the reduction of calcium [145,149,150]. Docking studies have shown that pregabalin's most stable conformation fits in a pocket where 12 interactions including 6 hydrogen bonds were found, including electrostatic interactions with Arg217. It has been shown experimentally that Arg217 is a gating charge carrier essential in pregabalin's action mechanism, opening the possibility of developing new leads to treat neuropathic pain through calcium channel modulation [149,151,152].

6. Conclusions

Pain and inflammation are conditions that affect the quality of life of a high percentage of the human population. Humans have been looking for ways to treat pain and inflammation since ancient times, finding the first treatments in herbal plants such as the willow bark. Salix species contains salicin, which is the main active ingredient helping reduce pain and inflammation in conjunction with other compounds such as flavonoids. Furthermore, this compound was used as starting material to produce one of the most famous anti-inflammatory drugs, aspirin. Although the main mechanisms of action of anti-inflammatory drugs are the COX pathways involving COX-1 and COX-2 enzymes, it is now known the inflammation and analgesic pathways are much more complex and include other proteins such as 5-LOX, FAAH, CAI, ion channels, γ-aminobutyric acid, serotonergic, adrenergic, opioid, and CB1/CB2 receptors that can be and are now tested as potential targets for pain management treatments. In this sense, new types of molecules including cannabinoids are being tested as potential pain treatments with excellent results. Efforts have been focused on designing more potent and safer drugs. In this sense, research has leaned toward the synthesis of cocrystals, COX-selective inhibitors (either COX-1- or COX-2-selective), and dual inhibitors. Thanks to the advances of the technology and through the elucidation of the biological targets and QSAR studies, interesting new molecules have been designed, taking advantage of the pocket size and properties in order to enhance selectivity and reduce side effects. Still, more studies need to be done to produce more effective treatments against more complex conditions including rheumatoid arthritis, neuropathic pain, and inflammation processes present in different types of cancer.

Author Contributions: Conceptualization, L.M. and S.A.C.; formal analysis, L.M. and S.A.C.; investigation, L.M. and S.A.C.; writing—original draft preparation, L.M. and S.A.C.; writing—review and editing, L.M. and S.A.C. Both authors have read and agreed to the published version of the manuscript.

Funding: This research received no external funding.

Institutional Review Board Statement: Not applicable.

Informed Consent Statement: Not applicable.

Data Availability Statement: Not applicable.

Conflicts of Interest: The authors declare no conflict of interest.

References

1. Ferrer, M.D.; Busquets-Cortés, C.; Capó, X.; Tejada, S.; Tur, J.A.; Pons, A.; Sureda, A. Cyclooxygenase-2 Inhibitors as a Therapeutic Target in Inflammatory Diseases. *Curr. Med. Chem.* **2018**, *26*, 3225–3241. [CrossRef]
2. Aggarwal, R.; Singh, G.; Kaushik, P.; Kaushik, D.; Paliwal, D.; Kumar, A. Molecular docking design and one-pot expeditious synthesis of novel 2,5-diarylpyrazolo [1,5-a] pyrimidin-7-amines as anti-inflammatory agents. *Eur. J. Med. Chem.* **2015**, *101*, 326–333. [CrossRef]
3. Shenvi, S.; Kiran, K.R.; Kumar, K.; Diwakar, L.; Reddy, G.C. Synthesis and biological evaluation of boswellic acid-NSAID hybrid molecules as anti-inflammatory and anti-arthritic agents. *Eur. J. Med. Chem.* **2015**, *98*, 170–178. [CrossRef]
4. Chan, K.L.; Cathomas, F.; Russo, S.J. Central and peripheral inflammation link metabolic syndrome and major depressive disorder. *Physiology* **2019**, *34*, 123–133. [CrossRef]
5. Collins, K.H.; Herzog, W.; MacDonald, G.Z.; Reimer, R.A.; Rios, J.L.; Smith, I.C.; Zernicke, R.F.; Hart, D.A. Obesity, metabolic syndrome, and musculoskeletal disease: Common inflammatory pathways suggest a central role for loss of muscle integrity. *Front. Physiol.* **2018**, *9*, 112. [CrossRef] [PubMed]
6. Frühbeck, G.; Catalán, V.; Rodríguez, A.; Ramírez, B.; Becerril, S.; Salvador, J.; Portincasa, P.; Colina, I.; Gómez-Ambrosi, J. Involvement of the leptin-adiponectin axis in inflammation and oxidative stress in the metabolic syndrome. *Sci. Rep.* **2017**, *7*, 6619. [CrossRef]

7. Lopez-Candales, A.; Hernández Burgos, P.M.; Hernandez-Suarez, D.F.; Harris, D. Linking Chronic Inflammation with Cardiovascular Disease: From Normal Aging to the Metabolic Syndrome. *J. Nat. Sci.* **2017**, *3*, e341. [PubMed]
8. Reddy, P.; Lent-Schochet, D.; Ramakrishnan, N.; McLaughlin, M.; Jialal, I. Metabolic syndrome is an inflammatory disorder: A conspiracy between adipose tissue and phagocytes. *Clin. Chim. Acta* **2019**, *496*, 35–44. [CrossRef]
9. Zanandrea, R.; Bonan, C.D.; Campos, M.M. Zebrafish as a model for inflammation and drug discovery. *Drug Discov. Today* **2020**, *25*, 2201–2211. [CrossRef] [PubMed]
10. Ngo, H.X.; Garneau-Tsodikova, S. What are the drugs of the future? *MedChemComm* **2018**, *9*, 757–758. [CrossRef] [PubMed]
11. Neha, K.; Wakode, S. Contemporary advances of cyclic molecules proposed for inflammation. *Eur. J. Med. Chem.* **2021**, *221*, 113493. [CrossRef] [PubMed]
12. Demay, J.; Halary, S.; Knittel-Obrecht, A.; Villa, P.; Duval, C.; Hamlaoui, S.; Roussel, T.; Yéprémian, C.; Reinhardt, A.; Bernard, C.; et al. Anti-inflammatory, antioxidant, and wound-healing properties of cyanobacteria from thermal mud of balaruc-les-bains, France: A multi-approach study. *Biomolecules* **2021**, *11*, 28. [CrossRef]
13. Chen, Z.-H.; Li, W.-S.; Zhang, Z.-Y.; Luo, H.; Wang, J.-R.; Zhang, H.-Y.; Zeng, Z.-R.; Chen, B.; Li, X.-W.; Guo, Y.-W. Sinusiaetone A, an Anti-inflammatory Norditerpenoid with a Bicyclo [11.3.0] hexadecane Nucleus from the Hainan Soft Coral Sinularia siaesensis. *Org. Lett.* **2021**. [CrossRef]
14. Shi, Z.; Dun, B.; Wei, Z.; Liu, C.; Tian, J.; Ren, G.; Yao, Y. Peptides Released from Extruded Adzuki Bean Protein through Simulated Gastrointestinal Digestion Exhibit Anti-inflammatory Activity. *J. Agric. Food Chem.* **2021**. [CrossRef]
15. Burgos, R.A.; Alarcón, P.; Quiroga, J.; Manosalva, C.; Hancke, J. Andrographolide, an Anti-Inflammatory Multitarget Drug: All Roads Lead to Cellular Metabolism. *Molecules* **2020**, *26*, 5. [CrossRef]
16. Tian, H.; Wei, L.; Yao, Y.; Zeng, Z.; Liang, X.; Zhu, H. Analysis of the Anti-Inflammatory and Analgesic Mechanism of Shiyifang Vinum Based on Network Pharmacology. *Evid. Based Complement. Altern. Med.* **2021**, *2021*, 8871276. [CrossRef] [PubMed]
17. Schmid, B.; Lüdtke, R.; Selbmann, H.K.; Kötter, I.; Tschirdewahn, B.; Schaffner, W.; Heide, L. Efficacy and tolerability of a standardized willow bark extract in patients with osteoarthritis: Randomized, placebo-controlled, double blind clinical trial. *Z. Rheumatol.* **2000**, *59*, 314–320. [CrossRef]
18. Desborough, M.J.R.; Keeling, D.M. The aspirin story—From willow to wonder drug. *Br. J. Haematol.* **2017**, *177*, 674–683. [CrossRef]
19. Wood, E.J. Aspirin: The remarkable story of a wonder drug. *Biochem. Mol. Biol. Educ.* **2006**, *34*, 459–460. [CrossRef]
20. Khayyal, M.T.; El-Ghazaly, M.A.; Abdallah, D.M.; Okpanyi, S.N.; Kelber, O.; Weiser, D. Mechanisms involved in the anti-inflammatory effect of a standardized willow bark extract. *Arzneim. Forsch. /Drug Res.* **2005**, *55*, 677–687. [CrossRef]
21. Chrubasik, S.; Eisenberg, E.; Balan, E.; Weinberger, T.; Luzzati, R.; Conradt, C. Treatment of low back pain exacerbations with willow bark extract: A randomized double-blind study. *Am. J. Med.* **2000**, *109*, 9–14. [CrossRef]
22. Vlachojannis, J.E.; Cameron, M.; Chrubasik, S. A systematic review on the effectiveness of willow bark for musculoskeletal pain. *Phyther. Res.* **2009**, *23*, 897–900. [CrossRef]
23. Shara, M.; Stohs, S.J. Efficacy and Safety of White Willow Bark (*Salix alba*) Extracts. *Phyther. Res.* **2015**, *29*, 1112–1116. [CrossRef] [PubMed]
24. Dwivedi, A.K.; Gurjar, V.; Kumar, S.; Singh, N. Molecular basis for nonspecificity of nonsteroidal anti-inflammatory drugs (NSAIDs). *Drug Discov. Today* **2015**, *20*, 863–873. [CrossRef] [PubMed]
25. Rouzer, C.A.; Marnett, L.J. Structural and Chemical Biology of the Interaction of Cyclooxygenase with Substrates and Non-Steroidal Anti-Inflammatory Drugs. *Chem. Rev.* **2020**, *120*, 7592–7641. [CrossRef] [PubMed]
26. Pannunzio, A.; Coluccia, M. Cyclooxygenase-1 (COX-1) and COX-1 inhibitors in cancer: A review of oncology and medicinal chemistry literature. *Pharmaceuticals* **2018**, *11*, 101. [CrossRef]
27. Lolli, M.L.; Cena, C.; Medana, C.; Lazzarato, L.; Morini, G.; Coruzzi, G.; Manarini, S.; Fruttero, R.; Gasco, A. A new class of ibuprofen derivatives with reduced gastrotoxicity. *J. Med. Chem.* **2001**, *44*, 3463–3468. [CrossRef] [PubMed]
28. Gudis, K.; Sakamoto, C. The role of cyclooxygenase in gastric mucosal protection. *Dig. Dis. Sci.* **2005**, *50*, S16–S23. [CrossRef] [PubMed]
29. Simopoulos, A.P. Omega-3 fatty acids in inflammation and autoimmune diseases. *J. Am. Coll. Nutr.* **2002**, *21*, 495–505. [CrossRef]
30. Vitale, P.; Tacconelli, S.; Perrone, M.G.; Malerba, P.; Simone, L.; Scilimati, A.; Lavecchia, A.; Dovizio, M.; Marcantoni, E.; Bruno, A.; et al. Synthesis, pharmacological characterization, and docking analysis of a novel family of diarylisoxazoles as highly selective cyclooxygenase-1 (COX-1) inhibitors. *J. Med. Chem.* **2013**, *56*, 4277–4299. [CrossRef] [PubMed]
31. Migliore, M.; Habrant, D.; Sasso, O.; Albani, C.; Bertozzi, S.M.; Armirotti, A.; Piomelli, D.; Scarpelli, R. Potent multitarget FAAH-COX inhibitors: Design and structure-activity relationship studies. *Eur. J. Med. Chem.* **2016**, *109*, 216–237. [CrossRef]
32. Huang, Z.; Velázquez, C.A.; Abdellatif, K.R.A.; Chowdhury, M.A.; Reisz, J.A.; Dumond, J.F.; King, S.B.; Knaus, E.E. Ethanesulfohydroxamic acid ester prodrugs of nonsteroidal anti-inflammatory drugs (NSAIDs): Synthesis, nitric oxide and nitroxyl release, cyclooxygenase inhibition, anti-inflammatory, and ulcerogenicity index studies. *J. Med. Chem.* **2011**, *54*, 1356–1364. [CrossRef]
33. Huang, Y.; Zhang, B.; Li, J.; Liu, H.; Zhang, Y.; Yang, Z.; Liu, W. Design, synthesis, biological evaluation and docking study of novel indole-2-amide as anti-inflammatory agents with dual inhibition of COX and 5-LOX. *Eur. J. Med. Chem.* **2019**, *180*, 41–50. [CrossRef]
34. Gedawy, E.M.; Kassab, A.E.; El Kerdawy, A.M. Design, synthesis and biological evaluation of novel pyrazole sulfonamide derivatives as dual COX-2/5-LOX inhibitors. *Eur. J. Med. Chem.* **2020**, *189*, 112066. [CrossRef]

35. Smith, W.L.; DeWitt, D.L.; Garavito, R.M. Cyclooxygenases: Structural, cellular, and molecular biology. *Annu. Rev. Biochem.* **2000**, *69*, 145–182. [CrossRef] [PubMed]
36. Kiefer, J.R.; Pawlitz, J.L.; Moreland, K.T.; Stegeman, R.A.; Hood, W.F.; Glerse, J.K.; Stevens, A.M.; Goodwin, D.C.; Rowlinson, S.W.; Marnett, L.J.; et al. Structural insights into the stereochemistry of the cyclooxygenase reaction. *Nature* **2000**, *405*, 97–101. [CrossRef] [PubMed]
37. Selinsky, B.S.; Gupta, K.; Sharkey, C.T.; Loll, P.J. Structural analysis of NSAID binding by prostaglandin H2 synthase: Time-dependent and time-independent inhibitors elicit identical enzyme conformations. *Biochemistry* **2001**, *40*, 5172–5180. [CrossRef]
38. Viegas, A.; Manso, J.; Corvo, M.C.; Marques, M.M.B.; Cabrita, E.J. Binding of ibuprofen, ketorolac, and diclofenac to COX-1 and COX-2 studied by saturation transfer difference NMR. *J. Med. Chem.* **2011**, *54*, 8555–8562. [CrossRef]
39. Uddin, M.J.; Elleman, A.V.; Ghebreselasie, K.; Daniel, C.K.; Crews, B.C.; Nance, K.D.; Huda, T.; Marnett, L.J. Design of fluorine-containing 3,4-diarylfuran-2(5H)-ones as selective COX-1 inhibitors. *ACS Med. Chem. Lett.* **2014**, *5*, 1254–1258. [CrossRef]
40. Bhardwaj, A.; Kaur, J.; Wuest, M.; Wuest, F. In situ click chemistry generation of cyclooxygenase-2 inhibitors. *Nat. Commun.* **2017**, *8*, 1–14. [CrossRef] [PubMed]
41. Desai, S.J.; Prickril, B.; Rasooly, A. Mechanisms of Phytonutrient Modulation of Cyclooxygenase-2 (COX-2) and Inflammation Related to Cancer. *Nutr. Cancer* **2018**, *70*, 350–375. [CrossRef] [PubMed]
42. Yamakawa, N.; Suemasu, S.; Matoyama, M.; Kimoto, A.; Takeda, M.; Tanaka, K.I.; Ishihara, T.; Katsu, T.; Okamoto, Y.; Otsuka, M.; et al. Properties and synthesis of 2-{2-fluoro (or bromo)-4-[(2-oxocyclopentyl) methyl]phenyl}propanoic acid: Nonsteroidal anti-inflammatory drugs with low membrane permeabilizing and gastric lesion-producing activities. *J. Med. Chem.* **2010**, *53*, 7879–7882. [CrossRef] [PubMed]
43. Smalley, W.E.; Ray, W.A.; Daugherty, J.R.; Griffin, M.R. Nonsteroidal anti-inflammatory drugs and the incidence of hospitalizations for peptic ulcer disease in elderly persons. *Am. J. Epidemiol.* **1995**, *141*, 539–545. [CrossRef] [PubMed]
44. Green, G.A. Understanding NSAIDs: From aspirin to COX-2. *Clin. Cornerstone* **2001**, *3*, 50–59. [CrossRef]
45. Radwan, M.F.; Dalby, K.N.; Kaoud, T.S. Propyphenazone-based analogues as prodrugs and selective cyclooxygenase-2 inhibitors. *ACS Med. Chem. Lett.* **2014**, *5*, 983–988. [CrossRef] [PubMed]
46. Bjarnason, I.; Scarpignato, C.; Holmgren, E.; Olszewski, M.; Rainsford, K.D.; Lanas, A. Mechanisms of Damage to the Gastrointestinal Tract from Nonsteroidal Anti-Inflammatory Drugs. *Gastroenterology* **2018**, *154*, 500–514. [CrossRef]
47. Reddy, M.V.R.; Billa, V.K.; Pallela, V.R.; Mallireddigari, M.R.; Boominathan, R.; Gabriel, J.L.; Reddy, E.P. Design, synthesis, and biological evaluation of 1-(4-sulfamylphenyl)-3-trifluoromethyl-5-indolyl pyrazolines as cyclooxygenase-2 (COX-2) and lipoxygenase (LOX) inhibitors. *Bioorg. Med. Chem.* **2008**, *16*, 3907–3916. [CrossRef]
48. Wired Release Global Ibuprofen Market Projected to Witness Robust Development by 2020–2026. Available online: https://www.pharmiweb.com/press-release/2020-01-17/ibuprofen-market-is-valued-at-2944-million-us-in-2020-is-expected-to-reach-4476-million-us-by-th (accessed on 14 February 2021).
49. Harrak, Y.; Casula, G.; Basset, J.; Rosell, G.; Plescia, S.; Raffa, D.; Cusimano, M.G.; Pouplana, R.; Pujol, M.D. Synthesis, anti-inflammatory activity, and in vitro antitumor effect of a novel class of cyclooxygenase inhibitors: 4-(Aryloyl)phenyl methyl sulfones. *J. Med. Chem.* **2010**, *53*, 6560–6571. [CrossRef] [PubMed]
50. Taher, E.S.; Ibrahim, T.S.; Fares, M.; AL-Mahmoudy, A.M.M.; Radwan, A.F.; Orabi, K.Y.; El-Sabbagh, O.I. Novel benzenesulfonamide and 1,2-benzisothiazol-3(2H)-one-1,1-dioxide derivatives as potential selective COX-2 inhibitors. *Eur. J. Med. Chem.* **2019**, *171*, 372–382. [CrossRef] [PubMed]
51. Onder, G.; Pellicciotti, F.; Gambassi, G.; Bernabei, R. NSAID-related psychiatric adverse events: Who is at risk? *Drugs* **2004**, *64*, 2619–2627. [CrossRef]
52. Machado, G.C.; Maher, C.G.; Ferreira, P.H.; Day, R.O.; Pinheiro, M.B.; Ferreira, M.L. Non-steroidal anti-inflammatory drugs for spinal pain: A systematic review and meta-analysis. *Ann. Rheum. Dis.* **2017**, *76*, 1269–1278. [CrossRef] [PubMed]
53. Cooper, C.; Chapurlat, R.; Al-Daghri, N.; Herrero-Beaumont, G.; Bruyère, O.; Rannou, F.; Roth, R.; Uebelhart, D.; Reginster, J.Y. Safety of Oral Non-Selective Non-Steroidal Anti-Inflammatory Drugs in Osteoarthritis: What Does the Literature Say? *Drugs Aging* **2019**, *36*, 15–24. [CrossRef] [PubMed]
54. Baigent, C.; Bhala, N.; Emberson, J.; Merhi, A.; Abramson, S.; Arber, N.; Baron, J.A.; Bombardier, C.; Cannon, C.; Farkouh, M.E.; et al. Vascular and upper gastrointestinal effects of non-steroidal anti-inflammatory drugs: Meta-analyses of individual participant data from randomised trials. *Lancet* **2013**, *382*, 769–779.
55. Nissen, S.E.; Yeomans, N.D.; Solomon, D.H.; Lüscher, T.F.; Libby, P.; Husni, M.E.; Graham, D.Y.; Borer, J.S.; Wisniewski, L.M.; Wolski, K.E.; et al. Cardiovascular Safety of Celecoxib, Naproxen, or Ibuprofen for Arthritis. *N. Engl. J. Med.* **2016**, *375*, 2519–2529. [CrossRef] [PubMed]
56. Grosser, T.; Ricciotti, E.; FitzGerald, G.A. The Cardiovascular Pharmacology of Nonsteroidal Anti-Inflammatory Drugs. *Trends Pharmacol. Sci.* **2017**, *38*, 733–748. [CrossRef]
57. Sisa, M.; Dvorakova, M.; Temml, V.; Jarosova, V.; Vanek, T.; Landa, P. Synthesis, inhibitory activity and in silico docking of dual COX/5-LOX inhibitors with quinone and resorcinol core. *Eur. J. Med. Chem.* **2020**, *204*, 112620. [CrossRef]
58. Fiorucci, S.; Meli, R.; Bucci, M.; Cirino, G. Dual inhibitors of cyclooxygenase and 5-lipoxygenase. A new avenue in anti-inflammatory therapy? *Biochem. Pharmacol.* **2001**, *62*, 1433–1438. [CrossRef]
59. Manju, S.L.; Ethiraj, K.R.; Elias, G. Safer anti-inflammatory therapy through dual COX-2/5-LOX inhibitors: A structure-based approach. *Eur. J. Pharm. Sci.* **2018**, *121*, 356–381.

60. Gaba, M.; Singh, D.; Singh, S.; Sharma, V.; Gaba, P. Synthesis and pharmacological evaluation of novel 5-substituted-1-(phenylsulfonyl)-2-methylbenzimidazole derivatives as anti-inflammatory and analgesic agents. *Eur. J. Med. Chem.* **2010**, *45*, 2245–2249. [CrossRef]
61. Lednicer, D. *The Organic Chemistry of Drug Synthesis*; John Wiley and Sons: Hoboken, NJ, USA, 2007; Volume 7, ISBN 9780470107508.
62. Varrassi, G.; Pergolizzi, J.V.; Dowling, P.; Paladini, A. Ibuprofen Safety at the Golden Anniversary: Are all NSAIDs the Same? A Narrative Review. *Adv. Ther.* **2020**, *37*, 61–82. [CrossRef]
63. Veber, D.F.; Johnson, S.R.; Cheng, H.Y.; Smith, B.R.; Ward, K.W.; Kopple, K.D. Molecular properties that influence the oral bioavailability of drug candidates. *J. Med. Chem.* **2002**, *45*, 2615–2623. [CrossRef] [PubMed]
64. Kleemiss, F.; Justies, A.; Duvinage, D.; Watermann, P.; Ehrke, E.; Sugimoto, K.; Fugel, M.; Malaspina, L.A.; Dittmer, A.; Kleemiss, T.; et al. Sila-Ibuprofen. *J. Med. Chem.* **2020**, *63*, 12614–12622. [CrossRef]
65. Franz, A.K.; Wilson, S.O. Organosilicon molecules with medicinal applications. *J. Med. Chem.* **2013**, *56*, 388–405. [CrossRef]
66. Sodano, F.; Lazzarato, L.; Rolando, B.; Spyrakis, F.; De Caro, C.; Magliocca, S.; Marabello, D.; Chegaev, K.; Gazzano, E.; Riganti, C.; et al. Paracetamol-Galactose Conjugate: A Novel Prodrug for an Old Analgesic Drug. *Mol. Pharm.* **2019**, *16*, 4181–4189. [CrossRef]
67. Lassalas, P.; Oukoloff, K.; Makani, V.; James, M.; Tran, V.; Yao, Y.; Huang, L.; Vijayendran, K.; Monti, L.; Trojanowski, J.Q.; et al. Evaluation of Oxetan-3-ol, Thietan-3-ol, and Derivatives Thereof as Bioisosteres of the Carboxylic Acid Functional Group. *ACS Med. Chem. Lett.* **2017**, *8*, 864–868. [CrossRef] [PubMed]
68. Ju, Z.; Su, M.; Hong, J.; La Kim, E.; Moon, H.R.; Chung, H.Y.; Kim, S.; Jung, J.H. Design of balanced COX inhibitors based on anti-inflammatory and/or COX-2 inhibitory ascidian metabolites. *Eur. J. Med. Chem.* **2019**, *180*, 86–98. [CrossRef] [PubMed]
69. Neves, B.J.; Braga, R.C.; Melo-Filho, C.C.; Moreira-Filho, J.T.; Muratov, E.N.; Andrade, C.H. QSAR-based virtual screening: Advances and applications in drug discovery. *Front. Pharmacol.* **2018**, *9*, 1275. [CrossRef] [PubMed]
70. Simeon, S.; Anuwongcharoen, N.; Shoombuatong, W.; Malik, A.A.; Prachayasittikul, V.; Wikberg, J.E.S.; Nantasenamat, C. Probing the origins of human acetylcholinesterase inhibition via QSAR modeling and molecular docking. *PeerJ* **2016**, *4*, e2322. [CrossRef] [PubMed]
71. Flores, M.C.; Márquez, E.A.; Mora, J.R. Molecular modeling studies of bromopyrrole alkaloids as potential antimalarial compounds: A DFT approach. *Med. Chem. Res.* **2018**, *27*, 844–856. [CrossRef]
72. Flores-Sumoza, M.; Alcázar, J.J.; Márquez, E.; Mora, J.R.; Lezama, J.; Puello, E. Classical QSAR and docking simulation of 4-pyridone derivatives for their antimalarial activity. *Molecules* **2018**, *23*, 3166. [CrossRef]
73. Imran, S.; Taha, M.; Ismail, N.H.; Kashif, S.M.; Rahim, F.; Jamil, W.; Hariono, M.; Yusuf, M.; Wahab, H. Synthesis of novel flavone hydrazones: In-vitro evaluation of α-glucosidase inhibition, QSAR analysis and docking studies. *Eur. J. Med. Chem.* **2015**, *105*, 156–170. [CrossRef] [PubMed]
74. Mora, J.R.; Márquez, E.A.; Calle, L. Computational molecular modelling of N-cinnamoyl and hydroxycinnamoyl amides as potential α-glucosidase inhibitors. *Med. Chem. Res.* **2018**, *27*, 2214–2223. [CrossRef]
75. Lakhlili, W.; Yasri, A.; Ibrahimi, A. Structure-activity relationships study of mTOR kinase inhibition using QSAR and structure-based drug design approaches. *OncoTargets Ther.* **2016**, *9*, 7345–7353. [CrossRef]
76. Meng, X.; Cui, L.; Song, F.; Luan, M.; Ji, J.; Si, H.; Duan, Y.; Zhai, H. 3D-QSAR and Molecular Docking Studies on Design Anti-Prostate Cancer Curcumin Analogues. *Curr. Comput. Aided. Drug Des.* **2018**, *16*, 245–256. [CrossRef]
77. Cabrera, N.; Mora, J.R.; Marquez, E.A. Computational Molecular Modeling of Pin1 Inhibition Activity of Quinazoline, Benzophenone, and Pyrimidine Derivatives. *J. Chem.* **2019**, *2019*, 2954250. [CrossRef]
78. Suvannang, N.; Preeyanon, L.; Malik, A.A.; Schaduangrat, N.; Shoombuatong, W.; Worachartcheewan, A.; Tantimongcolwat, T.; Nantasenamat, C. Probing the origin of estrogen receptor alpha inhibition: Via large-scale QSAR study. *RSC Adv.* **2018**, *8*, 11344–11356. [CrossRef]
79. Misaka, E.; Yamaguchi, T.; Iizuka, Y. Anti-inflammatory, analgesic and antipyretic activities of sodium 2-[4-(2-oxocyclopentan-1-ylmethyl) phenyl] propionate dihydrate (CS-600). *Pharmacometrics* **1981**, *21*, 753–771.
80. Kawano, S.; Tsuji, S.; Hayashi, N.; Takei, Y.; Nagano, K.; Fusamoto, H.; Kamada, T. Effects of loxoprofen sodium, a newly synthesized non-steroidal anti-inflammatory drug, and indomethacin on gastric mucosal haemodynamics in the human. *J. Gastroenterol. Hepatol.* **1995**, *10*, 81–85. [CrossRef]
81. Yamakawa, N.; Suemasu, S.; Okamoto, Y.; Tanaka, K.I.; Ishihara, T.; Asano, T.; Miyata, K.; Otsuka, M.; Mizushima, T. Synthesis and biological evaluation of derivatives of 2-{2-fluoro-4-[(2-oxocyclopentyl)methyl]phenyl}propanoic acid: Nonsteroidal anti-inflammatory drugs with low gastric ulcerogenic activity. *J. Med. Chem.* **2012**, *55*, 5143–5150. [CrossRef] [PubMed]
82. Hsu, N.; Cai, D.; Damodaran, K.; Gomez, R.F.; Keck, J.G.; Laborde, E.; Lum, R.T.; Macke, T.J.; Martin, G.; Schow, S.R.; et al. Novel cyclooxygenase-1 inhibitors discovered using affinity fingerprints. *J. Med. Chem.* **2004**, *47*, 4875–4880. [CrossRef] [PubMed]
83. Shen, T.Y. Perspectives in Nonsteroidal Anti-inflammatory Agents. *Angew. Chem. Int. Ed. Engl.* **1972**, *11*, 460–472. [CrossRef]
84. Curtis, E.; Fuggle, N.; Shaw, S.; Spooner, L.; Ntani, G.; Parsons, C.; Corp, N.; Honvo, G.; Baird, J.; Maggi, S.; et al. Safety of Cyclooxygenase-2 Inhibitors in Osteoarthritis: Outcomes of a Systematic Review and Meta-Analysis. *Drugs Aging* **2019**, *36*, 25–44. [CrossRef]
85. Black, W.C.; Brideau, C.; Chan, C.C.; Charleson, S.; Chauret, N.; Claveau, D.; Ethier, D.; Gordon, R.; Greig, G.; Guay, J.; et al. 2,3-diarylcyclopentenones as orally active, highly selective cyclooxygenase-2 inhibitors. *J. Med. Chem.* **1999**, *42*, 1274–1281. [CrossRef] [PubMed]

86. Khanna, I.K.; Yu, Y.; Huff, R.M.; Weier, R.M.; Xu, X.; Koszyk, F.J.; Collins, P.W.; Cogburn, J.N.; Isakson, P.C.; Koboldt, C.M.; et al. Selective cyclooxygenase-2 inhibitors: Heteroaryl modified 1,2-diarylimidazoles are potent, orally active antiinflammatory agents. *J. Med. Chem.* **2000**, *43*, 3168–3185. [CrossRef]
87. Leblanc, Y.; Gauthier, J.Y.; Ethier, D.; Guay, J.; Mancini, J.; Riendeau, D.; Tagari, P.; Vickers, P.; Wong, E.; Prasit, P. Synthesis and biological evaluation of 2,3-diarylthiophenes as selective Cox-2 and Cox-1 inhibitors. *Bioorg. Med. Chem. Lett.* **1995**, *5*, 2123–2128. [CrossRef]
88. Talley, J.J.; Bertenshaw, S.R.; Brown, D.L.; Carter, J.S.; Graneto, M.J.; Koboldt, C.M.; Masferrer, J.L.; Norman, B.H.; Rogier, D.J.; Zweifel, B.S.; et al. 4,5-diaryloxazole inhibitors of cyclooxygenase-2 (COX-2). *Med. Res. Rev.* **1999**, *19*, 199–208. [CrossRef]
89. Penning, T.D.; Talley, J.J.; Bertenshaw, S.R.; Carter, J.S.; Collins, P.W.; Docter, S.; Graneto, M.J.; Lee, L.F.; Malecha, J.W.; Miyashiro, J.M.; et al. Synthesis and biological evaluation of the 1,5-diarylpyrazole class of cyclooxygenase-2 inhibitors: Identification of 4-[5-(4-methylphenyl)- 3(trifluoromethyl)-1h-pyrazol-1-yl]benzenesulfonamide (sc-58635, celecoxib). *J. Med. Chem.* **1997**, *40*, 1347–1365. [CrossRef]
90. Nicoll-Griffith, D.A.; Yergey, J.A.; Trimble, L.A.; Silva, J.M.; Li, C.; Chauret, N.; Gauthier, J.Y.; Grimm, E.; Léger, S.; Roy, P.; et al. Synthesis, characterization, and activity of metabolites derived from the cyclooxygenase-2 inhibitor rofecoxib (MK-0966, Vioxx(TM)). *Bioorg. Med. Chem. Lett.* **2000**, *10*, 2683–2686. [CrossRef]
91. Ahmed, E.M.; Kassab, A.E.; El-Malah, A.A.; Hassan, M.S.A. Synthesis and biological evaluation of pyridazinone derivatives as selective COX-2 inhibitors and potential anti-inflammatory agents. *Eur. J. Med. Chem.* **2019**, *171*, 25–37. [CrossRef]
92. Palomer, A.; Cabré, F.; Pascual, J.; Campos, J.; Trujillo, M.A.; Entrena, A.; Gallo, M.A.; García, L.; Mauleón, D.; Espinosa, A. Identification of novel cyclooxygenase-2 selective inhibitors using pharmacophore models. *J. Med. Chem.* **2002**, *45*, 1402–1411. [CrossRef] [PubMed]
93. Kalgutkar, A.S.; Rowlinson, S.W.; Crews, B.C.; Marnett, L.J. Amide derivatives of meclofenamic acid as selective cyclooxygenase-2 inhibitors. *Bioorg. Med. Chem. Lett.* **2002**, *12*, 521–524. [CrossRef]
94. Kalgutkar, A.S.; Marnett, A.B.; Crews, B.C.; Remmel, R.P.; Marnett, L.J. Ester and amide derivatives of the nonsteroidal antiinflammatory drug, indomethacin, as selective cyclooxygenase-2 inhibitors. *J. Med. Chem.* **2000**, *43*, 2860–2870. [CrossRef] [PubMed]
95. Uddin, M.J.; Xu, S.; Crews, B.C.; Aleem, A.M.; Ghebreselasie, K.; Banerjee, S.; Marnett, L.J. Harmaline Analogs as Substrate-Selective Cyclooxygenase-2 Inhibitors. *ACS Med. Chem. Lett.* **2020**, *11*, 1881–1885. [CrossRef]
96. Prusakiewicz, J.J.; Duggan, K.C.; Rouzer, C.A.; Marnett, L.J. Differential sensitivity and mechanism of inhibition of COX-2 oxygenation of arachidonic acid and 2-arachidonoylglycerol by ibuprofen and mefenamic acid. *Biochemistry* **2009**, *48*, 7353–7355. [CrossRef]
97. Windsor, M.A.; Hermanson, D.J.; Kingsley, P.J.; Xu, S.; Crews, B.C.; Ho, W.; Keenan, C.M.; Banerjee, S.; Sharkey, K.A.; Marnett, L.J. Substrate-selective inhibition of cyclooxygenase-2: Development and evaluation of achiral profen probes. *ACS Med. Chem. Lett.* **2012**, *3*, 759–763. [CrossRef] [PubMed]
98. Thipparaboina, R.; Kumar, D.; Chavan, R.B.; Shastri, N.R. Multidrug co-crystals: Towards the development of effective therapeutic hybrids. *Drug Discov. Today* **2016**, *21*, 481–490. [CrossRef] [PubMed]
99. Godman, B.; McCabe, H.; Leong, T.D.; Mueller, D.; Martin, A.P.; Hoxha, I.; Mwita, J.C.; Rwegerera, G.M.; Massele, A.; de Oliveira Costa, J.; et al. Fixed dose drug combinations–are they pharmacoeconomically sound? Findings and implications especially for lower- and middle-income countries. *Exp. Rev. Pharmacoecon. Outcomes Res.* **2020**, *20*, 1–26. [CrossRef]
100. Nascimento, A.L.C.S.; Fernandes, R.P.; Charpentier, M.D.; ter Horst, J.H.; Caires, F.J.; Chorilli, M. Co-crystals of non-steroidal anti-inflammatory drugs (NSAIDs): Insight toward formation, methods, and drug enhancement. *Particuology* **2021**, *58*, 227–241. [CrossRef]
101. Kumari, P.; Singh, P.; Kaur, J.; Bhatti, R. Design, Synthesis, and Activity Evaluation of Stereoconfigured Tartarate Derivatives as Potential Anti-inflammatory Agents In Vitro and In Vivo. *J. Med. Chem.* **2021**. [CrossRef]
102. Abdelazeem, A.H.; El-Din, A.G.S.; Arab, H.H.; El-Saadi, M.T.; El-Moghazy, S.M.; Amin, N.H. Design, synthesis and anti-inflammatory/analgesic evaluation of novel di-substituted urea derivatives bearing diaryl-1,2,4-triazole with dual COX-2/sEH inhibitory activities. *J. Mol. Struct.* **2021**, *1240*, 130565. [CrossRef]
103. Liu, K.; Li, D.; Zheng, W.; Shi, M.; Chen, Y.; Tang, M.; Yang, T.; Zhao, M.; Deng, D.; Zhang, C.; et al. Discovery, Optimization, and Evaluation of Quinazolinone Derivatives with Novel Linkers as Orally Efficacious Phosphoinositide-3-Kinase Delta Inhibitors for Treatment of Inflammatory Diseases. *J. Med. Chem.* **2021**. [CrossRef]
104. Melagraki, G.; Afantitis, A.; Igglessi-Markopoulou, O.; Detsi, A.; Koufaki, M.; Kontogiorgis, C.; Hadjipavlou-Litina, D.J. Synthesis and evaluation of the antioxidant and anti-inflammatory activity of novel coumarin-3-aminoamides and their alpha-lipoic acid adducts. *Eur. J. Med. Chem.* **2009**, *44*, 3020–3026. [CrossRef]
105. Abdellatif, K.R.A.; Chowdhury, M.A.; Dong, Y.; Das, D.; Yu, G.; Velázquez, C.A.; Suresh, M.R.; Knaus, E.E. Dinitroglyceryl and diazen-1-ium-1,2-diolated nitric oxide donor ester prodrugs of aspirin, indomethacin and ibuprofen: Synthesis, biological evaluation and nitric oxide release studies. *Bioorg. Med. Chem. Lett.* **2009**, *19*, 3014–3018. [CrossRef] [PubMed]
106. Chowdhury, M.A.; Abdellatif, K.R.A.; Dong, Y.; Das, D.; Suresh, M.R.; Knaus, E.E. Synthesis of celecoxib analogues possessing a N-difluoromethyl-1,2- dihydropyrid-2-one 5-lipoxygenase pharmacophore: Biological evaluation as dual inhibitors of cyclooxygenases and 5-lipoxygenase with anti-inflammatory activity. *J. Med. Chem.* **2009**, *52*, 1525–1529. [CrossRef] [PubMed]

107. Al-Ostoot, F.H.; Zabiulla; Salah, S.; Khanum, S.A. Recent investigations into synthesis and pharmacological activities of phenoxy acetamide and its derivatives (chalcone, indole and quinoline) as possible therapeutic candidates. *J. Iran. Chem. Soc.* **2021**. [CrossRef]
108. Jacob, Í.T.T.; Gomes, F.O.S.; de Miranda, M.D.S.; de Almeida, S.M.V.; da Cruz-Filho, I.J.; Peixoto, C.A.; da Silva, T.G.; Moreira, D.R.M.; de Melo, C.M.L.; de Oliveira, J.F.; et al. Anti-inflammatory activity of novel thiosemicarbazone compounds indole-based as COX inhibitors. *Pharmacol. Rep.* **2021**, *73*, 907–925. [CrossRef]
109. Qian, S.; Huang, Y.; Li, J.; Zhang, Y.; Zhang, B.; Jin, F. Synthesis and Anti-proliferative Activity of Indole-2-amide Derivatives as Cyclooxygenase-2/5-lipoxygenase (COX-2/5-LOX) Dual Inhibitors. *Chin. J. Org. Chem.* **2021**, *41*, 1631–1638. [CrossRef]
110. Ghanim, A.M.; Rezq, S.; Ibrahim, T.S.; Romero, D.G.; Kothayer, H. Novel 1,2,4-triazine-quinoline hybrids: The privileged scaffolds as potent multi-target inhibitors of LPS-induced inflammatory response via dual COX-2 and 15-LOX inhibition. *Eur. J. Med. Chem.* **2021**, *219*, 113457. [CrossRef]
111. Maghraby, M.T.E.; Abou-Ghadir, O.M.F.; Abdel-Moty, S.G.; Ali, A.Y.; Salem, O.I.A. Novel class of benzimidazole-thiazole hybrids: The privileged scaffolds of potent anti-inflammatory activity with dual inhibition of cyclooxygenase and 15-lipoxygenase enzymes. *Bioorg. Med. Chem.* **2020**, *28*, 115403. [CrossRef]
112. Berrino, E.; Carradori, S.; Angeli, A.; Carta, F.; Supuran, C.T.; Guglielmi, P.; Coletti, C.; Paciotti, R.; Schweikl, H.; Maestrelli, F.; et al. Dual carbonic anhydrase ix/xii inhibitors and carbon monoxide releasing molecules modulate LPS-mediated inflammation in mouse macrophages. *Antioxidants* **2021**, *10*, 56. [CrossRef]
113. Akgul, O.; Di Cesare Mannelli, L.; Vullo, D.; Angeli, A.; Ghelardini, C.; Bartolucci, G.; Alfawaz Altamimi, A.S.; Scozzafava, A.; Supuran, C.T.; Carta, F. Discovery of Novel Nonsteroidal Anti-Inflammatory Drugs and Carbonic Anhydrase Inhibitors Hybrids (NSAIDs-CAIs) for the Management of Rheumatoid Arthritis. *J. Med. Chem.* **2018**, *61*, 4961–4977. [CrossRef]
114. Bua, S.; Di Cesare Mannelli, L.; Vullo, D.; Ghelardini, C.; Bartolucci, G.; Scozzafava, A.; Supuran, C.T.; Carta, F. Design and Synthesis of Novel Nonsteroidal Anti-Inflammatory Drugs and Carbonic Anhydrase Inhibitors Hybrids (NSAIDs-CAIs) for the Treatment of Rheumatoid Arthritis. *J. Med. Chem.* **2017**, *60*, 1159–1170. [CrossRef]
115. Myasoedova, E.; Crowson, C.S.; Kremers, H.M.; Therneau, T.M.; Gabriel, S.E. Is the incidence of rheumatoid arthritis rising? Results from Olmsted County, Minnesota, 1955-2007. *Arthritis Rheum.* **2010**, *62*, 1576–1582. [CrossRef]
116. Placha, D.; Jampilek, J. Chronic inflammatory diseases, anti-inflammatory agents and their delivery nanosystems. *Pharmaceutics* **2021**, *13*, 64. [CrossRef] [PubMed]
117. Abdel-Aziz, A.A.M.; Angeli, A.; El-Azab, A.S.; Hammouda, M.E.A.; El-Sherbeny, M.A.; Supuran, C.T. Synthesis and anti-inflammatory activity of sulfonamides and carboxylates incorporating trimellitimides: Dual cyclooxygenase/carbonic anhydrase inhibitory actions. *Bioorg. Chem.* **2019**, *84*, 260–268. [CrossRef]
118. Vučković, S.; Savić Vujović, K.; Srebro, D.; Medić, B.; Ilic-Mostic, T. Prevention of Renal Complications Induced by Non- Steroidal Anti-Inflammatory Drugs. *Curr. Med. Chem.* **2016**, *23*, 1953–1964.
119. Vučković, S.; Srebro, D.; Vujović, K.S.; Vučetić, Č.; Prostran, M. Cannabinoids and pain: New insights from old molecules. *Front. Pharmacol.* **2018**, *9*, 1259. [CrossRef]
120. Baron, E.P. Medicinal Properties of Cannabinoids, Terpenes, and Flavonoids in Cannabis, and Benefits in Migraine, Headache, and Pain: An Update on Current Evidence and Cannabis Science. *Headache* **2018**, *58*, 1139–1186. [CrossRef] [PubMed]
121. Meng, H.; Johnston, B.; Englesakis, M.; Moulin, D.E.; Bhatia, A. Selective Cannabinoids for Chronic Neuropathic Pain: A Systematic Review and Meta-Analysis. *Anesth. Analg.* **2017**, *125*, 1638–1652. [CrossRef]
122. Khurana, L.; Mackie, K.; Piomelli, D.; Kendall, D.A. Modulation of CB1 cannabinoid receptor by allosteric ligands: Pharmacology and therapeutic opportunities. *Neuropharmacology* **2017**, *124*, 3–12. [CrossRef]
123. Slivicki, R.A.; Xu, Z.; Kulkarni, P.M.; Pertwee, R.G.; Mackie, K.; Thakur, G.A.; Hohmann, A.G. Positive Allosteric Modulation of Cannabinoid Receptor Type 1 Suppresses Pathological Pain Without Producing Tolerance or Dependence. *Biol. Psychiatry* **2018**, *84*, 722–733. [CrossRef]
124. Starowicz, K.; Finn, D.P. Cannabinoids and Pain: Sites and Mechanisms of Action. In *Advances in Pharmacology*; Academic Press Inc.: Cambridge, MA, USA, 2017; Volume 80, pp. 437–475.
125. Bai, R.; Yao, C.; Zhong, Z.; Ge, J.; Bai, Z.; Ye, X.; Xie, T.; Xie, Y. Discovery of natural anti-inflammatory alkaloids: Potential leads for the drug discovery for the treatment of inflammation. *Eur. J. Med. Chem.* **2021**, *213*, 113165. [CrossRef]
126. Andre, C.M.; Hausman, J.F.; Guerriero, G. Cannabis sativa: The plant of the thousand and one molecules. *Front. Plant Sci.* **2016**, *7*, 1–17. [CrossRef]
127. Friedman, D.; French, J.A.; Maccarrone, M. Safety, efficacy, and mechanisms of action of cannabinoids in neurological disorders. *Lancet Neurol.* **2019**, *18*, 504–512. [CrossRef]
128. Rahn, E.J.; Hohmann, A.G. Cannabinoids as Pharmacotherapies for Neuropathic Pain: From the Bench to the Bedside. *Neurotherapeutics* **2009**, *6*, 713–737. [CrossRef]
129. Hill, K.P.; Palastro, M.D.; Johnson, B.; Ditre, J.W. Cannabis and Pain: A Clinical Review. *Cannabis Cannabinoid Res.* **2017**, *2*, 96–104. [CrossRef] [PubMed]
130. O'Sullivan, S.E. An update on PPAR activation by cannabinoids. *Br. J. Pharmacol.* **2016**, *173*, 1899–1910. [CrossRef] [PubMed]
131. Morales, P.; Hurst, D.P.; Reggio, P.H. Molecular Targets of the Phytocannabinoids: A Complex Picture. *Prog. Chem. Org. Nat. Prod.* **2017**, *103*, 103–131.

132. Bakas, T.; van Nieuwenhuijzen, P.S.; Devenish, S.O.; McGregor, I.S.; Arnold, J.C.; Chebib, M. The direct actions of cannabidiol and 2-arachidonoyl glycerol at GABAA receptors. *Pharmacol. Res.* **2017**, *119*, 358–370. [CrossRef] [PubMed]
133. ElSohly, M.A.; Radwan, M.M.; Gul, W.; Chandra, S.; Galal, A. Phytochemistry of *Cannabis sativa* L. *Prog. Chem. Org. Nat. Prod.* **2017**, *103*, 1–36.
134. Romero-Sandoval, E.A.; Kolano, A.L.; Alvarado-Vázquez, P.A. Cannabis and Cannabinoids for Chronic Pain. *Curr. Rheumatol. Rep.* **2017**, *19*, 67. [CrossRef]
135. Laprairie, R.B.; Bagher, A.M.; Kelly, M.E.M.; Denovan-Wright, E.M. Cannabidiol is a negative allosteric modulator of the cannabinoid CB1 receptor. *Br. J. Pharmacol.* **2015**, *172*, 4790–4805. [CrossRef] [PubMed]
136. Stockings, E.; Campbell, G.; Hall, W.D.; Nielsen, S.; Zagic, D.; Rahman, R.; Murnion, B.; Farrell, M.; Weier, M.; Degenhardt, L. Cannabis and cannabinoids for the treatment of people with chronic noncancer pain conditions: A systematic review and meta-analysis of controlled and observational studies. *Pain* **2018**, *159*, 1932–1954. [CrossRef]
137. Stott, C.G.; White, L.; Wright, S.; Wilbraham, D.; Guy, G.W. A phase i study to assess the effect of food on the single dose bioavailability of the THC/CBD oromucosal spray. *Eur. J. Clin. Pharmacol.* **2013**, *69*, 825–834. [CrossRef] [PubMed]
138. Abrams, D.I.; Guzman, M. Cannabis in Cancer Care. *Clin. Pharmacol. Ther.* **2015**, *97*, 575–586. [CrossRef] [PubMed]
139. Kiran Vemuri, V.; Makriyannis, A. Medicinal Chemistry of Cannabinoids. *Clin. Pharmacol. Ther.* **2015**, *97*, 553–558. [CrossRef] [PubMed]
140. Lynch, M.E.; Ware, M.A. Cannabinoids for the Treatment of Chronic Non-Cancer Pain: An Updated Systematic Review of Randomized Controlled Trials. *J. Neuroimmune Pharmacol.* **2015**, *10*, 293–301. [CrossRef] [PubMed]
141. Field, M.J.; Cox, P.J.; Stott, E.; Melrose, H.; Offord, J.; Su, T.Z.; Bramwell, S.; Corradini, L.; England, S.; Winks, J.; et al. Identification of the α2-δ-1 subunit of voltage-calcium calcium channels as a molecular target for pain mediating the analgesic actions of pregabalin. *Proc. Natl. Acad. Sci. USA* **2006**, *103*, 17537–17542. [CrossRef]
142. Moore, R.A.; Straube, S.; Wiffen, P.J.; Derry, S.; McQuay, H.J.; Moore, M. Pregabalin for acute and chronic pain in adults. *Cochrane Database Syst. Rev.* **2009**. [CrossRef]
143. Bouhassira, D.; Lantéri-Minet, M.; Attal, N.; Laurent, B.; Touboul, C. Prevalence of chronic pain with neuropathic characteristics in the general population. *Pain* **2008**, *136*, 380–387. [CrossRef]
144. Dworkin, R.H.; Kirkpatrick, P. Pregabalin. *Nat. Rev. Drug Discov.* **2005**, *4*, 455–456. [CrossRef] [PubMed]
145. Gajraj, N.M. Pregabalin: Its pharmacology and use in pain management. *Anesth. Analg.* **2007**, *105*, 1805–1815. [CrossRef] [PubMed]
146. Taylor, C.P. Mechanisms of analgesia by gabapentin and pregabalin—Calcium channel α2-δ [Cavα2-δ] ligands. *Pain* **2009**, *142*, 13–16. [CrossRef] [PubMed]
147. Kavoussi, R. Pregabalin: From molecule to medicine. *Eur. Neuropsychopharmacol.* **2006**, *16*, S128–S133. [CrossRef] [PubMed]
148. Sills, G.J. The mechanisms of action of gabapentin and pregabalin. *Curr. Opin. Pharmacol.* **2006**, *6*, 108–113. [CrossRef] [PubMed]
149. Micó, J.-A.; Prieto, R. Elucidating the Mechanism of Action of Pregabalin. *CNS Drugs* **2012**, *26*, 637–648. [CrossRef] [PubMed]
150. Ben-Menachem, E. Pregabalin pharmacology and its relevance to clinical practice. *Epilepsia* **2004**, *45*, 13–18. [CrossRef]
151. Armstrong, C.T.; Mason, P.E.; Anderson, J.L.R.; Dempsey, C.E. Arginine side chain interactions and the role of arginine as a gating charge carrier in voltage sensitive ion channels. *Sci. Rep.* **2016**, *6*, 21759. [CrossRef]
152. Taylor, C.P.; Angelotti, T.; Fauman, E. Pharmacology and mechanism of action of pregabalin: The calcium channel α2-δ (alpha2-delta) subunit as a target for antiepileptic drug discovery. *Epilepsy Res.* **2007**, *73*, 137–150. [CrossRef]

Perspective

Drug Design: Where We Are and Future Prospects

Giuseppe Zagotto [1,*] and Marco Bortoli [2]

[1] Department of Pharmaceutical Sciences, University of Padova, Via Marzolo 5, 35131 Padova, Italy
[2] Institute of Computational Chemistry and Catalysis (IQCC) and Department of Chemistry, Faculty of Sciences, University of Girona, C/M. A. Capmany 69, 17003 Girona, Spain; marco.bortoli@udg.edu
* Correspondence: giuseppe.zagotto@unipd.it

Abstract: Medicinal chemistry is facing new challenges in approaching precision medicine. Several powerful new tools or improvements of already used tools are now available to medicinal chemists to help in the process of drug discovery, from a hit molecule to a clinically used drug. Among the new tools, the possibility of considering folding intermediates or the catalytic process of a protein as a target for discovering new hits has emerged. In addition, machine learning is a new valuable approach helping medicinal chemists to discover new hits. Other abilities, ranging from the better understanding of the time evolution of biochemical processes to the comprehension of the biological meaning of the data originated from genetic analyses, are on their way to progress further in the drug discovery field toward improved patient care. In this sense, the new approaches to the delivery of drugs targeted to the central nervous system, together with the advancements in understanding the metabolic pathways for a growing number of drugs and relating them to the genetic characteristics of patients, constitute important progress in the field.

Keywords: drug discovery; precision medicine; pharmacodynamics; pharmacokinetics

1. Introduction

The fascinating path of drug discovery shares many features with a very complex and multidimensional maze, in which the medicinal chemist starts from the chemical space, more than 10^{60} small drug-like molecules of which only about 10^8 of have been synthesized so far [1], and has to find the way to the drug at the center of the maze. A further intriguing property of this maze is that the walls and the center keep moving with time. Obviously, the maze is the body, with its barriers and transporters, and the center is the target site for a drug, which we call a receptor in its extensive sense. Very often, for sake of simplicity, the properties of a drug are grouped in pharmacodynamics and pharmacokinetics, but it must be remembered that a drug is a single entity, comprising both groups. The pharmacodynamic and pharmacokinetic properties stem from the chemistry of the drug molecule; the chemistry of the body, which is also made up by molecules; and the chemistry of the water that is interacting with both. All these molecules are not fixed in place, but continuously move. Particularly important are the movements happening when a molecule interacts with the target or off-target sites, leading to a biological effect. The understanding of body–drug interactions is a very complex problem where the properties of the molecules, which the medicinal chemist can know either from fundamental chemistry rules or from empirical observations of complex macromolecules and the biology of living organisms, can help to find the right way to the drug. A substantial simplification that is often made by medicinal chemists is to forget about the time course: in other words, to consider both the small molecules and the macromolecules as fixed. In this short perspective on the drug discovery process, a static situation is considered first, then some considerations on how to possibly account for the chemical motions are made. Moreover, the discussion is limited mainly to chemical entities complying with Lipinski's rule of five (Ro5), even if there are new trends looking at chemical entities beyond the Ro5 (bRo5) [2]. This enlargement of the chemical space from Ro5 to bRo5 molecules may reflect the enlargement of the target space, up to now often limited at the inhibition of enzymatic action antagonizing the action of signal transmission in the body, to the modulation of macromolecule behavior as it happens in protein–protein or protein–nucleic acid interaction. This enlargement of the target space also requires new tools to manage the absorption, distribution, metabolism, and excretion (ADME) and toxicity of the new active pharmaceutical ingredients. In this short perspective, the attention is focused only on active pharmaceutical ingredients and not on excipients, drug delivery, or dosing regimens. Furthermore, pathways leading to "me-too drugs", even if very important for the pharmaceutical industry, are not considered.

2. The Lead Discovery

2.1. Target Selection and Validation: Possible Expansion of Chemical Space

When looking for a new drug, the first choice to be made, almost always, is the selection of the target disease [3]. This decision is based mostly on economic considerations but can also be made following the discovery of a new biochemical pathway leading to a pathological state, or the finding of an important biomarker. Sometimes, it is the observation of the properties of a substance or metabolite that sparks the interest to better investigate a topic. Recently, many pharmaceutical companies have been searching in universities or in contract research organizations (CROs) [4,5] for new biochemical pathways or new chemical entities to treat diseases. The selection of the target disease is a basic choice, since the overall failure rate in drug discovery is very high, over 96%, including a 90% failure rate during clinical development, and costs are massive [6]. Once the target disease is defined, the next step toward the discovery of a new drug is the selection and validation of the biological target: a protein [7], a nucleic acid [8], or a different biochemical structure critical to the development of the disease, that can be characterized and is druggable [2]. During the last thirty years, a tremendous effort has been devoted to the selection of the correct target, with the widespread use of genome-wide association studies [9]. Even if this approach is increasingly showing problems [10], it remains a fundamental investigation tool for target selection, and in the last several years, since the chemical space available has expanded with the massive introduction of monoclonal antibodies, there are many new opportunities, such as the challenging task of the delivery of monoclonal antibodies to the central nervous system.

2.2. From Hit to Lead: Structure-Guided Drug Design and Beyond

Once the target is selected and validated, there are many possible pathways leading to a chemical entity able to bind the target. The selection of chemical structures able to bind the selected target is usually conducted by screening large libraries of molecules, available in-house or from outsourcing, against the selected biological target [5]. The molecule collection may be made up of virtual or real chemical entities. Real chemical entities, until the second half of the last century, were obtained by a separate synthesis of every single molecule or by isolating them from natural sources. After Merrifield's discovery of solid phase synthesis (Figure 1) [11], the development of computers able to manipulate large datasets, and the possibility of performing biochemical and biological experiments using very small amounts of compounds, combinatorial chemistry was introduced to help and speed up the drug discovery and optimization process [12].

Combinatorial chemistry started in the mid-1980s with the synthesis of hundreds of thousands of peptides on solid supports in parallel or with the split-and-mix methods (Figure 2). Lam et al. [13] introduced the one-bead, one-compound combinatorial peptide libraries, which first allowed only one bond to be formed: the peptidic amide bond. The following step in the evolution of solid phase synthesis was to make the phosphoroester bond connecting the nucleotides using mainly phosphoramidite chemistry [14,15]. In 1992, Bunin and Ellman reported the first example of a small-molecule combinatorial library [16] and started the era of the synthesis of libraries of small drug-like molecules satisfying the Ro5.

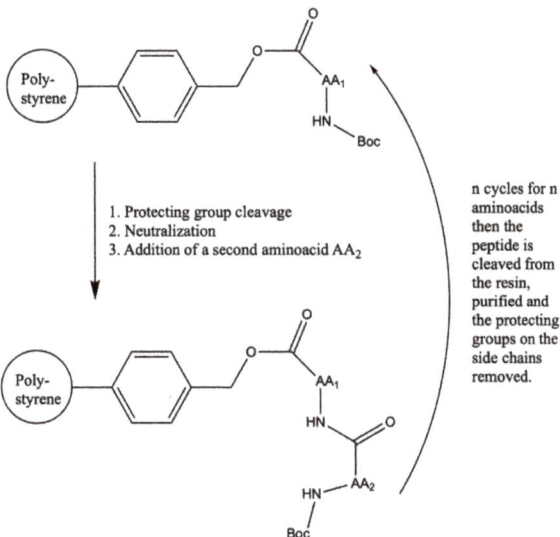

Figure 1. Merrifield peptide synthesis on a polystyrene support.

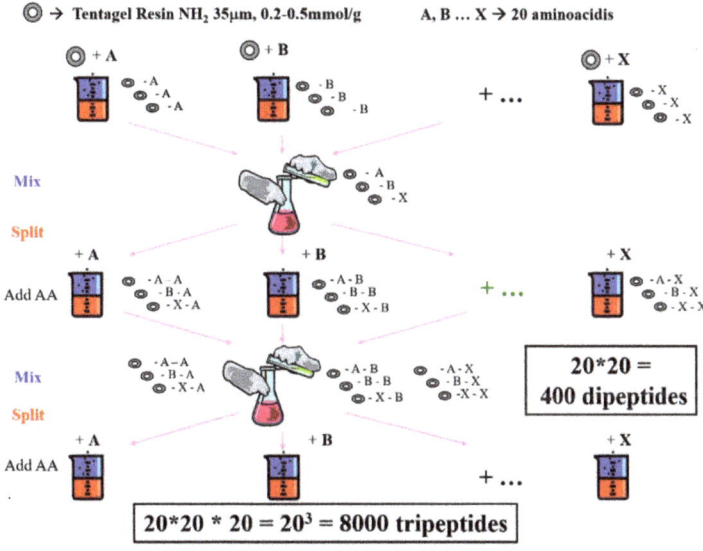

Figure 2. Schematic illustration of split–and–mix combinatorial synthesis.

The combinatorial synthesis of drug-like compounds was strongly pushed on by the pharmaceutical industry around the turn of the century, and so were the analytical tools, with respect to the chemical characterization of very small quantities of substance, the decoding of chemical libraries, and the biochemical assays. The high-throughput era had begun, and someone referred to this as an accelerated evolution in the search of new active compounds: instead of using millions of years to select molecules able to protect life against predators or perform precise biochemical tasks, scientists could obtain the same result in a period that was very short in comparison. At the end of the last century, libraries of oligopeptides, oligonucleotides, and drug-like small molecules [17] were routinely prepared on automated synthesizers, providing pure substances in a rapid and efficient manner. However, the solid phase synthesis of oligosaccharide libraries was not yet feasible due to either the presence of functional groups of similar reactivity on each saccharide monomer, or to the fact that a new stereogenic center is created each time a glycosidic bond is formed [18]. Another consideration that slowed the development of the solid phase synthesis of oligosaccharide libraries was that while proteins and nucleic acids are genetically encoded structures, polysaccharides are not. This made polysaccharides ill-defined and unappealing targets for many investigators and pharmaceutical managers, as carbohydrates are considered particularly important only in few signal transduction pathways [19] and vaccines [20].

The availability of a very large number of compounds from combinatorial synthesis, in-house libraries, robotics, high-throughput screening methods, and fast structure determination constitutes a great help in the drug discovery process. Moreover, computers and software able to store, organize, and manage a huge, and continuously growing, amount of data are available to the pharmaceutical field. Despite this, we need something else to improve and speed up the pharmacodynamics in drug discovery when a validated target is established. No recipe is available for this, but taking into consideration the time evolution of chemical processes, instead of the static snapshots of the target structure as determined by X-ray crystallography, NMR, or cryo-electron microscopy, can help the medicinal chemist. To better understand the target structure behavior while it is performing its biological task, we must extrapolate the time course from many structure determinations, often crystallographic or NMR. The determination of the time course for a biochemical process, which is fascinating, although very challenging, will allow us to understand how the signal is managed by the validated target structures such as proteins, nucleic acids, or other biochemical players. More accurate structural data and improved chemistry software will allow a better look at the structure and its changes with time, environment, and regulator molecules. A recent example that explicitly considers the time evolution of a target molecule is the PPI-FIT method, which involves the targeting of intermediates along the path of protein folding (Figure 3) [21]. These structures are regarded as the druggable targets because they present binding pockets not present in the protein's final structure. The drug-intermediate interaction should stabilize the complex, thus preventing the protein from reaching its native conformation. The method employs computer simulations together with experimental techniques, and supports the idea that folding intermediate targeting could represent a useful way to regulate protein levels. Regarding the crystallographic support to the drug-discovery process, it was recently reported that a detailed understanding of the interactions between drugs and their targets is crucial to developing the best possible therapeutic agents, and that structure-based drug design still relies on the availability of high-resolution structures obtained primarily through X-ray crystallography [22]. Working on a single crystal is marginally useful to understanding the enzyme movements during the catalytic process and to plan possible molecular structures interacting or interfering with different conformational states of the enzyme. At the moment, it is possible to combine different crystal snapshots to have an idea of the enzyme conformational changes during the catalytic process, as it was performed for the ubiquitous enzymes α-D-phosphohexomutases [23].

To characterize the various enzyme conformations involved in the isomerization of 1-phospho to 6-phosphohexoses, 15 high-resolution crystal structures of the phosphoglucomutase enzyme while performing the isomerization of glucose 1-phosphate to glucose 6-phosphate were obtained. Glucose 1,6-bisphosphate undergoes a 180° reorientation between the two phosphoryl transfer steps of the reaction. The enzyme with the phosphoserine bound to a Mg^{2+} ion has the same conformation at the beginning of the catalytic process, when it is bound to the substrate glucose 1-phosphate, and at the end of it, when it is bound to the product glucose 6-phosphate. During the reorientation of the sugar, when the catalytic serine is in the dephosphorylated state and bound to the glucose 1,6-bisphosphate intermediate, the enzyme has a different structure. In the future, the structure of such intermediates of the enzymes may suggest new drug molecules eventually able to trap, in these intermediate conformations, even the enzymes that are currently not druggable. It is also possible to use an in silico methodology combining a classic and quantum mechanics approach [24] to better understand the catalytic path, as is performed on the selenoenzyme glutathione peroxidase in the reduction reaction of hydrogen peroxides and organic hydroperoxides by glutathione. NMR [25] and EPR [26] measurements can also feed data to molecular in silico calculations to determine the evolution of a protein with time, although limited to the active site or oligonucleotide structures. At the moment, dealing with the changes of structures with time in protein–protein interaction, as in GPCR receptors and the intracellular effector proteins or in a protein regulator [27], or with protein-oligonucleotides binding, as in transcription regulators, is much more complicated, but very appealing [28].

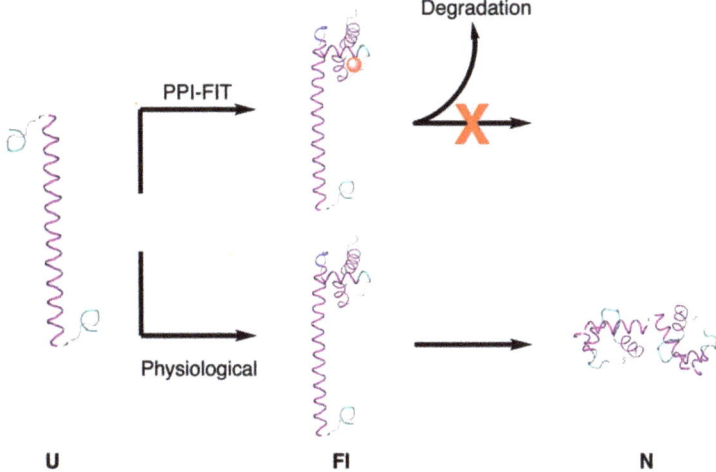

Figure 3. Schematic representation of the PPI-FIT approach to protein regulation. U = unfolded; FI = folding intermediate; N = native. The red sphere represents the drug molecule.

2.3. Speeding up Screening and Design: Artificial Intelligence in Drug Discovery

In addition, artificial intelligence (AI) is finding its way in helping the process of speeding up drug discovery [29] with the design of improved experiments and more sophisticated machine learning (ML) algorithms to better understand the behavior of the target structure when performing its biological task. The increasing volume of available data has given a strong impulse to computer-aided drug design, with the latest developments focused on the applications of deep learning (DL) [30,31]. These methods take advantage of the already known concept of artificial neural networks and, due to the augmented performance of calculators, increase their complexity, reaching a much improved performance compared to other ML algorithms [32–34]. Moreover, their application reaches to not only the molecular discovery process of drug design (as in structure-activity predictions [35] or de novo design [36]), but also the synthetic (or retrosynthetic) route [37,38] and formulation design [39–41], and takes steps to also encompass fields that, while still pertaining to the drug discovery process, lie outside of wet laboratory activity, such as product quality assurance, marketing, and clinical trial management [29,30]. Other recent studies that benchmarked DL against other machine learning algorithms for properties prediction, using large biomolecular datasets comprising hundreds of thousands of compounds, consistently showed that deep neural networks are the best performing approach [42,43]. In addition to properties prediction or screening, DL has been employed in de novo design. As an example, a particular neural network was designed with the aim of transforming a set of molecular structures of known properties into a continuous representation of a molecular structure that could be exploited to maximize a desired property, and then reversibly transformed into an optimized molecular structure expressing such desired property [44]. With this machinery, novel structures were proposed that showed potential specific anticancer properties [45] and a predicted activity against dopamine receptor type 2 [46]. Analogous approaches employing the power of DL have been used to develop tools for the design of a molecule that can adapt best to a given 3D protein pocket [47] or that can display a particular desired property [48]. Moreover, DL has been integrated with more traditional computer techniques to decrease the computational cost without losing their predictive power. For instance, a DL-driven quantum mechanical approach was employed to efficiently calculate electronic wavefunctions of possible drug candidates [49], and the application of a neural network trained on MD simulations showed that the calculation of free energies of transfer of 1500 small molecules is possible with small errors [50]. Finally, DL techniques can also complement the drug discovery process, shedding light on the fundamental interactions that take place in the human body at a molecular level and on their disruption at the onset of disease. On the other hand, the lack of a large amount of high-quality data, required to train the algorithms successfully, is one of the main drawbacks of these methods. For example, the atomistic structure of many proteins, which is essential to understand their mechanism of action, is still not known. Again, DL has proven to be effective in these areas, as demonstrated by the successful development of the AlphaFold method [51] and its extension, ColabFold [52], two of the most promising structure prediction algorithms that, starting from an amino acid sequence, can predict the 3D folded structure of a protein with an accuracy competing with experimental structures [51]. Another feature that renders the obtained data sometimes hard to interpret but, more importantly, provides no insight into the underlying biochemical mechanism, is the fact that DL algorithms operate as a black box [35,53]. Nevertheless, the clear knowledge of the molecular cause of a pathological condition combined with the ability to obtain through AI-driven methods an effective and efficient compound without severe side effects in a very short time can impart a strong impulse to successful drug development. Moreover, as these techniques continue to develop, treatment possibilities increase, opening new possible choices to fight pathological conditions. Again, DL has proven useful in aiding the fine tailoring of the best treatment choice based on the analysis of patient data such as life history, previous diagnostics, and manifested symptoms [53].

2.4. One Size Does Not Fit All: From General to Precision Medicine

The availability of large collections of molecules, the development of a large number of microscale analytical tools, the genome-wide association studies, and the simple and fast methods for the detection of target genes having a single-nucleotide polymorphisms took modern medicinal chemistry to the precision medicine era. The early steps of precision medicine were taken in the oncology field. The personalized therapies of the anticancer drugs, along with the identification of tumor-specific targets, were in part due to the general cytotoxicity and, as a consequence, the severe side effects of existing one-size-fits-all cancer drugs [9]. Examples are the molecularly targeted cancer therapies, such as small-molecule kinase inhibitors blocking the incorrect signaling of tumor cells from the intracellular side of a growth factor receptor protein, and monoclonal antibodies that often stop the same signal from outside the cell membrane. An early application of this was the epidermal growth factor receptor (EGFR), abnormally activated in cancer, against which the two classes of anti-EGFR agents, monoclonal antibodies and low-molecular-weight tyrosine kinase inhibitors, showed antitumor activity in patients. It was also reported that the kinase inhibitor gefitinib (Figure 4) and the monoclonal antibody cetuximab share complementary mechanisms of action on EGFR and that a combined EGFR targeting is a clinically exploitable strategy [54].

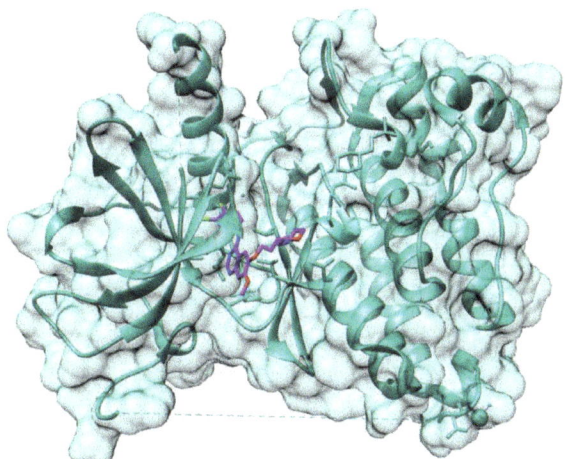

Figure 4. Gefitinib in complex with EGFR (PDB ID: 4WKQ; the image was obtained using UCSF Chimera, San Francisco, CA, USA).

Many dysregulated pathways are now characterized, and new targets, proteins, and polynucleotides are attracting medicinal chemists. Among the new targets are not only the classical receptors, but many enzymes that can be inhibited by binding the small molecule to them, as in the case of the BCL-2 inhibitor venetoclax currently on the market [55], or by hitting a regulator protein [56]. To better understand the mechanism of action of drugs and to progress in the field of pharmacodynamics and precision medicine, we need to know the different conformations that the targets, proteins, or polynucleotides, assume in their energy minima during their functioning within the natural environment.

3. Pharmacokinetics

Pharmacokinetics, i.e., what is happening in the body to the drug molecule before and after the interaction with the target, is often divided in absorption, distribution, metabolism, and excretion (ADME). Many factors can influence the individual response to pharmaceutical compounds, among which genomic differences, gut microbiome, sex, nutrition, age, stress, and health status are included. They can impact drug absorption and distribution, the metabolic profile, with the drug–drug and drug–food interactions, and the toxicity in an individual. As for molecular design, computer simulations based on artificial intelligence help with the recognition of toxicity of the administered drug candidate. For example, an algorithm based on DL correctly predicted the toxicity of drug compounds in the data set with an accuracy of over 80% in almost all instances and was the Tox21 Data Challenge winner [33]; a similar approach was employed to study the possible epoxidation sites of drug candidates, obtaining a detailed picture on the likeliness of a molecule to be epoxidized and its consequent toxicity due to the structural modification [57]. On the experimental side, the advances made in gene sequencing, mainly using next-generation sequencing technologies [58] for pharmacogenomic studies and in the chemical analysis of metabolites, in particular by HPLC-MS, allows the better characterization the individuals and move toward what is commonly defined as precision medicine, not only as far as the target selection in a pathological state, but also for the pharmacokinetic effects. Precision medicine, which is defined as the capacity to prescribe the most effective treatment with the fewest adverse effects to a patient [59], applies principally to medical diagnostic, prescribing, and prevention [60], and is progressing very fast. A main problem to be resolved for precision medicine is the development of effective therapies targeted at the central nervous system (CNS). This is due to the failure to achieve therapeutically relevant concentrations in the CNS, due to the presence of the blood–brain barrier and to the strong neuronal interconnection between the different brain regions, as in the case of the dopaminergic effects of morphine and its derivatives targeted to the opioid receptors. A very important and challenging therapeutic area is that of brain tumors. Some of the approaches explored to address this challenge include blood–brain barrier disruption and drug modifications to enhance CNS permeability; unfortunately, neither approach has proven successful. Another approach is to deliver therapeutics locoregionally, directly into the tumor mass and the surrounding tumor-infiltrated brain parenchyma. The most widely used method for direct brain delivery is convection-enhanced delivery (CED), whereby specially designed catheters are introduced into target tissue, and the infusate is delivered slowly over a prolonged period of time. CED enables the delivery of conventional, nano-, bio-, gene, and even cellular therapies [61–73]. Hopefully, in the near future, it will be possible to deliver more small molecules in a therapeutic useful concentration and antibodies to the CNS.

Progress in precision medicine in the pharmacokinetic field is also increasing due to the improved use of experimental data on metabolic reactions and to the fact that the collection of DNA samples from clinical trial participants to perform pharmacogenomic studies has become standard practice for most pharmaceutical companies [74]. The analysis of single-nucleotide polymorphisms (SNPs) is rapidly growing, in particular for genetic variants that alter the activity of drug metabolizing enzymes and drug transporters.

As far as the experimental data is concerned, the massive work performed to gain information on metabolic pathways and the relative metabolizing enzymes of clinically used drugs to better understand their therapeutic effect is central to understanding the therapeutic drug properties, as well as the drug–drug and drug–food interactions. The metabolism of opioids, also considering their low clinical dosage, always attracted the attention of many investigators [75]. The developments in the pharmacokinetics of opioids is considered as a case study to briefly show the role of metabolism as a predictor of the clinical response and side effects of opioid analgesics, keeping the opioid crisis in mind [76]. The important side effects are due to the neuronal connectivity between the reward, dopaminergic, and opioid regions, as well as to the respiratory depression in the CNS, while many other side effects, e.g., constipation, are derived from the interaction with the peripheral opioid receptors. The common metabolic phase I reactions of opioids are dealkylations, O-dealkylation being CYP2D6-mediated, while N-dealkylation is CYP3A4-mediated, and redox reactions (e.g., for oxycodone and methadone); for phase II, glucuronation at positions three and six of the morphine nucleus and on reduced keto groups or dealkylated ethers, is the most important (Figure 5).

CYP2D6-mediated O-dealkylation of morphine 3-methoxy derivatives, such as codeine, and tramadol, are required to generate the phenolic OH group important for binding to a histidine of the opioid receptor. CYP2D6 is highly polymorphic, and the expression of different variants results in several phenotypes. The implementation of pharmacogenetics-based codeine prescribing that accounts for the CYP2D6 metabolizer status was described in a recent work [77] and is an example of precision medicine. Genome-wide association studies and candidate gene findings suggest that genetic approaches may help in choosing the most appropriate opioid and its dosage, while preventing adverse drug reactions [78].

Beyond the experimental data on metabolic enzymes and transporters, it is also possible to examine the genetic variants that alter the activity of enzymes or transporters and to use this information in ADME and toxicity studies [74,79]. Pharmacogenomic studies provide a growing list of clinically relevant markers that could be used to improve patient care [80], but such information is still not widely used in clinical practice. The difficulty of translating the pharmacogenomic information into ADME and toxicity studies during clinical phases was examined [81], but the basic reason is that the drug response is often highly complex, resulting from the interaction of many influencing factors. In the future, this approach will be a very useful tool for helping in the drug discovery process and in personalized medicine.

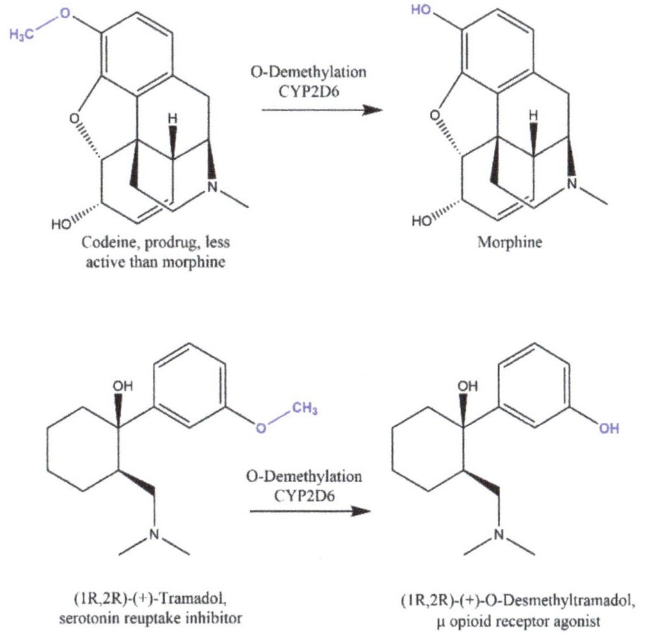

Figure 5. Examples of metabolic pathways for which CYP2D6 polymorphism is important.

4. Conclusions

The maze of the drug discovery process is still very complex and challenging, even when only considering the small-molecule approach and no other promising approaches, such as those involving monoclonal antibodies or polynucleotides. Precision medicine, from drug discovery to the bedside, is the main concern nowadays. New powerful tools are made available almost every day, but medicinal chemists are still looking in every direction, from natural products [82] to sophisticated modeling [83], in search of new drug candidates complying with the new targets emerging from precision medicine needs. To further progress in the medicinal chemistry field, we need, in addition to new targets, a more accurate description of their different conformations and possibly of the evolution of the target structure with time during the biological process. This, combined with the knowledge of the genetic variants of the targets, will lead to an increased number and precision of the "magic bullets" that are drugs, and allow the progress of precision medicine.

Author Contributions: Conceptualization, G.Z.; writing–original draft preparation: G.Z., M.B.; writing–review and editing, G.Z. and M.B. All authors have read and agreed to the published version of the manuscript.

Funding: This research received no external funding.

Institutional Review Board Statement: Not applicable.

Informed Consent Statement: Not applicable.

Data Availability Statement: Not applicable.

Conflicts of Interest: The authors declare no conflict of interest.

References

1. Druchok, M.; Yarish, D.; Gurbych, O.; Maksymenko, M. Toward efficient generation, correction, and properties control of unique drug-like structures. *J. Comput. Chem.* **2021**, *42*, 746–760. [CrossRef]
2. Yang, W.; Gadgil, P.; Krishnamurthy, V.R.; Landis, M.; Mallick, P.; Patel, D.; Patel, P.J.; Reid, D.L.; Sanchez-Felix, M. The Evolving Druggability and Developability Space: Chemically Modified New Modalities and Emerging Small Molecules. *AAPS J.* **2020**, *22*, 21. [CrossRef]
3. Roses, A.D.; Burns, D.K.; Chissoe, S.; Middleton, L.; Jean, P.S. Keynote review: Disease-specific target selection: A critical first step down the right road. *Drug Discov. Today* **2005**, *10*, 177–189. [CrossRef]
4. Steadman, V.A. Drug Discovery: Collaborations between Contract Research Organizations and the Pharmaceutical Industry. *ACS Med. Chem. Lett.* **2018**, *9*, 581–583. [CrossRef] [PubMed]
5. Decorte, B.L. Evolving Outsourcing Landscape in Pharma R&D: Different Collaborative Models and Factors to Consider When Choosing a Contract Research Organization. *J. Med. Chem.* **2020**, *63*, 11362–11367. [CrossRef]
6. Hingorani, A.D.; Kuan, V.; Finan, C.; Kruger, F.A.; Gaulton, A.; Chopade, S.; Sofat, R.; MacAllister, R.J.; Overington, J.P.; Hemingway, H.; et al. Improving the odds of drug development success through human genomics: Modelling study. *Sci. Rep.* **2019**, *9*, 1–25. [CrossRef] [PubMed]
7. Knowles, J.; Gromo, G. Target selection in drug discovery. *Nat. Rev. Drug Discov.* **2003**, *2*, 63–69. [CrossRef] [PubMed]
8. Roberts, T.C.; Langer, R.; Wood, M.J.A. Advances in oligonucleotide drug delivery. *Nat. Rev. Drug Discov.* **2020**, *19*, 673–694. [CrossRef] [PubMed]
9. Dugger, S.A.; Platt, A.; Goldstein, D.B. Drug development in the era of precision medicine. *Nat. Rev. Drug Discov.* **2018**, *17*, 183–196. [CrossRef]
10. Boyle, E.A.; Li, Y.I.; Pritchard, J.K. An Expanded View of Complex Traits: From Polygenic to Omnigenic. *Cell* **2017**, *169*, 1177–1186. [CrossRef] [PubMed]
11. Merrifield, R.B. Solid Phase Peptide Synthesis. I. The Synthesis of a Tetrapeptide. *J. Am. Chem. Soc.* **1963**, *85*, 2149–2154. [CrossRef]
12. Liu, R.; Li, X.; Lam, K.S. Combinatorial chemistry in drug discovery. *Curr. Opin. Chem. Biol.* **2017**, *38*, 117–126. [CrossRef] [PubMed]
13. Lam, K.S.; Salmon, S.E.; Hersh, E.M.; Hruby, V.J.; Kazmeierski, W.M.; Knapp, R.J. A new type of synthetic peptide library for identifying ligand-binding activity. *Nature* **1991**, *354*, 82–84. [CrossRef] [PubMed]
14. Matteucci, M.D.; Caruthers, M.H. Synthesis of Deoxyoligonucleotides on a Polymer Support. *J. Am. Chem. Soc.* **1981**, *103*, 3185–3191. [CrossRef]
15. Marshall, W.S.; Boymel, J.L. Oligonucleotide synthesis as a tool in drug discovery research. *Drug Discov. Today* **1998**, *3*, 34–42. [CrossRef]
16. Bunin, B.A.; Ellman, J.A. A General and Expedient Method for the Solid-Phase Synthesis of 1,4-Benzodiazepine Derivatives. *J. Am. Chem. Soc.* **1992**, *114*, 10997–10998. [CrossRef]
17. Nicolaou, K.C.; Xiao, X.-Y.; Parandoosh, Z.; Senyei, A.; Nova, M.P. Radiofrequency Encoded Combinatorial Chemistry. *Angew. Chem. Int. Ed. Engl.* **1995**, *34*, 2289–2291. [CrossRef]
18. Seeberger, P.H.; Haase, W.C. Solid-phase oligosaccharide synthesis and combinatorial carbohydrate libraries. *Chem. Rev.* **2000**, *100*, 4349–4393. [CrossRef]
19. Furukawa, K.; Ohkawa, Y.; Yamauchi, Y.; Hamamura, K.; Ohmi, Y.; Furukawa, K. Fine tuning of cell signals by glycosylation. *J. Biochem.* **2012**, *151*, 573–578. [CrossRef]
20. Seeberger, P.H. Discovery of Semi- and Fully-Synthetic Carbohydrate Vaccines against Bacterial Infections Using a Medicinal Chemistry Approach. *Chem. Rev.* **2021**, *121*, 3598–3626. [CrossRef]
21. Spagnolli, G.; Massignan, T.; Astolfi, A.; Biggi, S.; Rigoli, M.; Brunelli, P.; Libergoli, M.; Ianeselli, A.; Orioli, S.; Boldrini, A.; et al. Pharmacological inactivation of the prion protein by targeting a folding intermediate. *Commun. Biol.* **2021**, *4*, 1–16. [CrossRef]
22. Mazzorana, M.; Shotton, E.J.; Hall, D.R. A comprehensive approach to X-ray crystallography for drug discovery at a synchrotron facility—The example of Diamond Light Source. *Drug Discov. Today Technol.* **2020**, in press. [CrossRef]
23. Stiers, K.M.; Graham, A.C.; Zhu, J.S.; Jakeman, D.L.; Nix, J.C.; Beamer, L.J. Structural and dynamical description of the enzymatic reaction of a phosphohexomutase. *Struct. Dyn.* **2019**, *6*, 024703. [CrossRef] [PubMed]
24. Bortoli, M.; Torsello, M.; Bickelhaupt, F.M.; Orian, L. Role of the Chalcogen (S, Se, Te) in the Oxidation Mechanism of the Glutathione Peroxidase Active Site. *ChemPhysChem* **2017**, *18*, 2990–2998. [CrossRef] [PubMed]
25. Gołowicz, D.; Kasprzak, P.; Orekhov, V.; Kazimierczuk, K. Fast time-resolved NMR with non-uniform sampling. *Prog. Nucl. Magn. Reson. Spectrosc.* **2020**, *116*, 40–55. [CrossRef] [PubMed]
26. Horitani, M.; Kusubayashi, K.; Oshima, K.; Yato, A.; Sugimoto, H.; Watanabe, K. X-ray Crystallography and Electron Paramagnetic Resonance Spectroscopy Reveal Active Site Rearrangement of Cold-Adapted Inorganic Pyrophosphatase. *Sci. Rep.* **2020**, *10*, 4368. [CrossRef]
27. Zonta, F.; Pagano, M.A.; Trentin, L.; Tibaldi, E.; Frezzato, F.; Trimarco, V.; Facco, M.; Zagotto, G.; Pavan, V.; Ribaudo, G.; et al. Lyn sustains oncogenic signaling in chronic lymphocytic leukemia by strengthening SET-mediated inhibition of PP2A. *Blood* **2015**, *125*, 3747–3755. [CrossRef]
28. Schreiber, G. CHAPTER 1: Protein-Protein Interaction Interfaces and their Functional Implications. *RSC Drug Discov. Ser.* **2021**, 1–24. [CrossRef]

29. Paul, D.; Sanap, G.; Shenoy, S.; Kalyane, D.; Kalia, K.; Tekade, R.K. Artificial intelligence in drug discovery and development. *Drug Discov. Today* **2021**, *26*, 80–93. [CrossRef]
30. Chen, H.; Engkvist, O.; Wang, Y.; Olivecrona, M.; Blaschke, T. The rise of deep learning in drug discovery. *Drug Discov. Today* **2018**, *23*, 1241–1250. [CrossRef]
31. Klambauer, G.; Hochreiter, S.; Rarey, M. Machine Learning in Drug Discovery. *J. Chem. Inf. Model.* **2019**, *59*, 945–946. [CrossRef]
32. Ma, J.; Sheridan, R.P.; Liaw, A.; Dahl, G.E.; Svetnik, V. Deep Neural Nets as a Method for Quantitative Structure–Activity Relationships. *J. Chem. Inf. Model.* **2015**, *55*, 263–274. [CrossRef] [PubMed]
33. Mayr, A.; Klambauer, G.; Unterthiner, T.; Hochreiter, S. DeepTox: Toxicity Prediction using Deep Learning. *Front. Environ. Sci.* **2016**, *3*, 80. [CrossRef]
34. Wainberg, M.; Merico, D.; Delong, A.; Frey, B.J. Deep learning in biomedicine. *Nat. Biotechnol.* **2018**, *36*, 829–838. [CrossRef] [PubMed]
35. Zhang, L.; Tan, J.; Han, D.; Zhu, H. From machine learning to deep learning: Progress in machine intelligence for rational drug discovery. *Drug Discov. Today* **2017**, *22*, 1680–1685. [CrossRef]
36. Mouchlis, V.D.; Afantitis, A.; Serra, A.; Fratello, M.; Papadiamantis, A.G.; Aidinis, V.; Lynch, I.; Greco, D.; Melagraki, G. Advances in de novo drug design: From conventional to machine learning methods. *Int. J. Mol. Sci.* **2021**, *22*, 1676. [CrossRef]
37. Segler, M.H.S.; Waller, M.P. Neural-Symbolic Machine Learning for Retrosynthesis and Reaction Prediction. *Chem.-Eur. J.* **2017**, *23*, 5966–5971. [CrossRef]
38. Coley, C.W.; Barzilay, R.; Jaakkola, T.S.; Green, W.H.; Jensen, K.F. Prediction of Organic Reaction Outcomes Using Machine Learning. *ACS Cent. Sci.* **2017**, *3*, 434–443. [CrossRef]
39. Guo, M.; Kalra, G.; Wilson, W.; Peng, Y.; Augsburger, L.L. A Prototype Intelligent Hybrid System for Hard Gelatin Capsule Formulation Development. *Pharm. Technol.* **2002**, *26*, 44–60. [CrossRef]
40. Mehta, C.H.; Narayan, R.; Nayak, U.Y. Computational modeling for formulation design. *Drug Discov. Today* **2019**, *24*, 781–788. [CrossRef]
41. Zhao, C.; Jain, A.; Hailemariam, L.; Suresh, P.; Akkisetty, P.; Joglekar, G.; Venkatasubramanian, V.; Reklaitis, G.V.; Morris, K.; Basu, P. Toward intelligent decision support for pharmaceutical product development. *J. Pharm. Innov.* **2006**, *1*, 23–35. [CrossRef]
42. Koutsoukas, A.; Monaghan, K.J.; Li, X.; Huan, J. Deep-learning: Investigating deep neural networks hyper-parameters and comparison of performance to shallow methods for modeling bioactivity data. *J. Cheminform.* **2017**, *9*, 42. [CrossRef]
43. Lenselink, E.B.; ten Dijke, N.; Bongers, B.; Papadatos, G.; van Vlijmen, H.W.T.; Kowalczyk, W.; IJzerman, A.P.; van Westen, G.J.P. Beyond the hype: Deep neural networks outperform established methods using a ChEMBL bioactivity benchmark set. *J. Cheminform.* **2017**, *9*, 45. [CrossRef] [PubMed]
44. Gómez-Bombarelli, R.; Wei, J.N.; Duvenaud, D.; Hernández-Lobato, J.M.; Sánchez-Lengeling, B.; Sheberla, D.; Aguilera-Iparraguirre, J.; Hirzel, T.D.; Adams, R.P.; Aspuru-Guzik, A. Automatic Chemical Design Using a Data-Driven Continuous Representation of Molecules. *ACS Cent. Sci.* **2018**, *4*, 268–276. [CrossRef]
45. Kadurin, A.; Nikolenko, S.; Khrabrov, K.; Aliper, A.; Zhavoronkov, A. druGAN: An Advanced Generative Adversarial Autoencoder Model for de Novo Generation of New Molecules with Desired Molecular Properties in Silico. *Mol. Pharm.* **2017**, *14*, 3098–3104. [CrossRef] [PubMed]
46. Blaschke, T.; Olivecrona, M.; Engkvist, O.; Bajorath, J.; Chen, H. Application of Generative Autoencoder in De Novo Molecular Design. *Mol. Inform.* **2018**, *37*, 1700123. [CrossRef] [PubMed]
47. Bai, Q.; Tan, S.; Xu, T.; Liu, H.; Huang, J.; Yao, X. MolAICal: A soft tool for 3D drug design of protein targets by artificial intelligence and classical algorithm. *Brief. Bioinform.* **2021**, *22*. [CrossRef] [PubMed]
48. Mariya, P.; Olexandr, I.; Alexander, T. Deep reinforcement learning for de novo drug design. *Sci. Adv.* **2021**, *4*, eaap7885. [CrossRef]
49. Schütt, K.T.; Gastegger, M.; Tkatchenko, A.; Müller, K.-R.; Maurer, R.J. Unifying machine learning and quantum chemistry with a deep neural network for molecular wavefunctions. *Nat. Commun.* **2019**, *10*, 5024. [CrossRef]
50. Bennett, W.F.D.; He, S.; Bilodeau, C.L.; Jones, D.; Sun, D.; Kim, H.; Allen, J.E.; Lightstone, F.C.; Ingólfsson, H.I. Predicting Small Molecule Transfer Free Energies by Combining Molecular Dynamics Simulations and Deep Learning. *J. Chem. Inf. Model.* **2020**, *60*, 5375–5381. [CrossRef]
51. Jumper, J.; Evans, R.; Pritzel, A.; Green, T.; Figurnov, M.; Ronneberger, O.; Tunyasuvunakool, K.; Bates, R.; Žídek, A.; Potapenko, A.; et al. Highly accurate protein structure prediction with AlphaFold. *Nature* **2021**, *596*, 583–589. [CrossRef]
52. Mirdita, M.; Ovchinnikov, S.; Steinegger, M. ColabFold-Making protein folding accessible to all. *bioRxiv* **2021**. [CrossRef]
53. Ching, T.; Himmelstein, D.S.; Beaulieu-Jones, B.K.; Kalinin, A.A.; Do, B.T.; Way, G.P.; Ferrero, E.; Agapow, P.-M.; Zietz, M.; Hoffman, M.M.; et al. Opportunities and obstacles for deep learning in biology and medicine. *J. R. Soc. Interface* **2018**, *15*, 20170387. [CrossRef]
54. Matar, P.; Rojo, F.; Cassia, R.; Moreno-Bueno, G.; Di Cosimo, S.; Tabernero, J.; Guzmán, M.; Rodriguez, S.; Arribas, J.; Palacios, J.; et al. Combined epidermal growth factor receptor targeting with the tyrosine kinase inhibitor Gefitinib (ZD1839) and the monoclonal antibody Cetuximab (IMC-C225): Superiority over single-agent receptor targeting. *Clin. Cancer Res.* **2004**, *10*, 6487–6501. [CrossRef]
55. Roberts, A.W. Therapeutic development and current uses of BCL-2 inhibition. *Hematol. Am. Soc. Hematol. Educ. Program Book* **2020**, *2020*, 1–9. [CrossRef] [PubMed]

56. Pagano, M.A.; Tibaldi, E.; Molino, P.; Frezzato, F.; Trimarco, V.; Facco, M.; Zagotto, G.; Ribaudo, G.; Leanza, L.; Peruzzo, R.; et al. Mitochondrial apoptosis is induced by Alkoxy phenyl-1-propanone derivatives through PP2A-mediated dephosphorylation of Bad and Foxo3A in CLL. *Leukemia* **2019**, *33*, 1148–1160. [CrossRef]
57. Hughes, T.B.; Miller, G.P.; Swamidass, S.J. Modeling Epoxidation of Drug-like Molecules with a Deep Machine Learning Network. *ACS Cent. Sci.* **2015**, *1*, 168–180. [CrossRef]
58. Behjati, S.; Tarpey, P.S. What is next generation sequencing? *Arch. Dis. Child. Educ. Pract. Ed.* **2013**, *98*, 236–238. [CrossRef]
59. Beger, R.D.; Schmidt, M.A.; Kaddurah-Daouk, R. Current concepts in pharmacometabolomics, biomarker discovery, and precision medicine. *Metabolites* **2020**, *10*, 129. [CrossRef] [PubMed]
60. Boccia, S.; Liu, J.; Demirkan, A.; van Duijn, C.; Mariani, M.; Castagna, C.; Pastorino, R.; Fiatal, S.; Pikó, P.; Ádány, R.; et al. Identification of Biomarkers for the Prevention of Chronic Disease. *SpringerBriefs Public Health* **2021**, 9–32. [CrossRef]
61. Upadhyayula, P.S.; Spinazzi, E.F.; Argenziano, M.G.; Canoll, P.; Bruce, J.N. Convection Enhanced Delivery of Topotecan for Gliomas: A Single-Center Experience. *Pharmaceutics* **2021**, *13*, 39. [CrossRef]
62. Wu, S.-K.; Tsai, C.-L.; Huang, Y.; Hynynen, K. Focused Ultrasound and Microbubbles-Mediated Drug Delivery to Brain Tumor. *Pharmaceutics* **2021**, *13*, 15. [CrossRef]
63. Griffith, J.I.; Rathi, S.; Zhang, W.; Zhang, W.; Drewes, L.R.; Sarkaria, J.N.; Elmquist, W.F. Addressing BBB Heterogeneity: A New Paradigm for Drug Delivery to Brain Tumors. *Pharmaceutics* **2020**, *12*, 1205. [CrossRef] [PubMed]
64. Zonta, N.; Cozza, G.; Gianoncelli, A.; Korb, O.; Exner, T.E.; Meggio, F.; Zagotto, G.; Moro, S. Scouting novel protein kinase A (PKA) inhibitors by using a consensus docking-based virtual screening approach. *Lett. Drug Des. Discov.* **2009**, *6*, 327–336. [CrossRef]
65. Straehla, J.P.; Warren, K.E. Pharmacokinetic Principles and Their Application to Central Nervous System Tumors. *Pharmaceutics* **2020**, *12*, 948. [CrossRef]
66. Tosi, U.; Souweidane, M. Convection Enhanced Delivery for Diffuse Intrinsic Pontine Glioma: Review of a Single Institution Experience. *Pharmaceutics* **2020**, *12*, 660. [CrossRef] [PubMed]
67. Molotkov, A.; Carberry, P.; Dolan, M.A.; Joseph, S.; Idumonyi, S.; Oya, S.; Castrillon, J.; Konofagou, E.E.; Doubrovin, M.; Lesser, G.J.; et al. Real-Time Positron Emission Tomography Evaluation of Topotecan Brain Kinetics after Ultrasound-Mediated Blood–Brain Barrier Permeability. *Pharmaceutics* **2021**, *13*, 405. [CrossRef]
68. Chatterjee, K.; Atay, N.; Abler, D.; Bhargava, S.; Sahoo, P.; Rockne, R.C.; Munson, J.M. Utilizing Dynamic Contrast-Enhanced Magnetic Resonance Imaging (DCE-MRI) to Analyze Interstitial Fluid Flow and Transport in Glioblastoma and the Surrounding Parenchyma in Human Patients. *Pharmaceutics* **2021**, *13*, 212. [CrossRef]
69. Sharabi, S.; Last, D.; Daniels, D.; Fabian, I.D.; Atrakchi, D.; Bresler, Y.; Liraz-Zaltsman, S.; Cooper, I.; Mardor, Y. Non-Invasive Low Pulsed Electrical Fields for Inducing BBB Disruption in Mice—Feasibility Demonstration. *Pharmaceutics* **2021**, *13*, 169. [CrossRef]
70. Nwagwu, C.D.; Immidisetti, A.V.; Bukanowska, G.; Vogelbaum, M.A.; Carbonell, A.-M. Convection-Enhanced Delivery of a First-in-Class Anti-β1 Integrin Antibody for the Treatment of High-Grade Glioma Utilizing Real-Time Imaging. *Pharmaceutics* **2021**, *13*, 40. [CrossRef]
71. Brady, M.; Raghavan, R.; Sampson, J. Determinants of Intraparenchymal Infusion Distributions: Modeling and Analyses of Human Glioblastoma Trials. *Pharmaceutics* **2020**, *12*, 895. [CrossRef] [PubMed]
72. Binaschi, M.; Zagotto, G.; Palumbo, M.; Zunino, F.; Farinosi, R.; Capranico, G. Irreversible and reversible topoisomerase II DNA cleavage stimulated by clerocidin: Sequence specificity and structural drug determinants. *Cancer Res.* **1997**, *57*, 1710–1716. [PubMed]
73. Mehta, J.N.; McRoberts, G.R.; Rylander, C.G. Controlled Catheter Movement Affects Dye Dispersal Volume in Agarose Gel Brain Phantoms. *Pharmaceutics* **2020**, *12*, 753. [CrossRef] [PubMed]
74. Bienfait, K.; Chhibber, A.; Marshall, J.C.; Armstrong, M.; Cox, C.; Shaw, P.M.; Paulding, C. Current challenges and opportunities for pharmacogenomics: Perspective of the Industry Pharmacogenomics Working Group (I-PWG). *Hum. Genet.* **2021**. [CrossRef] [PubMed]
75. Peiró, A.M. Pharmacogenetics in Pain Treatment. *Adv. Pharmacol.* **2018**, *83*, 247–273. [CrossRef]
76. DeWeerdt, S. Tracing the US opioid crisis to its roots. *Nature* **2019**, *573*, S10–S12. [CrossRef]
77. Gammal, R.S.; Crews, K.R.; Haidar, C.E.; Hoffman, J.M.; Baker, D.K.; Barker, P.J.; Estepp, J.H.; Pei, D.; Broeckel, U.; Wang, W.; et al. Pharmacogenetics for Safe Codeine Use in Sickle Cell Disease. *Pediatrics* **2016**, *138*, e20153479. [CrossRef]
78. Benjeddou, M.; Peiró, A.M. Pharmacogenomics and prescription opioid use. *Pharmacogenomics* **2021**, *22*, 235–245. [CrossRef]
79. Tremaine, L.; Delmonte, T.; Francke, S. The role of ADME pharmacogenomics in early clinical trials: Perspective of the of the Industry Pharmacogenomics Working Group (I-PWG). *Pharmacogenomics* **2015**, *16*, 2055–2067. [CrossRef]
80. Relling, M.V.; Klein, T.E.; Gammal, R.S.; Whirl-Carrillo, M.; Hoffman, J.M.; Caudle, K.E. The Clinical Pharmacogenetics Implementation Consortium: 10 Years Later. *Clin. Pharmacol. Ther.* **2020**, *107*, 171–175. [CrossRef]
81. Chenoweth, M.J.; Giacomini, K.M.; Pirmohamed, M.; Hill, S.L.; van Schaik, R.H.N.; Schwab, M.; Shuldiner, A.R.; Relling, M.V.; Tyndale, R.F. Global Pharmacogenomics Within Precision Medicine: Challenges and Opportunities. *Clin. Pharmacol. Ther.* **2020**, *107*, 57–61. [CrossRef] [PubMed]

82. Atanasov, A.G.; Zotchev, S.B.; Dirsch, V.M.; Orhan, I.E.; Banach, M.; Rollinger, J.M.; Barreca, D.; Weckwerth, W.; Bauer, R.; Bayer, E.A.; et al. Natural products in drug discovery: Advances and opportunities. *Nat. Rev. Drug Discov.* **2021**, *20*, 200–216. [CrossRef] [PubMed]
83. Caballero, J. The latest automated docking technologies for novel drug discovery. *Expert Opin. Drug Discov.* **2021**, *16*, 625–645. [CrossRef] [PubMed]

MDPI
St. Alban-Anlage 66
4052 Basel
Switzerland
Tel. +41 61 683 77 34
Fax +41 61 302 89 18
www.mdpi.com

Molecules Editorial Office
E-mail: molecules@mdpi.com
www.mdpi.com/journal/molecules

www.ingramcontent.com/pod-product-compliance
Lightning Source LLC
LaVergne TN
LVHW070421100526
838202LV00014B/1498